Melanie Phillips is one of the best-known columnists writing in national newspapers in Britain. Her column, originally published in the *Guardian* and subsequently in the *Observer*, is renowned for its fearless and penetrating critique of the social issues of her time. Through writing that is both accessible and authoritative, she has demonstrated time and again an ability to spot developing social trends and to challenge from within the shibboleths of the liberal establishment. Awarded the Orwell Prize for journalism in 1996, she is married with two teenage children and lives in West London.

'The instinctive reaction of those within the education establishment — and I include myself as a member — will be to dismiss Melanie Phillips's book as deranged rantings from the far right. Since she is already the columnist most educators love to hate, this seems all too likely. To do so would be a grave error . . . To dismiss the book would be to miss the critical importance of some of her arguments'
Professor Michael Barber, *Guardian*

'This is crap by anybody's standards'
Professor Ted Wragg, *Independent*

'There is a cult of phoney egalitarianism around, and Phillips rustles up some choice specimens of its sentimentalist absurdities . . . A tunnel-vision text which discredits itself by hyperbole, monomania and reckless generalisation'
Professor Terry Eagleton, *Observer*

'The most frightening book of the decade . . . [I] recommend it to every parent, grandparent and teacher who cares about the education of the next generation'
Daily Express

'[A] farrago of ignorance and inaccuracy . . . A series of prejudices decked out with press cuttings, gossip and privately circulated papers . . . As I contemplated this hybrid creature [Phillips] – half stereotypical *Guardian* woman, half average *Daily Mail* Sloane – I felt an irresistible urge to take myself off to the Garrick'
Colin McCabe, *New Statesman*

'A blistering indictment of Britain's education system . . . It crystallises the arguments about education, family and crime, and makes a most convincing claim that the politics of both Left and Right have between them produced a selfish society with rotting values both in the classroom and the home. Ideological fads are exploded and a programme for survival offered in this book which at least every parent, if not every concerned citizen, should read'
Charles Osborne, *Sunday Telegraph*

MELANIE PHILLIPS

ALL MUST HAVE PRIZES

timewarner
paperbacks

A *Time Warner* Paperback

First published in Great Britain in 1996
by Little, Brown and Company
This edition published in 1998 by Warner Books
Reprinted by Time Warner Paperbacks in 2002

A CIP catalogue record for this book
is available from the British Library.

ISBN: 0 7515 2274 0

Typeset in Plantin by
Palimpsest Book Production Limited,
Polmont, Stirlingshire
Printed and bound in Great Britain
by Clays Ltd, St Ives plc

Time Warner Paperbacks
An imprint of
Time Warner Books UK
Brettenham House
Lancaster Place
London WC2E 7EN

www.TimeWarnerBooks.co.uk

For Gabriel and Abigail, and their friends

CONTENTS

There was no 'One, two three, and away,' but they began running when they liked, and left off when they liked, so that it was not easy to know when the race was over. However, when they had been running half an hour or so, and were quite dry again, the Dodo suddenly called out 'The race is over!' and they all crowded round it, panting, and asking, 'But who has won?'

This question the Dodo could not answer without a great deal of thought, and it sat for a long time with one finger pressed upon its forehead (the position in which you usually see Shakespeare, in the pictures of him) while the rest waited in silence. At last the Dodo said, '*Everybody* has won, and all must have prizes.'

Alice's Adventures in Wonderland, Lewis Carroll

PREFACE TO THE PAPERBACK EDITION

Since this book was first published in 1996, there has been a change of government. Labour's landslide victory on 1 May 1997 demonstrated not merely the public's disaffection with the Conservatives but a profound desire for cultural change. There was deep anxiety about what was perceived as a climate of pervasive selfishness and greed, and an absence of any sense of a shared national project. People wanted to believe once again that there was indeed such a thing as society. They approved of Tony Blair because he talked about restoring duty and responsibility, notions which people realised had been devalued during the past eighteen years of Conservative rule. In particular, Blair spoke to their deepest anxieties about children and family life. He underlined the importance of stable families; he indicated he would ensure that antisocial individuals faced the consequences of their behaviour; and above all he promised that he would haul Britain's failing education system out of the mire into which it had fallen.

His pledges on education were central to his programme. 'Education, education, education' wasn't just a pious sound-bite. A reformed education system lay at the heart of Labour's proposals to restore full employment for young people and end the culture of welfare dependency. Verbally, the signals could not have been clearer. Labour's Education Secretary David Blunkett employed ferocious rhetoric to ram home the message that henceforth there would be 'zero tolerance' of bad teaching. The Schools Minister Stephen Byers, who was given special responsibility for raising standards, published a list of failing schools and declared that they would henceforth be set on a fast track to improvement. The new government accepted the evidence – denied consistently by the education establishment – that Britain had huge problems with basic literacy and numeracy. It denounced the progressive ideologies favoured by that establishment, and made clear its preference for traditional teaching methods. The Education Department set up a new Standards and Effectiveness Unit and a new Standards Task Force. A White Paper, 'Excellence in Schools', showered initiatives like confetti. Schools would henceforth be set targets, which would be monitored by local education authorities, which would themselves have to meet targets which would in turn be monitored. In the cause of raising education standards, the first few months of the new government were characterised by frenetic hyperactivity – at least on the level of rhetoric.

Yet the more noise that was being made, the more scepticism seemed in order. There was a philosophical hole at the heart of Labour's education policy. For all its initiatives, the White Paper was devoid of any discussion about the content of education, any vision of what it was or what its purposes might be. The endorsement of traditional teaching methods was a significant and courageous shift for Labour. Yet the measures being proposed were essentially managerialist and thus did not get to grips with the hard issues at the heart of what had gone wrong. No one was asking the deeper questions. What was a bad teacher? What were the driving ideas behind failing educational practices – and to what extent were these ideas embedded not just within schools but within the whole education system, right up to the universities?

Targets may sound reassuring; but it was hard to see how these would succeed in practice, since the education authorities charged with monitoring standards were also charged with putting those problems right under the threat of being penalised themselves. They thus seemed to have every incentive to ensure school targets were not too demanding. Moreover, it was the local education authorities that had done so much damage to schools in the first place. The government emphasised its determination to crack down on inadequate authorities. Very early on it made an example of the London borough of Hackney, revealed by Ofsted to be catastrophically failing the schools in its governance. Yet for all the sound and fury, there was precious little sign that anyone knew how to put the situation right. The 'hit squad' the government appointed to sort out Hackney met fierce resistance from within the very council structure that had so badly failed. More generally, ministers seemed not to recognise that local education authorities were the problem and not, as they continued to believe, the solution.

Labour's education reform strategy appeared to be largely based on exhortation and noisy announcements that it would no longer put up with the second-rate. Many of its initiatives seemed to be little other than gimmicks – linking schools to the Internet, for example, or setting up a 'sneak-line' for pupils to complain about poor teachers. But on important matters, there emerged alarming signs of a failure to grasp the nettle. Proposals to introduce compulsory grammar tests for fourteen-year-olds to meet widespread concerns about pupils' poor grasp of language were delayed, in an apparent concession to militant English teachers who, as this book outlines, have led the way in sloppy educational thinking. The introduction of the compulsory curriculum for teacher training, deemed essential to counter the pedagogic vacuum in such institutions, was also delayed. Defenders of failed practice, or those who denied the scale of the problems, were being appointed to the education quangos. At the same time, academics with a robust and realistic appreciation of what had gone wrong were being turned down on the grounds that they couldn't be relied upon to toe the government line.

The government seemed not to grasp the essential truth that the public interest would only be served by bringing in people who would speak up if their policies were wrong. There seemed to be an unshakeable belief that education ministers were right, and so any advice to the contrary had to be excluded from the reform process. Such unparalleled central control helped explain the prominent role being restored to education authorities, which under the Conservatives had lost a lot of their powers over schools. Labour ministers believed they could control them and ensure they would do what the government wanted. Yet as the experiences related in this book indicate, it was wishful thinking to imagine they could even control their own civil servants. Tony Blair himself seemed to be in danger of discovering too late, as did Mrs Thatcher before him, that the Education Department had a life of its own and would frustrate an agenda with which it disagreed.

This was because a substantial ideological gulf separated Blair from most of the ministers, advisers and civil servants at the Education Department. Blair himself was acknowledged to be a meritocrat. He was seen to have little time for the doctrine of egalitarianism. He stood not for equality of outcome but for equality of opportunity, a completely different matter. Yet despite the fact that David Blunkett himself said that the old aspiration towards equality of outcomes was now defunct, the education policy over which he was presiding was promoting that very shibboleth. Its galvanising principle was that everyone had the 'right' to achieve equally, and had to be dealt with in exactly the same way as everyone else. Thus selection by ability was taboo. The policy was hostile to diversity and antagonistic towards excellence.

Indeed, its architects seemed not to understand what excellence was. Their managerialist approach appeared to assume that simply by kicking enough teachers, Britain would arrive painlessly at the nirvana of 'excellence for all', the title of one of their pre-election education documents. The fact that excellence cannot be achieved by everyone, since by definition excelling means doing better than others, was implicitly denied by the policies which began to unfold. New Labour still believed in prizes for all. Three examples stood out in particular: the acceptance of the fundamental principles behind Sir Ron Dearing's proposals for higher education, the proposed assault on Oxbridge, and the threat to grant-maintained schools.

Right from the start, ambiguity had shrouded Labour's policy for grant-maintained schools. It somehow maintained that they would be brought back under local authority control and at the same time enjoy no less independence than at present. These schools were hated by the Education Department and by many, if not most, local authorities for precisely the reasons they were so attractive. It was hard for third-rate bureaucrats to control them and foist upon them the failing ideologies which had crippled the school system. Their independence led them to adopt policies whose popularity with parents showed up the deficiencies in other schools and destroyed the usual alibis for poor practice. It was always very unlikely that Blair would wish to damage grant-maintained schools. After all, it was to one of them, the Catholic London Oratory, that he had decided to send his two sons. Yet his education policy was on course to destroy the very qualities which made such schools so popular. Restoring to local authorities the power to interfere with their running threatened to strangle their independence and initiative within the deadly embrace of meddlesome officials. It appeared inconceivable that Blair wanted this to happen. It was all too likely that he didn't realise this was what would happen, because no one had told him.

He also appeared not to realise that the Dearing proposals on higher education ran counter to his own meritocratic philosophy. Dearing had been commissioned by the previous Conservative government to square a circle, to give more and more people access to higher education while finding the means to pay for such continuous expansion. His proposals for student fees attracted most attention and controversy. But the significance of his report was far more profound and destructive. If implemented, it would simply destroy the universities. As the philosophy professor Anthony O'Hear pointed out in a scathing critique (*The Dearing Report on Higher Education*; Centre for Policy Studies), Dearing effectively demolished the liberal ideal of education for its own sake, insisting instead that research and scholarship had to be related to 'the wider needs of society'. The fruits of that philosophy were already all too plain. Students could take university degrees in Golf Course Studies, Real Time Studies, Accommodation Management, Exhibition Design, Hospitality Studies, Creative Therapies and many other similar absurdities. These simply aren't academic disciplines. But Dearing didn't appear to understand what academic study was for. He was impervious to the idea that physics, maths, literature or history should be studied for their own sakes, because they enlightened people about what the world was, what they were in that world and what they might become. Dearing believed instead that scholarship must be related to wealth creation and that much specific learning would become obsolete. As O'Hear despairingly protested: 'Does Dearing really think that Plato and Aeschylus are obsolete, that Aristotle should be forgotten, that Giotto and Titian have been superseded by Damien Hirst . . . that in the days of *The X-Files*, the Millennium Dome and the Internet we have no need to study Newton or Einstein?'

Dearing wanted to raise the proportion of young people in higher education from 33 per cent to 45 per cent. It was hard to understand the establishment's continuing obsession with getting more and more young people to university when British numbers were already way above European norms. The result of this obsession was that the vocational, on-the-job skills training which was what most young people really needed was being sacrificed for bogus 'academic' qualifications. There was no way these numbers could be accommodated in higher education without destroying its pursuit of excellence. Indeed, the relentless pursuit of the chimera of equality was already destroying the universities. Vice-chancellors and professors spoke privately of the collapse of standards as they were forced to take more and more students who had no aptitude for university education. They spoke of the ubiquity of the 2:1 degree under the pressure to produce results identical to everyone else; of the collapse of research standards under the pressure of the market; of their view that soon students would have to gain a masters degree or above to reach the standard once represented by a first degree. Dearing threatened to push this to its conclusion, aided and abetted by an Education Department determined to dumb down A-levels by relentless modularisation in the drive to produce a spurious equivalence between A-levels and Britain's inadequate vocational qualifications.

Equivalence was the name of the game. That was what lay behind the threat to Oxford and Cambridge posed by the government's plan to deprive them of college fees. Their protest that this would mean the end of the excellent tutorial system was a two-edged strategy which would only succeed if the egalitarians were defeated. Excellence, it seemed, was not to be permitted when not everyone else could be as excellent. Oxbridge would have to become the same as everyone else. The logical outcome of that thinking was that a degree in maths or history from Oxford or Cambridge would become the 'equivalent' of a degree in cosmetics or sport from an 'equivalent' university. This seemed to be the apotheosis of the old Labour egalitarian agenda, but with one crucial difference. Now students being dumbed down would be forced to pay for the privilege.

There were signs during the early months of the new government of a power struggle behind the scenes in education, as the Prime Minister tried to steer policy in the direction he wanted rather than where the education establishment was trying to take it. In particular, when he realised the perverse consequences looming for grant-maintained schools, he took steps to head them off. However, he would not succeed in reforming the system unless he came to understand that there was a deeper problem with his whole approach to government. The great issues of education and family life, on which he had gained such approval from the public for his apparent identification with traditional values, were riven by an ideological chasm. People were either on one side or the other. It was not possible to oppose selection and diversity in education and at the same time espouse meritocratic ideals; it was not possible to express indifference towards adultery or approval for cohabitation without undermining married family life. The Prime Minister's problem, however, was that he wanted to straddle that chasm for fear of alienating a significant section of the population. He wanted to take everybody with him. He was reluctant to make the hard choices.

In education, this was symbolised by Blunkett's appointment of Tim Brighouse and Chris Woodhead as joint vice-chairmen of his Standards Task Force. Brighouse was Birmingham's chief education officer, a hero to teachers for whom he acted as cheerleader and champion, a staunch defender of the municipal education authority as the principal catalyst for reform, and a critic of Ofsted. Woodhead, the Chief Inspector of Schools and head of Ofsted, had been demonised by the education establishment for his unflinching criticisms of failing teaching practices, his disapproval of education authorities and his determination to promote traditional teaching approaches. There was no way that these two arch-opponents could jointly fashion an education reform. One of them had to win. As two diametrically opposing strands emerged in education policy-making during those early months, it was anyone's guess how this struggle would resolve itself.

All the signs were, however, that it wasn't just the Labour government that wanted to duck the hardest choices over the great cultural questions of the day. The new Conservative leader William Hague raised eyebrows when, at his 1997 party conference, he endorsed Michael Portillo's new-found enthusiasm for non-judgementalism in family life. Suddenly, the traditional family found itself deserted. No one in public life seemed prepared to acknowledge that intact, stable families could only be supported by preferring and according privilege to marriage over un-marriage. Instead of choosing which side of the chasm on which to stand, public figures seemed to believe that tolerance now required an abdication of moral judgement. What was necessary instead was 'compassion' or 'giving'. But true compassion requires judgements to be made so that harm can be prevented. The case against Woodhead was that he made teachers feel bad about themselves. It was apparently more important that teachers should feel good about themselves than that their pupils were properly educated. How, though, can it be the sign of a 'caring' society to abandon children to illiteracy and low expectations? How can it be compassionate to be indifferent to the increase in never-married mothers struggling alone and unsupported to bring up their children; or to the distress and damage done to children by the fragmentation of family life; or to the dishonesty and breaches of trust involved in adultery?

Feelings had become more important than facts, a development illustrated in a startling manner by the public reaction to the death of Diana, Princess of Wales. That episode was deeply disturbing, not least for revealing the extent to which emotion had come to displace reason, and a culture of narcissistic individualism now masqueraded as 'the giving age'. There was no doubt that the millions who took to the streets with their flowers and their tears were expressing a need to connect with something beyond themselves, to share in some collective enterprise. It was a disturbing mass demonstration of the loneliness of disconnected individuals. But it was also a demonstration of mass self-centredness. People appeared to be grieving not so much for the Princess as for themselves. Indeed, the crowds grew resentful that the Royal Family wasn't acknowledging the public's grief. Moreover, they concluded that its members couldn't themselves be grieving, simply because they weren't displaying their emotions in public. The crowds seemed to believe therefore that their own feelings were more valid than those of the bereaved. Yet such astonishing arrogance and selfishness were deemed to be evidence of a new cultural 'compassion'. Sentimentality was mistaken for authentic grief, which was accordingly devalued. An alarming flight from reason was suddenly detectable in a sight not seen before in people's lifetimes: an English mob in the grip of powerful emotion, which at one point threatened to turn ugly as it demanded due acknowledgement.

Just as disturbing was the object of all this public emotion. The Princess was an iconic figure, not merely because she reflected in her own life the emotional hardships with which many ordinary people could identify but – crucially – because of the manner in which she appeared to have transcended them. The messages she seemed to embody were that women could make it triumphantly on their own without those faithless, dishonourable and emotionally illiterate creatures called men; and that the damage and distress wreaked upon children by their parents' divorce could be soothed away by a hug. The Princess thus epitomised an age desperate for both freedom and comfort – but with no price-tag attached. Her death exposed people's deep yearning for someone to come along and solve the crushing problems of the age, but with no pain or effort required of themselves. People felt keenly the increasing failure of mediating institutions such as the family to provide fixed points of stability and attachment. They wanted to be part of something beyond themselves, while simultaneously retreating into their narcissistic pursuit of individual fulfilment. That was surely why the Princess, obsessed with making herself whole and with repairing her own body, searching constantly for someone to love her and give her a sense of identity, yet at the same time touching and hugging others in an even worse plight than herself, was such a potent symbol. She gave self-centredness a veneer of moral righteousness.

This was an insidious message. The damage caused by the collapse of attachments to cultural traditions or within families cannot be hugged away. This social fragmentation has been fuelled by people's reluctance to place limits on their individualism and put others first. Soon after this book was first published, the contemporary moral confusion was crystallised in a series of highly publicised disasters and problems. A headmaster, Philip Lawrence, was murdered by schoolboys. Severe problems with disruptive children at the Ridings School in Halifax and Manton junior school in Nottingham highlighted the malign confluence of grossly inadequate parenting and weak and ineffectual teaching. The core issue was the erosion of adult authority over children. A forum set up by the School Curriculum and Assessment Authority published a framework of moral values for teachers to pass on to children because SCAA was so worried that neither teachers nor young people believed in the legitimacy of absolute moral rules. There was public uncertainty and confusion in an unaccustomed moral debate. Were some children really uneducable and uncontrollable? Where should the lines be drawn between public and private codes of conduct? Were teachers or parents to blame for children's bad behaviour?

It seemed to me that the remarkable public reaction to *All Must Have Prizes* reflected the deep cultural fissures and contradictions which emerged during that moral debate and during the aftermath of the death of the Princess of Wales. After all, my book had suggested that the explosive issues of education and family required hard choices to be made. The book had provoked such passions not merely because the education world didn't like its failings held up for public scrutiny (although that was certainly part of the explanation). The fury illuminated the culture war being waged over how we viewed ourselves and how we should live with each other, the values upon which our society should be organised and, crucially, the balance we should strike between individualism and duty.

It is easy for people to call for a more decent society. The difficulty lies in facing up to the clear choices in personal behaviour and political organisation without which that decent society will remain a chimera: the move, as the political theorist David Selbourne has argued in his book *The Principle of Duty*, from our present culture of dutiless rights to one in which duty comes to the fore.

In schools, attitudes among teachers are slowly changing as they begin to question hitherto unchallenged ideologies and realise there may be better ways of helping children to learn. Just as the moral debate remains superficial, however, the absence of a coherent knowledge-based ethic to challenge educational orthodoxies has rendered much ostensible concern over low standards shallow and opportunistic. In the universities, people who have made careers out of promoting or conniving at these 'child-centred' fallacies have been digging into their trenches. In the media, dominant cultural nihilists reach instantaneously for the weapons of mockery and insult. They hurl at their opponents accusations of moral authoritarianism; as if this were the only alternative to nihilism and social anarchy.

It is this intense and polarised context which explains the reaction to this book. It simply hit a nerve. Among parents and many teachers, the reaction was enthusiastic. 'I could have written every word of this myself,' one primary head told me. 'I recognise all of this as true,' said another. Yet another told me that as a result of reading the book, he was now rethinking the way he ran his school. Although many teachers expressed strong disagreement, others wrote or spoke appreciatively and courteously. Among parents, there was passionate support. Several recorded their bewilderment at their children's lack of progress at school and regaled me with horror stories. They now understood rather better, they said, why these things had happened. And university professors wrote or rang with yet more disturbing evidence, and to say how much they agreed with what the book had said.

These reactions from teachers and parents were in sharp contrast with many of the published reviews and articles about the book. Obviously, a polemical argument can be expected to draw a fierce response. What was more surprising, however, was the nature of the personalised attacks which seemed designed to shut down the debate rather than engage with it. Assertion took the place of argument; misrepresentation replaced reasoned disagreement. There was also uncomprehending outrage: one interviewer from the education press was aghast that anyone should think that prizes should not be awarded to everyone. He was quite unable to grasp that this would mean a prize was no longer a prize, or that there was a difference between a prize and a reward.

A common complaint among these critics was that the book 'bore no relation to reality'. Others claimed I had exaggerated or distorted the evidence, or used caricature, or relied on gossip and unreliable anecdote. Yet if this were really the case, what did this say about all those teachers, university professors, education psychologists, inspectors, examiners, parents and pupils who had provided me with this evidence? Did their experience therefore also 'bear no relation to reality'? What about the research literature or academic texts that I had quoted? Were they also unreal? And what about those many teachers who said the book described perfectly the world they worked in? In fact, the loudest protests were made not by teachers but by teacher educators, people who had made careers out of or turned a blind eye to the quack theories which have made the job of teaching particularly difficult.

Indeed, the claim that the book 'bore no relation to reality' surely suggested precisely the denial of evidence about children's educational needs that I had noted in the book itself. I felt an example of this took place at the debate staged by the *Observer* to mark the book's publication. From the panel the Labour peer Baroness Blackstone, who would go on to become an Education Minister in the new Labour government, dismissed the book as sloppy, mendacious and bearing no relation to reality. Even though a university German lecturer who had contributed to the book told the audience that the standard of his own undergraduates was often little short of catastrophic, the baroness said she found my evidence of appalling standards in German very hard to believe. Such an exchange appeared to demonstrate that for some people there may be simply no evidence, however authoritative, which can ever shake their unassailable conviction that alarm over the progressive approach to education is the product of an over-ripe imagination.

The phenomenon of denial took an even more unsettling form. I was accused of misrepresenting the meaning of certain texts. This is a most serious charge and one which no reputable author or journalist can afford to take lightly. Accordingly, I carefully re-read the contentious texts to ensure I had not been unfair. In every case, I felt that I had not misrepresented them. These disagreements seemed closely related to a yet stranger feature of the debate. A number of teachers and educationists to whom I spoke after the book was published declared they agreed with much of what I was saying. Yet they then went on to uphold assumptions and practices which embodied the opposite of what I believed. What all this seemed to show was considerable confusion about the educational principles which underpinned the current orthodoxy.

This moral and intellectual confusion has continued into our new 'age of giving'. Re-moralising the country means that the fundamentals of education and family have to be put right. The difficulty is that common sense has become politicised, and what was once obvious has been falsely portrayed as an ideological agenda by those who camouflage their own prejudices. Such critics wrap themselves in the mantle of liberalism and claim that the analysis in this book represents a new social authoritarianism. These critics, though, are not liberals. They are libertarians. Consequently, teachers, parents and the public at large have been bamboozled. As one reviewer remarked, the word 'liberalism' has been used by our élites as a 'Trojan horse for malign forces'.

Ordinary teachers and parents who know what's right and want to do the best for the children in their charge must be given back the authority and the self-confidence to do so. Parenthood is the most important job any of us do in our lives; teaching is the most noble and essential role anyone can play in society. Teachers and parents must be given real support and encouragement, based not on a sentimental and ultimately self-defeating utopianism which will cause them to fail, but on a realistic appraisal of human nature. That's what this book is about.

Melanie Phillips
November 1997

INTRODUCTION

In the late 1980s, not long after I had started writing a column for the *Guardian*, I interviewed a young black community worker in Tottenham, an economically deprived part of north London. We were talking about relations between black people and the police when he suddenly startled me with a passionate outburst about education. Black children, he said, were getting a rotten deal in the schools: they were being treated like rubbish. Working as I did for the *Guardian*, where writers and readers routinely condemned the government for its neglect of education, I was used to the complaint that schoolchildren were being short-changed. But what made me sit up and take notice was that this young man was not blaming Margaret Thatcher's government for this state of affairs. He despised those white liberals who wanted to give local councils more money to spend on education. He blamed instead an ideology which meant that the schools were refusing to give young black people like himself the education they needed, the knowledge of maths, language and history which would put them on the same level as white, middle-class children and enable them to compete fairly for a secure place in the world. At the *Guardian*, the prevailing view was that independent schools were a principal source of inequality. But this young man from a highly marginalised community wanted black children in Tottenham to be educated in private schools where he thought they would be taught properly and free of ideology; and he bitterly attacked the local Labour council for blocking the planning permission which would have allowed this to happen. Black youngsters like himself, he said, had been cheated by a liberal culture on a guilt trip.

To me, this was liberalism betrayed. But it was by no means my first indication that something peculiar was happening within the liberal culture of which I considered myself a part. From my earliest

Guardian columns and later when I started writing a column for the *Observer*, I had been engulfed by a tide of venomous fury unleashed by the education establishment whenever I tried to explore the phenomenon of persistent low education standards and expectations. These critics didn't just appear to be reacting irrationally. They were denying clearly demonstrable evidence that harm was being done to children and, even more chillingly, they were themselves advocating doctrines which were inimical to educational values. At the same time, I was receiving a steady stream of letters from teachers writing clandestinely with accounts of how they had to conceal their orderly and successful teaching methods for fear of being victimised. And I was shocked by the passion and desperation with which they urged me to continue to write about such matters.

I encountered even more hostile reactions when I started to explore the breakdown in traditional family life and its effects on children and society. Once again, I discovered that liberal society was denying or misrepresenting evidence of the harm being done to children by the fragmentation of the family. Astonishing and alarming as I found all this, I was even more perplexed by the reaction to what I was writing, not just by the liberal establishment but by a number of friends, colleagues and *Guardian* and *Observer* readers. It rapidly became clear to me that it was almost impossible to discuss this issue in the way other topics were debated. Family breakdown and irregular relationships were already so entrenched within society that more and more apparently 'neutral' commentators were personally compromised. Ostensibly arguing about policy, my critics almost invariably were disguising expressions of their own personal pain, defiance or guilt. Like the educationists, these furious family theorists accused me of having journeyed from the political left to the right, of becoming not only a 'reactionary' but a 'moraliser', the term of abuse which has become for the left what the insult 'do-gooder' has long been for the right.

But was it true? Had I changed my position? I still cared about the issues which had always concerned me: justice and liberty, official lies and the abuse of power and the fate of the most vulnerable and dispossessed. None of those concerns had changed at all. What had changed, however, was my perspective on them. I began to understand that so-called liberal culture no longer embodied tolerance but had become profoundly intolerant, and that its ostensible aim of helping the disadvantaged had become so twisted that it was trapping them instead in that disadvantage. Instead of helping the oppressed, it had itself become oppressive. Why had morality become such a dirty word? Surely, a society without a moral sense couldn't claim to be civilised? And why did the people who supported my opinions themselves feel so beleaguered and fearful of saying in public what they thought? How could there be so much intellectual timidity in a society whose commitment to tolerance and free speech underpinned every public thought and deed? My concern about abuses of power shifted into a different focus. It was no longer just the government which needed invigilating, but its opponents too.

Gradually, it became obvious to me that these explosive divisions of opinion represented nothing less than a culture war between opposing mind-sets. They weren't merely battles over transient and local political issues. Instead, they represented a struggle for the power to define such fundamentals as our personal and national identity, our relationship with each other and with our children, the balance to be struck between rights and responsibilities, and the link between private behaviour and public morality. In short, this was a battle for the very soul of society itself, raging in particular through issues such as education, family structure and crime and punishment.

What also became clear was the frightening gap that had opened up between the public on the one hand and the political and other élites, including the universities and the media, on the other. These people somehow managed to avoid engaging with the issues that so troubled the public; they appeared instead to be intent on pursuing their own arcane and introverted agenda. The result was an alarming level of public cynicism and frustration with the democratic process. Coupled with the rampant individualism which was coursing through society and which characterised political discourse on both left and right, it all seemed to be flashing a set of warning lights about the safety of our liberal democracy.

The other notable feature about this culture war was that it made utterly redundant the old divisions between 'left' and 'right' in politics. If these terms had ever possessed any meaning, that had now disappeared. Since 1979, the Conservative party had not been in the business of conserving; instead, it had presided over a state of permanent revolution. People became as eclectic in their choice of stances on individual issues as they were in their choice of morning newspaper, building up a portfolio of attitudes which defied simple categorisation. They were in favour of expansionary economics and traditional family policies, for example; or in favour of libertarian sexual codes of behaviour and environmental conservation.

Alongside this emphasis on individual choice, there was increasing public disenchantment. More alarmingly, children's distress and disorder were also on the increase. Crime levels, truancy, standards of illiteracy and innumeracy, rates of family breakdown, more and more young people running away from home, schools in a state of permanent crisis: all seemed to testify to some fundamental disorder that went way beyond the utilitarian economic diagnoses and prescriptions which passed for political discourse. It seemed to me that one reason why all these crucial problems were falling outside the political process was that the politics of both the left and the right were heavily implicated in their creation. Both sides of the political divide had conspired to create the 'me-society' of atomised and alienated individuals, a society that saw freedom not as a means but as an end in itself – and then didn't know what to do with it.

This book is an attempt to unravel some of these trends. It seeks to discover how, in an age of unparalleled scientific progress and material prosperity, we have de-moralised our society and put at risk some of our most priceless assets. It provides a detailed analysis of the ways in which we have devalued our children's education: a process not merely of intellectual collapse but of a moral disintegration which now threatens the survival of our culture. This book attempts to show how the doctrines of individualism and sameness have eroded social bonds. It shows how politicians of both left and right, despite their theatrical confrontations, have colluded to create a society indifferent to the plight of others, from neglected children to unemployment. It demonstrates how the crippling of education and the fragmentation of family relationships have further eroded those social bonds and led to widespread crime and disorder. The book concludes with a blueprint for a new type of politics – one which steers a path between the extremes of the state on the one hand and the unfettered market on the other, one which shifts the emphasis from rights to duties. It is a blueprint which encourages us once again to take responsibility for our actions and for each other rather than to pursue our own interests in the economic or social sphere with little heed of personal or social consequences.

The book is an attempt to break the logjam of British politics. It is also a milestone on a personal journey.

In researching this book, I have been helped by a large number of people who have provided me with invaluable information, advice and wisdom. Since a number of them must remain anonymous, it would be invidious to single out anyone for special thanks. But they know who they are, and they have my deepest gratitude and admiration. I am also particularly grateful to Alan Samson at Little, Brown and to Peter Robinson at Curtis Brown for their unflagging enthusiasm, support and good advice. My greatest debt of gratitude, however, must be to my family: to Abigail and Gabriel, for enduring the bad temper and worse cooking of a distracted mother writing a book about good parenting; and above all to Joshua, who read it.

Chapter One

THE DE-EDUCATION OF BRITAIN

Standards Sliding

The University of Oxford is one of the most prestigious universities in the world. Its undergraduates are some of the brightest in Britain and their A-level grades are among the highest. One might imagine, therefore, that these students had been educated.

In December 1993, however, Dr Richard Sheppard, a tutor in German at Oxford, expressed his despair at the standard of knowledge among undergraduates studying for their degrees. Students, he observed, could now only speak pidgin German. One candidate from an independent school didn't know about passive verb forms or past participles, and had never learnt about the genders of German nouns or plural forms. One student spent the three years of his degree unable to learn that the word order in German was not a moveable feast and that word endings were not optional extras. At another university, he reported, a second year student with a B grade at A-level had demonstrated *after* a syntax revision course that he still didn't know the genders of common nouns, the basic rules of German orthography and punctuation, noun and adjectival endings, common strong verb forms, the passive, how to translate 'on' followed by a date from the calendar, what common prepositions took what cases and the position of verbs in subordinate clauses: 'the stuff', lamented Sheppard, 'of O-levels'. Nor was the general knowledge of undergraduates any better. At interviews with prospective candidates from 1991–93, he had had to explain who Homer was twice, what Buchenwald was three times and why Good Friday was not without significance for western civilisation.[1]

Dr Sheppard was not alone in his perceptions. The degree in English at the University of East Anglia enjoys a deservedly high reputation. Its undergraduates have A or B grades at A-level under their belt. Yet, when they arrive, they are presented with a booklet by the university. Entitled *A Guide to Essay Writing, Oral Presentation, Grammar, Punctuation and Related Matters*, it assumes a level of ignorance of the most basic techniques of essay writing and indeed of the English language itself. This is quite remarkable when you consider that all these students have just excelled in public examinations which have required them, it might be assumed, to write such essays to the most rigorous of standards.

The booklet advises them, for example, that essay writing involves 'information gathering, marshalling bodies of data, discussing alternative ways in which issues can be viewed and much else besides'. It suggests to the students that they plan their work. In case this particular advice throws them by its unfamiliarity, it carefully spells out that such a plan can take the form of either notes or diagram. It advises them that an essay must have a beginning, an argument and an end. It tells them, as if for the first time, what a paragraph is. It teaches them how to use the colon, the semicolon and the apostrophe. It shows them how to avoid falling into error with the preposition, the pronoun and the full stop. And it helpfully provides a list of words they are likely to misspell or confuse, including 'their and there', 'eligible and illegible', 'persecute and prosecute', 'who's and whose,' 'separate', 'similar' and 'seize'.[2]

Much of this would once have been taught to schoolchildren at primary school, or certainly by the first couple of years of secondary school. Yet now it cannot be assumed that university degree students know it. What kind of undergraduates are these who need to be taught such elementary points? How could they have achieved their high grades at A-level, the 'gold standard' of the British school examination system? If English degree students don't even know what an essay is, or the rudiments of grammar, punctuation and spelling, what must the educational standards of young people with less academic potential be like?

It must be said that this problem is not particular to students at the University of East Anglia. The phenomenon is universal. So bad have things got that a few years ago a proposal for an A-level syllabus suggested that students who had difficulties in writing essays should present their material in the form of paintings, dramatic performance, music or other media.[3] The proposal was turned down; but the fact that it was made at all indicates that something very disturbing is taking place in the upper reaches of education.

Many university dons are in a state of utter despair about the low levels of knowledge presented by the young people turning up to read for their degrees. Alarm about education standards is by no means confined, however, to the universities. It extends throughout the education system and is reinforced almost with every official report that is published. There is evidence that something more significant and far-reaching has been happening than any politician has yet managed to identify, let alone address – and that it extends far beyond education.

There is now a yawning gap between the standards reached by British schoolchildren and their counterparts in Europe or Japan. A report by the National Institute of Economic and Social Research, published in 1995, found that the bottom 40 per cent of English 13-year-olds lagged two years behind their German counterparts, while fewer than 30 per cent of Britain's workforce had vocational qualifications, compared to more than 60 per cent in Germany.[4] In 1990–91, the proportion of 16-year-olds gaining the equivalent of three GCSE grades between A and C in maths, the national language and one science was 62 per cent in Germany, 66 per cent in France, 50 per cent in Japan and 27 per cent in England. In 1990, the proportion of young people obtaining a comparable school qualification at 18 was 68 per cent in Germany, 48 per cent in France, 80 per cent in Japan and 29 per cent in England.[5]

There is constant concern about standards of literacy and numeracy among the population at large. Employers consistently complain that many job applicants cannot even write a correctly spelled and punctuated letter. How can it be, in a modern, advanced economy which has had universal education since 1944, that so many of its citizens, including those who have been 'well educated', are so illiterate?

What's more, the problems appear to have been getting steadily worse. In 1995, secondary school head teachers reported a sharp and accelerating drop in standards, particularly in reading, among 11-year-old pupils coming up from the primary schools. Their survey of nearly 500 state and independent schools found that standards of reading, spelling, comprehension and literacy had fallen continuously between 1991 and 1994. 'We are shocked,' said Dr John Dunford, president of the Secondary Heads Association. 'It's clear that literacy standards in primary schools are falling.'[6] The finger was pointed at failures by the primary schools, which tried to defend themselves by blaming the extra workload imposed by the National Curriculum. But this excuse seemed implausible. As Kath Brooke, head of Garth Hill comprehensive school in Berkshire commented: 'Our verbal reasoning score has dropped over the past 10 years from 104 to 93. The decline started long before the National Curriculum was introduced.'[7]

Certainly, the National Curriculum tests have revealed significant problems in the primary schools, particularly in the more senior classes. While a core of about 20 per cent of 7-year-olds have failed to reach the standard for their age since the first of these tests for their age group was held in 1991, 11-year-olds have fared worse. The first tests for 11-year-olds held in 1995 revealed that more than half were not up to scratch in English or maths.[8] The poor results for 11-year-olds produced a flood of excuses from primary teachers. The government drew comfort from the fact that the results at the ages of 7 and 14 had been better. But a report by the School Curriculum and Assessment Authority suggested that the malaise in education was very much deeper.

The report revealed that some teachers who marked the 1995 English tests for 14-year-olds had such a poor knowledge of Shakespeare's plays themselves that they gave pupils the wrong grades. Some markers were confused about who were Montagues and who were Capulets in *Romeo and Juliet*, marking wrong answers as right. Some gave pupils marks for naming figures of speech which were patent gobbledegook. One marker, for example, ticked one pupil's script when she wrote that Romeo used 'Simples and Metfords' instead of similes and metaphors.[9] How can it possibly have come about that current or former English teachers can themselves be so ignorant of Shakespeare's plays – which, if they've forgotten, they only have to re-read before marking any scripts – and so indifferent to rank gibberish? What can be happening to British society?

Whatever reservations might be expressed about the curriculum tests, which have after all taken time to bed down, the evidence is overwhelming that standards throughout the education system, from infant classes right through to degree level, give cause for the most intense concern. Statistics are an erratic guide to any deterioration, since changes in criteria or collection methods make comparisons notoriously unreliable. What can be said with confidence, however, is that the standard of knowledge of many thousands of schoolchildren, adults and even teachers is lamentable. Reports by the national education inspectors identify about 30 per cent of lessons as unsatisfactory. But even 'satisfactory', by the inspectors' standards, is actually pretty mediocre. The unpalatable fact is that about two-thirds of British schools simply aren't good enough. There *are* good schools and good teachers; but they are fighting a cultural tide that washes well beyond the shores of their schools.

The rot sets in at primary school level and runs through the system. In 1990, a survey of 500 pupils in Cheshire primary schools found that one fifth of them had difficulty forming their letters and writing legibly. But the difficulty wasn't confined to the children. Rhona Stainthorp, lecturer in education at Reading University, said: 'Some of our students have never been taught cursive script at school themselves, and they are very uncertain about it. We laid on handwriting workshops this term, but nobody turned up.'[10] When even trainee teachers can't do joined-up writing, clearly something profound and disturbing has happened.

Most of the political controversy over education concerns the performance of children in state maintained schools. Every time evidence is produced that this performance isn't good enough, however, its validity is denied or it is dismissed as irrelevant, or – if it is accepted – the blame is always shifted onto someone else. The government blames the teachers, the teachers blame the parents and the parents blame the television. Some people blame it all upon the independent schools for existing. All that is left tends to be confusion, an irritation at political bad-mouthing and a vague feeling that things aren't right.

There is also an impression that the problems, such as they are, are concentrated at the lower ability end of the population. The top end of the British ability range has always maintained the highest standards. No problems there; or so people fondly believe. So the middle classes play the system as they always have done, moving house to be in the catchment areas of good schools where there are greater numbers of high ability children. This means that their children are likely to get good enough exam grades to go onto higher education and thus sidestep the difficulties that beset British schools. The assumption is that such a predictable progression is proof in itself that the problems with education are localised – bad teachers, bad schools, bad parents – while reliable benchmarks of quality persist at the end of the process: the high A-level grades, the university degree. If children can just get through to those sunlit uplands, people believe, then all will be well.

But this complacency is grossly misplaced. The picture thus painted is simply not true. The difficulties in British education cannot now be sidestepped by anyone – neither by those who play the state system to their own advantage, nor by those who use the independent schools. Higher education is in a state of meltdown. What is happening is nothing less than a corruption of the very nature of education itself. It has spread throughout the system and has swept along with it Britain's brightest and best, the highest achievers and their university teachers. It is not just that university students can no longer spell or punctuate. In some subjects taught at degree level, the gaps in students' knowledge are so fundamental that university dons who can scarcely believe the evidence of their own eyes have had to dilute or extend their degree courses just to cope.

Languages

John Gordon, a German specialist and the admissions tutor in modern languages at the University of East Anglia, said in 1995 that his own subject had been particularly badly hit. Although other subjects had also been affected, there had been an inexorable decline in German since 1989.

> It's clear that people whose German is quite incompetent are being awarded As, Bs and Cs at A-level. There's absolutely no question about that. We've not been hit as badly as maths and physics, but worse than French or Spanish; German is the only widely taught modern language with a case system

and where the word order is very different from English, and above all these students lack knowledge of grammar. But there's also unease in history, where opinion is divided over declining standards. And in physics, students just don't have the maths, so it's commonplace for the first year or so to be dominated by remedial maths and physics.

In modern languages the standard of linguistic proficiency in an average 2:1 first degree now is often no more than that of an A-level-plus of 20 years ago. At the preliminary exams in French, they're just getting them to read French in very short poems and stories. Twenty years ago they were reading many major authors in German. Now they have to read them in translation. Most are quite incapable of writing a competent essay. Universities have to a large extent become the secondary schools' remedial sector.

We're finding this applies to students from both state and independent schools. We are doing fairly widespread remedial grammar teaching. A few years ago, this was unknown. People always made mistakes but what is happening now is grotesque. We always had grammar classes but five to ten years ago we used to be dealing largely with questions of style. Now we're teaching first year undergraduates things like 'the teacher gave the child a book', things that should have been taught in their first year of German.[11]

Gordon set out the situation starkly in an article entitled 'Kannitverstan' – an ironic reference in pidgin German to the students' lack of comprehension. In it, he gave details of a test given in the first week of the 1994 autumn term to first year students who were doing German as their main subject or as part of their degree. Of these students, 37 out of 43 had A and B grades at A-level. The sentences they were given to translate were very simple. These were the correct scores: 'I like to drink Chinese tea': 7 out of 43; 'I prefer to drink strong coffee': 4 out of 43; 'The train you came on was 20 minutes late': 10 out of 43; 'It was not even tried': 8 out of 43; 'They can be helped': 8 out of 43; 'Her father is a doctor': 30 out of 43; 'Our friend was injured by a falling stone': 8 out of 43; 'The people we met yesterday were quite friendly': 22 out of 43; 'Have you ever been to Germany?': 9 out of 43; 'The teacher gave the pupil the book': 8 out of 43.[12]

On the last sentence, Gordon commented: 'This sentence was once widely used to introduce learners to the use of the indirect object. Until recently the use of this sentence in a test after the first year or so of German would have been regarded as some kind of jest; and its inclusion in any sort of post A-level test at university or anywhere else would have been quite unthinkable: it would have been regarded as an insult to the students.'[13]

He concluded that there was no point in using A-level German as a criterion for admission to university German degree programmes. The exam boards were simply doling out a placebo.

It is very much open to question how many of the students have any systematic knowledge of the language at all: and haphazard, disorganised knowledge – whether of a language or any other complex phenomenon – is worthless.[14]

Another tutor in German, this time at Oxford, confirmed the bleak assessment. The paper on prose comprehension and composition in the University's German first year examination in 1993 revealed that half the undergraduates had little control over the language.

Errors were made with astonishing consistency in declension, cases, word order and number. Many students were incapable of making any correct use of the German given to them in the passage. I do a diagnostic test on new undergraduates every year. The results are getting worse year by year, even though most of these candidates arrive with A grades. Five years ago the independent schools were free from this, but no longer. I'm beginning to see the distinction crumble. The worst performance this year was from one of the major public schools whose German teacher didn't believe in grammar.[15]

The retreat from grammar is a phenomenon that underpins much of the crisis in modern language teaching. It stems from the belief that children pick up the codes of language by a kind of osmosis, and that to teach them those codes explicitly can actually do them harm. Its jargon term is the 'communicative' or 'direct' method of teaching. Originating in Britain in teachers' perception that schoolchildren used to be taught literary exercises but not how to speak foreign languages, it has now achieved the status of an obsession. The result is that teachers who want to teach grammar often feel so intimidated by the virulent opposition they encounter that they resort to doing it in secret, in their own time or not at all. 'We're hearing more and more that teachers don't believe in grammar or that grammar lessons had to be held in lunchtimes so the students never went,' said the Oxford tutor. 'Fewer and fewer teachers can teach it anyway. I've had to explain what a pronoun is to undergraduates, and they often don't know the difference between an adjective and an adverb.'[16]

Another lecturer commented: 'What's depressing talking to teachers is to hear how their ethos doesn't allow them to do the things they believe in. Teachers say to us: "I'd love to insist the children learn their vocabulary or set aside time to teach them formal grammar but I can't, I dare not, and if I go for another job I dare not say this is what I believe should be taught because I simply won't then get the job."'[17]

The lengths to which this anti-grammar obsession is being taken and the reasoning behind that ideology are explained in greater detail later in the book. What is notable, however, is not just the extraordinary degree of damage that has been done by this flight from language teaching. It is also that language teachers are so committed to the anti-grammar approach that they are denying the evidence of their own eyes. Irrationality has taken over. Derek McCulloch, a German tutor at Surrey University, was howled down when he attempted to lay such evidence before his colleagues. His outrage was heightened by the fact that he himself was once a supporter of the approach which appeared to give the spoken language a higher priority.

Now, *anything* that passes for communication is considered good. There's a 'good enough' philosophy in the schools. My students can't understand German word order. They don't understand who is doing what to whom in a sentence. For years I've been giving them a Heinrich Böll short story which starts: 'I was standing in the harbour looking at the gulls when a policeman noticed my face.' One after another, these students write in German: 'My face noticed the policeman.' They can't see the grammatical difference between 'He has a bad teacher' and 'He is a bad teacher'. In 1993, 31 out of 36 first year students couldn't write the latter sentence in German.[18]

McCulloch conducted a test on his own students and found that out of more than 40 students with A and B grades at A-level, hardly any could translate correctly the phrase: 'Please close the window.' All but two found that translating: 'The train she came on was late' was quite beyond them. A further test was devised using twelve sentences from everyday life, such as: 'My friend was bitten by her father's dog'; 'Is his wife a German?' 'I don't believe him'; 'Because of the bad weather I stayed at home'; 'I don't like Austrian coffee'; 'I prefer English tea'. The test was taken by 570 students, of whom the overwhelming majority didn't get even half of the sentences right and several got as many as 10 out of 12 wrong. Fewer than one out of 30 knew the German word for Austrian and hardly more than one out of 100 spelt it correctly or got the adjectival ending right. And more than half didn't know when to use *du* or *Sie*, the singular/informal and plural/formal forms of the pronoun 'you'. They were quite unable to extrapolate general rules from individual phrases.

Phrases seem to exist only in their own right. One group revealed that hardly any had been taught the perfect tense of verbs such as 'to be able' or 'to want'. And the subjunctive was played down as somewhere between an unnecessary complication and total irrelevance. The students say accuracy doesn't matter as long as you can be understood. Teachers appear to be working round the clock, finding materials, videoing TV programmes and so on. But the result is students who have discussed urban pollution but do not know the plural of *Stadt*; students who can list the political parties but do not know the gender of *Partei*; students who have discussed *Gleichberechtigung* and *Emanzipation* but cannot produce *Frauen* as the plural of the word for woman.

What's more, added the Oxford tutor, these students were even less able to speak the language than ever before, even though this was the ostensible point of this method of teaching. 'It's not my impression that these children are now any better at actually speaking these languages,' he said. 'If they speak them well, they've usually either got a foreign parent or have spent time in that country. The fact is they are very unsure how to string words together.'

McCulloch laid out these concerns at a conference of university teachers of German held in Britain in 1995. He was pulled apart by his colleagues. 'When I said all this, I was charged with espousing a right-wing agenda and accused of being dangerous!' But how can it be that university professors and lecturers, faced with a disaster of this magnitude, appear neither to know about it nor seem unduly concerned if they do? The answer appears to be a dispiriting mixture of ivory-tower ignorance and complacency, along with a spirit of defeatist time-serving and a dash of ideological zeal.

'A lot of university teachers don't know what's happening,' said the Oxford tutor. 'They can't really believe it. When I tell my German friends, they say surely this must be an English joke. The direct method was prominent on the continent between the wars and in the 1950s but then they threw it out because it hadn't worked.'

The greatest tensions are between the language teachers themselves and the linguisticians, or teachers of language education. 'The linguisticians say that students aren't bad at grammar, only morphology [the study of the forms of words], as if morphology were not part of grammar; and even if their whole diagnosis was true, which it is not,' said McCulloch.

Of the language dons themselves, the Oxford tutor said that some were too grand to do much teaching and therefore assumed their students were reading foreign literary works in the original, which they were not; a few were blinded by ideological zeal; and the great majority were keeping their heads down, adapting as best they could to the new situation they found themselves confronting, adjusting their degree courses accordingly and looking forward to their retirement.

'This ideology appeared to offer better results very quickly,' said the Oxford don. 'The old top-down system was imposed by the universities and worked to benefit the few. Now it has been designed to cater for the less able. And no-one wants to offend the teachers. The right-wing press goes in for a lot of teacher-bashing. We all really want to be liberal. We all have difficulty with the fact that something set up to be liberal is working against itself.'

Here in a nutshell is one of the driving impulses behind the thinking that has caused educational standards across the board to implode. At some point in the last few decades, the educational world came to agree that its overriding priority was to make children feel good about themselves: none of them should feel inferior to anyone else or a failure. At the same time, such people came to believe that children from relatively impoverished backgrounds, who unarguably started at a clear disadvantage, were somehow incapable of learning what other, more forward, children could learn. There was, of course, not a shred of evidence for such a belief. What disadvantaged children needed above all was more structured teaching, greater attention paid to those elementary rules of language or of arithmetic and a heavier emphasis on order. These were all features which were second nature to those children from more favoured homes but which tended to be lacking in their own.

But the educational world, heavily influenced by other profound currents of thinking which all conspired to undermine every form of external authority (and which are examined in more detail later), decided in its wisdom that disadvantaged children simply couldn't learn those 'difficult' things. Moreover, since it now held that no child was allowed to do better than any other, it decided that *no* child would learn them. Thus was created an examination system – the GCSE – which was structured so that many more children would be able to pass it. The only way the education establishment could attempt to square the circle it had set itself was by lowering the standard. And the teaching of individual subjects was similarly refined in such a way that it appeared no child could fail to cope with them. So no rules of language, for example, were to be taught; after all, if no rules were taught, no child could fail to learn them. Nothing was to be difficult; everything in the education garden was to be fun. The uncomfortable truth that little of value is achieved without effort, in education as elsewhere, was decried as a form of child abuse.

But circles, of course, cannot be squared and the attempt was accordingly misguided. It claimed to elevate equality to the highest virtue. But what it actually imposed, through the most doctrinaire means possible, was an equality of ignorance and under-achievement. It meant not only that *every* child was to be equally uneducated, not only that the brightest were stripped both of knowledge and of those challenges that would stretch them, but that the most disadvantaged children, those very children for whose benefit this was all ostensibly conceived, were left worst off of all – effectively abandoned and trapped in disadvantage by their grievous lack of a proper education. It was, indeed, a liberal ideal that was now in the process of destroying itself.

Mathematics

The subject that has suffered perhaps most unequivocally from the collapse of the authority of rules is mathematics. Maths teaching in Britain has effectively been deconstructed. A fundamental shift in emphasis away from knowledge transmitted by the teacher to skills and process 'discovered' by the child has undermined the fundamental premises of mathematics itself. The absolutes of exactness and proof on which the subject is based have been replaced by approximation, guesswork and context. The result has been disastrous and the effects have been felt in university maths departments up and down the land. Students, say their tutors, know far less than they used to, are less confident at handling simple mathematical expressions and need a great deal of spoon-feeding. Many departments have had to alter their teaching as a result. Cambridge University simplified its maths syllabus, traditionally the most difficult in the country, because of reduced knowledge by candidates. 'Cambridge can no longer expect the same mathematical knowledge and level of technical skill from undergraduates arriving in future years,' said a university spokesman.[19]

One maths teacher at the City of London independent school wrote of his horror at the decline in standards set in train by the GCSE exam. 'We now have a system in which children are asked to work out 17.2×0.0051 (calculators allowed). I remember doing sums like this when I was nine (calculators not invented). Time-consuming coursework has excluded topics such as formal geometry, calculus and harder algebraic manipulation. The written examination is now so easy that it has become necessary to emasculate A-level.'[20]

Easy for some, maybe. Others would claim that at GCSE at any rate, maths teaching has become seriously incoherent. Children are being expected to master higher order procedures in maths before showing they have properly absorbed less advanced material. The new approach has produced a bizarre mishmash of the very easy and the absurdly difficult. The harder algebra may have disappeared, but in its place have come statistics, set theory and probability. One of the key problems for children in maths is that, as in so many other areas of the curriculum, they are being expected to run before they can walk. They are suffering from a problem of disorder caused by the abolition of structure, imposed upon them by an adult world which appears to have forgotten that they are children and have an absolute need for structures if they are to thrive – a phenomenon which, as we shall see later, now extends well beyond the educational arena.

Like the modern linguists, the mathematicians were remarkably slow in bringing their anxieties to public attention. When they did so, however, they produced a devastating account of a qualitative change that had sent maths teaching into a spiral of decline. An authoritative report in 1995 by the London Mathematical Society and others identified three main problems: a serious lack of ability to undertake numerical and algebraic calculation with fluency and accuracy; a marked decline in analytical powers when faced with simple problems requiring more than one step; and a changed perception of what mathematics was, particularly the place of precision and proof.[21]

The emphasis on the practical application of maths and the obsession with presenting problems 'in context' – in other words, relating them to real life situations – had denigrated the primary importance of maths as a training for the mind. Schools, said the report, had shifted from teaching core techniques to time-consuming activities such as investigations, problem-solving and data surveys. Many of these were poorly focused and could obscure the underlying maths. The importance attached to process failed to recognise that to gain genuine understanding it was necessary first to achieve 'robust technical fluency', reducing familiar laborious processes to automatic mental routines to allow a progression to new unfamiliar ideas. Children were no longer being taught properly arithmetic, fractions, ratios, algebraic technique and basic geometry. Instead, they were much more dependent on calculators and computers.

Evidence that many pupils couldn't solve problems involving fractions, decimals, ratio, proportion or algebra, it said, was being interpreted as meaning that such topics were 'too hard' for most pupils. But this ignored the fact that most of these techniques were fundamental and could not be neglected, and that many countries where children started school later than they did in England somehow managed to teach these topics effectively to far larger proportions of their pupils.

At a meeting of the Conservative backbench education committee in 1995, Peter Saunders, professor of maths at King's College, London said: 'I don't see how you can teach skills and operations without facts and techniques. How can you operate with nothing to operate on? How do you teach any mathematics when the students lack the fluency in technique and the knowledge of the previous levels to follow what you are saying? It's hard to learn fractions if you aren't confident with arithmetic, it's hard to learn algebra if you aren't good at fractions and it's hard to learn calculus if you are still uncomfortable with algebra. It's also hard to navigate your car through a strange town if you haven't yet learned to change gears without thinking about it.'[22]

Despite this alarming evidence of the de-skilling of Britain, however, the people who taught mathematics teachers remained unabashed and defiant. As in so many subject areas, the retreat from knowledge into subjectivity had been driven along by the educationists in the universities, people who taught not maths itself but maths education. Like the linguisticians who fought with the modern languages professors, the teachers of maths education set themselves at odds with the lecturers and professors of maths itself. The educationists' absolute horror of 'rote leaning', repetition and memory work meant that they were fundamentally opposed to the very techniques which were essential for children to achieve mathematical fluency.

Margaret Brown, professor of maths education at King's College, London, and chairman of the Joint Mathematical Council, is the doyenne of the 'maths must be easy and fun' school of thought. She claimed:

> There is considerable evidence that those who give up mathematics do so because they perceive it as 'hard' or 'boring', that is, having little significance for them. Any recipe for

raising standards must not make the subject even 'harder' and more 'boring' or even fewer students will continue with it, and hence even lower standards will pertain . . . Ignorant diagnoses, such as that the national curriculum omits multiplication tables, algebraic manipulation and the process of proof, or that teachers have been brainwashed by trendy educationists into ignoring basic skills while spending their time on sloppy-minded investigations, are wide of the mark.[23]

But they weren't wide of the mark at all. Peter Saunders explained the chasm of view between those who taught maths and those who taught teachers to teach maths.

The problem isn't just a difference of view about how to teach maths. It's a fundamental disagreement about what the subject actually is. It's the maths education people who created the problem and won't admit it. They think maths is about investigation and we think it's about proof. They think proof, or axioms, isn't very important. Proof is what makes maths different from science. You should know the difference between knowing something and proving it. When challenged, the maths education people say most children never understood proof anyway. And now I've interviewed someone for a post as a maths education lecturer who had never even *heard* of a proof.[24]

Now, said Saunders, children lacked manipulative skills in algebra:

They're taught a bit of algebra but maths is like football: they've got to practise. There's a feeling that if you ask a kid to do something, that's enough, that practice is drudgery. There's this fear of boring the children. But that's a challenge for the teacher. The other thing they can't do is solve problems. The way the work is set out they're never asked to solve a problem. Euclidean geometry meant if you couldn't see the answer straight away you had to play with it for a bit and you could see what would happen. But they are not learning this. Exams now require pupils to show this and show that and they get marks for everything they can do. Our students throw equals signs around like no-one's business. They have no idea any more what they mean.

The reason for that, suggested Tony Gardiner, a maths lecturer at Birmingham University, was that one of the things that had gone dramatically wrong in the teaching of maths was the abuse of the calculator. Maths, he noted, rested on two premises: the basic language of the algebra of expressions, and the fact that the objects of this language and the methods used to analyse and transform them were in principle absolutely exact. But those who grew up using a calculator without developing the necessary mathematical instincts interpreted all expressions as potential recipes, not as algebraic expressions to be simplified and understood.

In mathematics, the = symbol conveys a moral message. It not only declares that the left-hand side is exactly equal to the right-hand side, but also imposes on the person who writes it the responsibility of being able to explain how one can transform one side (strictly according to the rules!) to the other side. In contrast, the equals button on the calculator is a completely different animal. Its nearest equivalent in everyday language is the magician's 'abracadabra'. Far from requiring the user to understand the connection between cause (= input) and effect (= output in the display), the equals button is almost guaranteed, like the magician's utterance, to focus attention on the effect and so distract the observer from looking for the true cause.[25]

For Tony Gardiner, maths was being reduced to an experimental bag of tricks, with calculators giving answers that were 'more or less correct' which pupils catastrophically mistook for mathematical behaviour. 'The calculator here is no longer an aid; instead it controls, obscures and distorts the meaning of the symbols and of the operations. The medium has become the message. In the space of less than a decade the key notion that mathematics is reliable and important because its calculations and methods are exact in principle has been thrown away. It will not be easy to recover.'[26] Yet under the National Curriculum calculators are compulsory in primary schools. In some European countries schoolchildren are not allowed to use them at all until they have mastered the mathematical concepts involved.

Gardiner argued that the main problem was a change in the understanding of education itself, from the concept of transmitting knowledge to teaching children 'how to learn'. 'Many teachers (and advisers, and HMIs [Her Majesty's Inspectors of Schools] and examiners, and educationalists) no longer understand that, or why, fluency in basic topics – such as numerical fluency, or fractions, or proportion, or congruence and similarity, or algebra – is fundamental to progress in elementary mathematics for all pupils . . . there has been a tendency to emphasise so-called "skills" as opposed to "knowledge" – ignoring the fact that "skills" can only be learned in a sufficiently rich context with a suitable knowledge base.'

The National Curriculum, he added, hadn't helped but had merely institutionalised the problem. It had

redefined the nature of proof, imposing as part of the compulsory national curriculum an educational sequence whereby students are encouraged, up to the age of 15 or 16, to believe that mathematical proof is based on a discredited and totally inappropriate kind of induction . . . The official curriculum then suggests (wrongly) that, around the age of 16, brighter pupils (less than 10 per cent) will suddenly begin to appreciate the superiority of deduction. In practice, there is very little prospect that most students who have been systematically deceived into thinking that mathematical proof is based on experimental evidence will ever make such a transition. The results are apparent to any observer. Our 18-year-olds no longer have any inkling of the fact that, if mathematical assertions are to have any value at all, statements must be formulated carefully and manipulated in precise, reliable, strictly defensible ways.

The first year university lecturer is thus confronted with a very genial anarchy. The students are pleasant enough; but they are mathematically amoral. The main problem is not that they cannot construct proofs; they do not understand what even the very simplest proofs are trying to achieve – what it is that makes them different from a random list of 'vaguely relevant' statements. They have no mental schemes that will allow them to apprehend what is significant in a simple proof presented to them. Basic errors can be repeatedly drawn to their attention without any perceptible change in their subsequent behaviour. They are reasonably intelligent, but have been deprived of the key components of the necessary mathematical (and educational) diet. Many are completely untroubled by contradictions, even with very simple material.[27]

The evidence provided by these subject specialists is devastating. Young people are arriving at university stripped of the basic knowledge of the subject they are to study at an advanced level. They display a risible standard of achievement and a failure to grasp the most elementary aspects of their chosen subject. And yet they have gained excellent grades in public examinations. Furthermore, every year an increasing number of young people gain high grades at both GCSE and A-level. Overall, the proportion of public exam successes continues to rise seamlessly, causing successive Education Secretaries to claim that education standards are improving all the time (despite the fact that those same Education Secretaries simultaneously lambaste teachers for failing their pupils). So how can education support these apparent contradictions? How can it be that university students whose grasp of German is pitiful have gained A and B grades at A-level? How is it that maths students are gaining A and B grades at A-level even though they no longer understand the meaning of an equals sign?

Examinations

The maths dons are in no doubt that public exam standards have slipped. Their report observed that there was 'clear evidence of grade dilution' in the sharp rises in the numbers getting GCSE grades A to C in the last decade. In their view, the mathematical knowledge required to get an A* grade at GCSE, they added, in no way corresponded to what was needed to get a good grade in its predecessor, the GCE additional maths paper.[28]

The modern linguists have also been attempting to unravel the puzzle. The Oxford German don quoted earlier was so troubled that he visited one of the exam boards to find out why undergraduates with high A-level grades were doing so badly in his tests.

> I looked at a comprehension that had been set from a newspaper article. I didn't understand the headline on the article. The exam board administrator said she didn't understand it either. So how could they include such a headline in the paper if there was no chance the students could understand it? The answer was they *weren't expected to understand it*. The board merely expected them to skim the text! The school teachers marking these A-levels are thinking 'there but for the grace of God go my own pupils'. I asked seven A-level boards how much attention they paid to syntax. Only one of them answered my question directly, and that one attached the lowest value to the acquisition of syntax and formal grammar.[29]

This don, like many others, was left mystified but highly suspicious of the exam boards' criteria and processes. There is now disturbing evidence, however, that the integrity of the public examinations has been corrupted by a deadly combination of educational ideology and crude market forces. It is a complex picture, involving a continuous decline in what the exam papers actually expect pupils to achieve along with a constant massaging up of the proportions of pupils gaining high grades – a sleight of hand connived at by the government which, for all its professed concern about educational standards, cannot afford to admit that they have fallen during its term of office.

David Horrocks is a lecturer in German at Keele University and the former chief examiner for the Oxford and Cambridge A-level board, which until it closed down was the most prestigious board of all because it was so heavily used by the highly academic independent schools. In 1990, however, Horrocks resigned his post in protest against the continuous manipulation of the content and grades of its German A-level exam to produce the false impression that more and more pupils were doing better and better when in fact they were producing 'a grotesquely comic pidgin German which in no sense deserves to be considered Advanced level work'.[30] He has painted a devastating picture of the pressures that drove him to resign.[31]

The demands being made of pupils by the boards have become less tight. I could produce a piece of translation done for an exam five years ago which could have been used at O-level 15 years ago. To some extent, that's been camouflaged by the boards making it appear that their demands are more varied and broader. For example, they can say that they've introduced 'listening comprehensions'. But one big thing that's gone from pretty well every board is translation from English into German. The reasons given for that disappearance are spurious: that it's not a realistic test and that pupils wouldn't be asked to do this in real life. But the real reason they've dropped it is that it's the acid test. It disguises whole areas of weakness, not just in grammar but also the fact that students can get away with a minimal vocabulary.

Marking has also been very important in camouflaging what was happening. The big change happened many years ago with pressure from the Schools Council [an early and less powerful curriculum advisory body] to introduce positive mark schemes rather than negative ones. This meant that in marking a translation, for example, you didn't deduct marks for every mistake but carved the piece up into chunks and marked each chunk positively. This had the effect of bunching the marks: you didn't lose as many marks and that had an enormous effect in bringing a lot more people into the pass grades. Most boards have very rigid mark schemes for language. These mark schemes are devised not to penalise ignorance and error. You are told you must ignore this or that mistake. If someone in a cluster of say three words not only gets them wrong but makes a fundamental error of an adjectival ending, that ending is immaterial now whereas it would once have counted as an error in its own right. Really fundamental things like genders and plurals no longer carry the weight they used to.

This was more or less imposed by fiat from outside. That was when I started to despair. It was imposed quite cynically to get more people passing the exams. The reason actually given was the usual psychological argument that it was damaging to pupils to concentrate on their mistakes. We were told in circulars this was 'best practice'. I tried to argue against it with these people, but unsuccessfully. It was a new orthodoxy, largely coming from the educationists in universities who didn't understand the subject.

This combination of the ideology of being kind to pupils and the political imperative of giving the appearance of higher educational standards received added impetus from the developing pressures of market forces. During the 1980s, as schools were given more autonomy in running their own affairs, the examination boards were sucked into a market that became more and more competitive. Schools started to shop around quite cynically for exam boards that would give their pupils a greater chance of success – in other words, for boards with lower standards. Later, when league tables were introduced by which parents could compare schools' exam results, that competition was to become cut-throat. So the boards responded in kind.

'Even the Oxford and Cambridge board was dropping its standards,' said Horrocks. 'They were in the market and they were losing candidates. I wrote to the secretary of the board saying I wasn't going to collaborate with giving A-level grades to pupils who weren't capable of more than pidgin German. I knew what I was talking about because I had seen the papers. I took part in a national exercise comparing A-level boards. There were utterly differing practices between them. There was a general pattern of a reduction in standards which was hidden away in the mark schemes. The pressure to have roughly the same percentage of pupils getting the same grades every year was very strong.' As a result, the boards put pressure on the examiners when they came to decide on the borderlines of the grades for any one year.

> There would be a fair amount of juggling to get the percentage that was satisfactory.
> People are under great pressure from the boards, the market, the educationists and from other subjects. If pupils see you can get a decent grade in a subject where there's hardly any disciplined learning, they will do that subject. We were constantly told this would happen and so we had to make these adaptations to keep our subject alive. Quite a few teachers would say this to us. But they knew children were coming to do A-level German who simply were not fitted to do it. In terms of their intelligence and potential, today's

undergraduates are as good as anyone; but they are not being given the tool kit.

Horrocks is not alone in his assessment. The London Mathematical Society's report observed that 'league tables encourage schools to move their examination entries to what are seen as "less demanding" examination boards – and the working group was supplied with compelling evidence that this was happening'.[32]

The exam boards themselves reject any such implication. They maintain that their standards are uniform, and they vehemently deny any deliberate manipulation of standards or results. George Turnbull, the Associated Examining Board's press officer, told the Channel Four *Dispatches* programme that there was 'no evidence to suggest' that some exams were easier than others.[33] However, research carried out by Professor David Burghes of Exeter University's Centre for Innovation in Mathematics Teaching revealed variations in standards between maths GCSEs set by two different boards which were sufficiently pronounced to 'give rise to real concerns about the fairness of the current system of having more than one exam group, and different ways of achieving a particular grade'. The authors were 'surprised, if not alarmed, about the clear inconsistencies in grading indicated here'.[34]

National Curriculum

Like the universities, it appears that the examination boards have been both reacting to and themselves contributing towards a drifting de-educational tide which has washed both upwards from the schools and downwards from the universities. This tide has left many good teachers stranded, unable to teach either what they believe their pupils should be taught or in the way they think they should be taught it. Their problems have not been eased by the introduction of the National Curriculum. One French teacher at a rural comprehensive school said both the curriculum and the GCSE were reinforcing the 'communicative' orthodoxy that in her view had all but destroyed children's ability to learn French properly. In particular, the emphasis in the curriculum upon teaching children French *in* French, the 'target' language, meant that most children simply never picked up the rudimentary codes to enable them to manipulate the language adequately.[35] Later, we shall see in detail how the National Curriculum, which was introduced in order to raise standards, has connived at their decline. But this teacher's account illustrates precisely why it is that after years of education in a foreign language, children may not have mastered even the basics.

> The curriculum tells us to teach only in the target language to make them develop the necessary skills; they have to listen, speak, read and write in that language. But you don't teach them the rudiments of the language and if you do teach them any grammar at all it will be after they've been exposed to these patterns. Then you may clarify these patterns for them but only in a very descriptive way. After three years, the children's performance has not improved and their competence is non-existent. They've not analysed any of the patterns to which they've been exposed, and the whole process means that halfway through the curriculum they've got such little understanding of the language they will go nowhere.

There's actually little chance these children will understand French language patterns since they have never been taught English sentence structure either. But they're told this doesn't matter. 'They've been told that learning a foreign language is simply about getting the gist. So we're teaching the children constantly to interpret the text, like look and see, interpreting the pictures on the page. It's treating language as a puzzle. There's no structure, no boundaries; the children just pick and mix.'

In other words, teaching a body of knowledge about the language has given way to a process in which pupils 'learn' by guesswork and approximation through a series of decontextualised, unconnected episodes. The notion of correctness or accuracy has disappeared. As long as they 'get the gist', that's good enough. The approach has all the integrity of a cheap ready-made furniture kit; it might enable you to construct a rather tacky wardrobe, but it doesn't give you the wherewithal to go away and make a chest of drawers from scratch. Just as the public exam system awards marks in 'chunks', so pupils are having to learn foreign languages in ready-made segments. This, of course, singularly fails to equip them to deal with the requirements of a language as they spontaneously arise. Despite its accompanying rhetoric of reaching out to the less able, this is nothing short of a radical method of dis-empowering *all* children.

Not only were pupils unable to understand parts of speech and sentence structure, said this French teacher, but they were being taught from a highly limited register of words and phrases which replicated their own culture and would enable them merely to get by as tourists. 'We're bringing up a generation of children who may be able to order a Coke in a French café but who can't deal with situations as they arise. The emphasis in the textbooks is on words found in restaurant menus, signs or posters. They are using words derived from English that the French actually forbid; or teenage expressions such as *moche* (naff), or *chouette* (brill). They are not being taught a vocabulary that extends their experience.

'There's heavy emphasis on cognates such as "pollution", "odeur", "déteste" which read like English words and so give a sense of instant success. The idea is that if a verb that's hard to understand is surrounded by cognates the child will understand the verb. But it doesn't work. The children have no idea of the different parts of speech or the order of the words. When asked a question in French, they don't just give the answer but they repeat the question indiscriminately because they don't know what it means.' A lot of the teaching takes place by matching French words to pictures in the textbook. This means that children may point to the correct French word, but they have no idea what that word actually means, only the 'general gist'. Thus if a child is faced with a picture of someone having a drink, and correctly points to the French verb *boire* (to drink) when asked what the verb means in the picture, the child is likely not to reply that the character is 'drinking' but to say instead that he's 'having a Coke'. The depth of these children's lack of comprehension, said the teacher, was illustrated by their inability even to understand the dictionary. 'I had one child who wrote: *Je vais prep college* or *Je mange prep cuisine*. She had looked up the dictionary to find the French for "to" or "at" and didn't realise "prep." was short for "preposition". She thought that was the French word for "to" or "at".'

The teacher pulled out the Midland Examining Group GCSE draft syllabus for 1995. Candidates for all grades were allowed to take dictionaries into the exam. The lowest grades would be marked for broad understanding of the content without any attention paid to the accuracy of the language used. It was all heavily non-verbal: candidates were to be tested by much linking of pictures to words and filling in of grids. Answers on the short texts that were included required little or no French. Even in the top grades, texts were to be marked for comprehension only: by ticking boxes, for example, or by very short responses.

The teacher's guide to the GCSE syllabus advised that pupils needed to be taught only the grammar they needed to communicate. It was quite explicit in its advice not to bother too much about accuracy. 'Note: the fact that the verb endings, *mange/manges*, *veux/veut* and *bois/boit* have the same sound is more relevant than their different spelling. Some learners will notice the changed ending but it need not be an issue.'[36]

She was horrified by what the National Curriculum and the GCSE courses between them were telling her to do. As far as she was concerned, she was being prevented from teaching children properly. She did her best to counter the trend but it was difficult. 'I teach grammar in secret. I don't tell anyone I'm doing it. My door is shut but it's got a window in it. I hate it if a colleague looks through the window when I've got the grammar on the overhead projector because I know my life will then be made a misery,' she said. What kind of an education system is it when a teacher is forced to teach the rudimentary codes of language in secret? What kind of a culture is it that appears to have turned knowledge itself into a thought-crime?

Parental Anxieties

An increasing number of parents is becoming alarmed by standards and expectations in their children's schools. Marion Reynolds, a parent from the Home Counties, believed that not enough was being expected of her eight-year-old daughter from the primary school she attended in Basingstoke. 'She simply can't sit still. They are all taught in groups, and work at their own pace on worksheets, manipulating answers to the questions. At certain times of the day they choose what to do. Maths isn't taught by any rules but by approximation. Subtraction involves a lot of guessing and rounding up and down. If the child does subtraction by a proper method the teachers go into paroxysms. They say the child has to have an overall concept in her mind of the shape of the numbers. But if only she had pages of sums to do she would just click through them. The theory is to get the child to understand, but they don't know how to do it. I haven't met a single parent, including those who send their children to private schools, who isn't outraged at the lack of whole class teaching. They all hate these informal methods.'[37]

Ms Reynolds's sister Helen Mackenzie lives in Scotland where, despite that country's history of reverence for education and academic excellence, she complained that her 16-year-old son had also suffered from extremely low standards of expectation and achievement. 'He was never taught prefixes or suffixes or how to understand the uses of words in a sentence. In the second year of secondary school he began to distinguish between nouns and adjectives. We had taught him because we had been appalled that he had got all the way through primary school with no grammar. All he was writing was just lots of stories about "how I feel". In modern languages, they were taught how to go into a café in France or Germany and order a Coke. It was absolutely disgusting. At 16 he was capable of doing French GCSE with the only written paper being a multiple choice. The idea that he would write a passage in French on a given choice of subjects was entirely optional.

'He has not been taught to think, to question or analyse. In English, which was a bit like a media studies course, he was given a free choice of what book to read. We were constantly astonished by the lack of understanding of their own language; given how they go on about communication, this is astonishing. Every six months we have to say to ourselves, don't worry, he'll survive; you have to numb yourself to it.'[38]

The collapse of the teaching of literacy and of numeracy is a matter of profound significance for our society. It tells us not merely about educational fashion but about some of our deepest human values. It is part of a 'child-centred' orthodoxy that purports to benefit children through an enhanced respect for their individuality. But it has the opposite effect. At the most immediate level, it deprives children of their elementary entitlement to gain some control over their own environment and handicaps them in their initiation into the adult world. But important as that is in itself, it also sends us many other disturbing messages. It tells us that we have thrown overboard the authority of external rules. We have decided that relevance, context and localised experience are what really matter. Instead of authority being located 'out there' in a body of knowledge handed down through centuries, we have repositioned it 'in here' within each child. In doing so, we have deprived those children of the structures through which human beings have traditionally made sense of the world. Instead, children are having to make it up for themselves as they go along. They are being abandoned to disorder, incoherence and flux.

This retreat from the rules of language and arithmetic is also helping to knock the moral stuffing out of our culture. It gives children the very clear message that there is no right or wrong; instead, everything is good enough as long as it is approximate and other people can 'get the gist'. A village primary school headmaster illustrated this moral relativism in his approach to mathematics. For him creativity was much more important than correctness. 'People are obsessed with the "right answer",' he said. 'If a child does a long division sum and gets the wrong answer that doesn't worry me. What worries me is if the child gets the right answer but doesn't know why that is a reasonable answer. We try to get across that exactness in maths is not the right answer but the process by which we play around with numbers.'[39] But this headmaster was wrong. That wasn't exactness at all.

Accuracy and correctness are not merely undervalued but now are positively disparaged as élitist. A fundamentalist egalitarianism has taken over in which rules are taboo because some people may break them. Since it is no longer permitted to have a hierarchy of right or wrong behaviour, everyone must be equally rule-less. It is, of course, both ironic and most telling that a society which has made equality into a fetish no longer even understands the meaning of the 'equals' sign in mathematics. Real equality, as Tony Gardiner observed, entails the responsibility of relating cause to effect. This has gone. In its place has come the something-for-nothing culture, achievement without effort, ready-made French phrases and long division on a calculator. It was to have been the educational no-pain, no-shame nirvana in which no-one would ever again be made to feel a failure. In reality, *all* are now failed; and those who are failed most grievously and disastrously by the collapse of educational authority are the children at the bottom of the social heap.

Chapter Two

THE RETREAT FROM TEACHING

Failing the Disadvantaged

Julie Armstrong has been teaching English at a further education college in Walthamstow, east London, for many years. Her students come overwhelmingly from poor backgrounds. They arrive wanting further qualifications; some need to re-take their GCSEs, while others want to study for English A-level. But the disastrous way they had been failed by their primary and secondary schools meant, she said, that most of them had serious difficulty in coping. They had been taught virtually nothing of the basics. Their teachers had rarely corrected their work, ignoring their mistakes and giving them the illusion of success. And because their teachers had helped them by correcting the coursework component of their GCSE exam before it was marked, none of them had any idea of exactly what they themselves could or couldn't do. The result was that, although they were bright children, most of them would fail. And their resulting sense of grievance was even more pronounced because the majority of them were black and thus already felt themselves to be at a disadvantage.

These students come to us knowing nothing. They come with their GCSE passes to do A-level literature without ever having been taught to structure an essay, with very poor literacy skills, with no knowledge of grammar, punctuation or paragraphing. They are almost incapable of independent thought; they have simply never been taught to think for themselves. They have no discipline because they've never had homework and they have very little general knowledge. Their first year here is pretty miserable. Quite a lot of them have no ability to do

A-level. But with four GCSE passes they consider themselves to be academic. They have hopes to go on to university. Some drop out in the first year. Most soldier on to the end of the two-year course. Many of them fail. All apply for university. Many have ideas of becoming barristers or psychologists. But they have no understanding of the requirements. Then they can't understand why they've done so poorly. What I feel even more angry about is the students who do have ability but who have been taught so badly it's impossible to raise them to the standards required for a degree course.

I really have no idea what they spend their time teaching them in the schools. History appears to be rarely taught at all. I always ask my students what historical knowledge they have because they need some to study literature. All they have at the age of 17 is some vague knowledge of the ancient Egyptians from primary school; then there's a huge gap and they will have a vague knowledge of the slave trade and apartheid and a little bit about Nazi Germany and that's it. They have no knowledge of geography at all.[1]

The students themselves were well aware of what they had missed and were angry. They were bright, but they knew now all too painfully what they had never been taught and what they suspected they would never now know.[2]

'Whatever we wrote, the teachers would give us good grades,' said one girl. 'I left school thinking I could do English at A-level. When I came here I found I couldn't do it.'

'All our essays were just descriptive writing,' said another, who had a B grade at GCSE. 'When we started here we had no knowledge about analysing anything. We just described what happened in the book.'

Not that they had previously read whole books. 'We weren't told we had to read any books,' said one student. ' All we got were extracts. For GCSE we did *Jane Eyre* and *Macbeth* but we only ever read extracts. We were never encouraged to read anything from start to finish.'

'It was a shock here to have to read a whole book in a period of time. I just lost motivation; I couldn't finish it,' said another.

The students reflected powerfully the ethos of low expectations and mutual lack of respect between teachers and pupils that had characterised their schooling. They complained about the informality of school, the assumption of equality between pupils and teachers; they said they had wanted instead structure and discipline and to have had a respect for their teachers which was in practice non-existent. 'None of the teachers seemed to know anything,' said one. And despite the rhetoric of creativity and empowerment that has accompanied the flight from structured teaching, these students believed they had been spoon-fed. 'The teachers always told us what to do, but never taught us how to think,' said one student.

They bitterly regretted their wholly inadequate grasp of British history. 'They teach you history but none of it is connected to the next stage,' said one. 'We didn't do any British history. But we need British history because these writers we're now studying reflect it.'

In Julie Armstrong's A-level class were two or three students who came from Nigeria. They had all been astonished at the standards in their British schools which they said had been far lower than in the country they had come from. 'We were never allowed to use calculators in school in Nigeria,' said one girl. 'There, we had exams every year and there was competition and school reports saying where we all were in the class. That's what motivated people. They all wanted competition. Here you feel you're a failure anyway.'

Aghast at her intake of students' poor grasp of language, Armstrong had strong-mindedly braved the disapproval of her colleagues and provided additional classes in formal grammar.

They enjoy it. I always explain why it's important that they learn grammar. I use the analogy of understanding the workings of a car engine. I understand from them that whatever work they've submitted in school, it's always been ticked. None of their spelling or punctuation was ever corrected. I resent my students embarking on research skills when they can't construct an essay. They've gone through their whole careers being told they write very good essays. The first essays they write for me are covered in red ink and they are horrified. I explain to them they won't pass their A-levels if they carry on like this. It's a little trying to start with but gradually they gain confidence.

I've also had to do a lot of history with them. They have no historical knowledge. They know nothing about Elizabethan England or the Industrial Revolution; and the eighteenth century is a complete blank. But as I've been teaching them history, they've become interested. They've started asking questions. They knew nothing about this before. I find it very frustrating because there isn't the time for all the remedial teaching that's necessary and we must finish the A-level texts. What is so sad is that now they have curiosity, and I know they will never get these answers again. Most are highly intelligent, but they are not going to progress because they don't know enough.

A number of them go to the new universities to read social sciences, but sociology reinforces their sense of oppression. It's readily absorbed by these students who see it as offering solutions to their problems. The psychological effect of this is disastrous. It creates a victim culture and it's very unhealthy. The Marxist rhetoric may have died down, but there's an intellectual vacuum.

Realising that the need for structured language teaching extended beyond her A-level students, Armstrong started evening classes called Academic English in which she taught grammar, comprehension, essay technique and analytical skills. The response was astonishing.

I've had A-level students, people in work who thought they might be in line for promotion, people whose employers sent them along; they were sitting on the window-sills because the room was so full. Quite a lot were working class people who came to me and said how worried they were about their own children's education, that they'd been to their children's schools and tried to speak to the teachers and got absolutely nowhere. This I found quite shocking. These were single parents on benefit or on very low pay who were saying how desperate they were, how they couldn't afford to send their children to private schools but maybe they could do so for part of their education and could I suggest at what age they should send them privately? But they couldn't possibly have afforded it. I don't think they had any idea what the fees would be, but they were absolutely desperate.

Armstrong movingly described her experiences in teaching GCSE English to a class of students who had been school drop-outs but were re-sitting the exam. It was an account which illustrated all too vividly the catastrophic combination of social disadvantage and the educational orthodoxy of egalitarianism.

I had to teach a class of GCSE re-sits. They told me they needed discipline and structure. They were contemptuous of the teachers who had tried to speak to them in their own language. I discovered the more structure they had, the better they responded: to be in class at certain times, to give in homework on time, to know that there were boundaries of behaviour. They said that's what they wanted because they had never had it. My impression was that they had been truanting at school and nothing was done about it and that all the teachers did was baby-sit. I've spoken to secondary school teachers and they've told me it would take at least 15 minutes to take the register and calm them all down, and they would be restless at the end of the lesson, so the actual teaching would be confined to about 20 minutes.

It was exhausting, because so much of my energy had to go into building a relationship with them. Quite often they would be missing because they had court appearances. There had to be an ongoing conversation about problems. They were quite offensive at the beginning because I was white and middle class and they didn't like the way I spoke. They couldn't sit still, but they liked being read to and they loved drama. They did an extract from Anouilh's *Becket*. There were two boys who were very difficult, and I was quite frightened of one of them. They volunteered to play Henry and Becket and did it absolutely brilliantly. It was a sensitive scene after Henry's exile. The performance brought tears to my eyes. And I thought, what a waste. What a terrible, terrible waste.

They would all wear personal stereos in the lessons. I used to have to mouth at them to take them off. I started to wonder whether the music increased their aggression. So I tried an experiment. I brought in some baroque music and played it to them and they all became totally quiet – and they were a very, very noisy class and never sat still – and only spoke to each other in whispers. I brought in other music, Bach and

Handel. What was amazing was that they stopped shouting and calmed down. They particularly liked Vivaldi. I came in one day and they'd written on the board: Vivaldi rules OK; and they had spelled Vivaldi correctly. But I only had them for a year.

No-one can doubt the difficulty of teaching children who are like this. Many of us might think that their teachers deserve a medal. And no-one can doubt that the problems presented by these children stem from a combination of circumstances extending way beyond the school gates. But school makes a difference. It can help improve matters, or it can help make them worse. One of the predictors of delinquency is, after all, low educational achievement or truancy.

The Inspectors' Views

The school inspectors have had harsh words to say about inner-city schools which are far too quick to blame their poor outcomes on the pupils and their home circumstances. A report in 1993 by the then Chief Inspector of Schools, Professor Stuart Sutherland, spelled out the effects of such unjustified low expectations.

In inner-city secondary schools, said the report,

teachers too readily over-generalised the extent and effect of pupils' disadvantage and their poorly developed basic skills to explain poor performance and depress expectations of improvement . . . Consequently, much of the work observed failed to tap the potential that pupils had for success. Only in very few lessons did pupils achieve what they were capable of; in only two schools was their achievement satisfactory or better for more than half the lessons seen, and even here there was some variation between subjects and year groups . . . Lessons rarely focused on clear learning goals that would enable pupils to be aware of what they were learning or the new skills they were gaining. Undemanding tasks and dull teaching failed to engage pupils' participation in lessons and their willing consent to learn . . . High standards were not always expected of them: some lessons sacrificed pace for good-natured relationships and teachers were, on occasion, unwilling to press pupils to complete work or to improve on a first attempt. Indeed, children rarely received the kind of feedback on their work that helped them to know what or how to improve.

The quality and standards of much of the work were 'inadequate and disturbing'. 'In the absence of accurate assessment teachers used more generalised estimates of pupils' abilities which led them in some schools to overestimate the effects of poverty and social disadvantage and to underestimate the potential ability of pupils.' In the primary schools, said the report, 'greater emphasis was needed on teaching pupils how to use their phonic skills to read more effectively and how to blend sounds in order to decode unfamiliar words . . . Standards of writing were correspondingly low . . . Too much use was made of worksheets requiring little beyond a word or gap-filling exercise, with the result that children acquired skills out of context and did not learn to apply these by writing independently . . . Standards in numeracy were unsatisfactory or poor'.[3]

This assessment painted a devastating picture of the way in which thousands of inner-city children were being failed by their schools. Two years later, the annual report of Sutherland's successor as Chief Inspector of Schools, Chris Woodhead, was no less damning. He observed: 'The issue of whether enough is currently expected of children who live in areas characterised by high levels of social and economic disadvantage is particularly pressing. Such children should not be assumed to be intrinsically less intelligent than their more fortunate peers.' Despite some good teaching, the overall situation was bleak. 'Less is expected of pupils in disadvantaged areas.'

Throughout the country, Woodhead found disturbing evidence of under-achievement. Standards of reading at age seven were unsatisfactory in one in 20 schools; in three in 20, pupils had difficulty with at least one major aspect of writing. 'These figures mean that a significant number of children are failing to master basic literacy skills,' he wrote. By age 11, reading and writing were bad in one in ten and one in four schools respectively, with one in eight poor in maths. By age 14, pupils were achieving poor standards of reading and writing in one fifth and one quarter of schools respectively. 'Such figures indicate that a substantial number of pupils have failed to master basic skills of punctuation and grammar and, more generally, that schools should be devoting more attention to the higher order skills of analysing, arguing a case, investigation, interpretation, reflection and the handling of evidence,' he wrote.[4]

In 1995, the Chief Inspector again painted a disturbing picture. Once more, Chris Woodhead paid tribute to the many good schools and good teachers around the country. But thousands of children were still being failed. Only one-third of primary schools and only two-fifths of secondary schools succeeded in meeting all the criteria set by Ofsted, the Office for Standards in Education. A further one-fifth of both primaries and secondaries met three criteria out of four. Ofsted found that nearly half the primary schools needed to improve the quality of their education, which meant that a huge number of children were moving up to their secondary schools under a serious disadvantage.[5] What's more, given the fact that a significant proportion of Ofsted's inspectors are sympathetic to the child-centred orthodoxy and share the resulting low expectations, their assessment of the scale of the problem was likely to be a serious underestimate.

In May 1996, however, Ofsted published its most controversial report to date. This was about reading standards in three London boroughs: Islington, Tower Hamlets and Southwark. It revealed that four out of five 7-year-olds were reading at a standard below their chronological age. Among 11-year-olds, three out of five were below their chronological reading age and two out of five were at least two years behind.[6] The report generated accusations that Woodhead had doctored the findings to highlight failure rather than success and to downplay the role played by social disadvantage rather than the inadequacy of the primary school teachers. In fact, although the emphasis of the report did change between drafts, this decision was taken by the drafting team of senior inspectors. It was these men who suggested to Woodhead that the scale and nature of the disaster in both the reading scores that had been recorded and in the teaching methods the inspectors had observed were so serious, they needed to be emphasised as strongly as possible.[7] In addition, contrary to the critics' claims social disadvantage *was* given prominence in the report. The fundamental facts, in any event, remained indisputable. Primary school reading standards in these highly disadvantaged areas were shockingly bad.

But it needn't be like this. Some schools in the most unpromising areas have transformed their pupils' chances by adopting the structured approach which is now so unfashionable. Sudbourne primary school, for example, in Brixton, south London, which was singled out for praise in the 1995 Ofsted annual report, had far exceeded the government's own norms in literacy and numeracy. This was despite the chronic social deprivation of the area: almost half its families were on welfare benefits and more than a fifth of the children spoke English only as a second language. The head teacher, Susan Scarsbrook, attributed the school's success to 'old-fashioned attitudes and teaching methods'. There were weekly tests and recitation of multiplication tables, bands of ability groups and school monitors. And crucially, the children were taught as a class. 'We go for whole-class teaching here because it focuses the children on what they should be doing,' said Mrs Scarsbrook. 'We do not believe child-centred teaching is as effective.'[8]

There are many other good primary and secondary schools around the country, staffed by good teachers. But there aren't enough of them; and they are good only because they set themselves against the culture that disapproves of their approach. They are run by strong-minded head teachers who don't give a fig for the prevailing 'child-centred' orthodoxy but teach children as they know they should be taught. Such head teachers are not in plentiful supply. Mrs Scarsbrook acknowledged that, in adopting her own approach, she had had to reassure teachers who were mindful of the prohibition against labelling children as 'better' or 'worse' than each other. Her remarks illustrated a crucial point: that the achievement of children in school is affected profoundly by not just the skills but also the attitudes towards education of their teachers and, most particularly, their head teacher.

It is often claimed that the reasons for such widespread educational disadvantage lie in the under-resourcing of schools and the large classes that many teachers have to teach. While it is undeniable that many schools are in a dismayingly dilapidated condition and that some classes are large, these explanations seem implausible. Only about 30 per cent of primary classes and 7 per cent of secondary classes contain more than thirty pupils. Moreover, research does not bear out the popular assumption that class size in itself makes a difference. It all depends on which teaching methods are used. Classes in the Far East, for example, often contain many more children than would be found in British schools and yet achieve higher standards because of the way they are taught. And under-resourcing can hardly be the reason for poor performance since there are many equally 'under-resourced' schools such as Sudbourne primary which achieve much higher standards than others. Indeed, it is the striking variation in standards among schools suffering the same under-resourcing and in areas of equal social disadvantage which indicate strongly that funding is not the root of the problem.

The problem is that so many teachers appear to lack those teaching skills highlighted by Mrs Scarsbrook, while displaying attitudes which are inimical to education. The result is that only about one-third of all lessons are good. In his 1994 report, Chris Woodhead expressed his frustration. Inspectors were making statements of the obvious all the time, and yet nothing appeared to change. 'Why, therefore, does the percentage of unsatisfactory lessons remain, year after year, so high? Why is it that expectations of what children can achieve remain too low in too many schools, in particular in too many schools serving areas of social disadvantage? Why is it that in too many primary schools "learning by doing" is preferred to "teaching by telling" to the point where sitting pupils down and telling them things becomes almost a marginal strategy?'[9]

The answer to that question lies in a cultural orthodoxy which is explained in greater depth during the rest of this book. But the evidence is that right from the time new teachers emerge from the teacher training colleges, they are singularly ill-equipped to teach children at all, let alone to meet the daunting challenges of teaching children from difficult backgrounds. A survey of new teachers by the education inspectors in 1987 found that among the new teachers studied, nearly four out of ten of their lessons were either good or excellent, three-quarters were satisfactory and one quarter unsatisfactory. Two-thirds of these new teachers were reasonably or well satisfied with their training but 'disturbing proportions' felt unprepared for important areas of teaching. 'Too many primary teachers felt less than adequately prepared for classroom management, the teaching of reading, teaching the more able and the under-five-year-olds and using audio-visual equipment, and too many secondary teachers felt under-prepared for teaching for public examinations,' it said.[10]

Teacher Training

Teacher training institutions, like everything else, vary in quality and attitudes. Many teachers believe the only training that is of much use is provided by actually doing the job. But all the evidence suggests that the vast majority of teacher training institutions subscribe to a doctrinaire, often highly politicised approach which strips new teachers of the authority they need to do that job. The prevailing philosophy is the child-centred, individualised approach in which learning has to be fun, no child must feel a failure and no rules must be taught – and through which society's prejudices in the fields of race or gender must be corrected. Many trainee teachers with robust, common-sense personalities shrug off such influences once they get into the classroom. But it is difficult to believe that others will not be influenced.

In 1995, a trainee secondary school teacher on a course in the west of England described the way she had been trained.[11]

The course wasn't incredibly rigorous. The calibre of the other students just wasn't good enough. They couldn't even pronounce the names of Shakespearean characters and they weren't corrected; it was just treated as a joke. We never got down to the nitty gritty. For example, we were told the government wanted us to teach grammar. I've never done grammar myself but I could learn it. The tutors told us they didn't approve of this and they refused to teach it to us. They just gave us some books and told us to go away and learn it on our own.

The trainees weren't just taught that child-centred learning was the only proper approach; their own course was run along similar lines, a kind of 'student-centred learning' which was the model for the approach they should take in the schools. So their work was never corrected even though standards were poor. 'The essays were very Mickey Mouse,' said this trainee. 'For instance, I might be told to prepare a lesson plan. But no-one quite knew what the curriculum provision meant. And I received no criticism of my work. We were seldom challenged in what we came up with; we'd be told: "that's a nice idea". They would ask us merely to record and analyse slightly our experiences in school, but that didn't challenge us to think beyond them.'

A typical assignment from this course illustrated how subjective this training was, how free from authoritative input and how heavily it relied on trainees just making it all up as they went along. Its brief read:

> This assignment is intended to help you to reflect on your own subject specialism and to begin to think about your own learning as a starting point for teaching your subject to young people. Negotiate with your subject tutor a starting point which will allow you to explore an aspect of your subject specialism which is relevant to the school curriculum. Work at this starting point at your own level. As you do this write a commentary on what you do. For example, you might comment on decisions you make, difficulties you encounter, feelings you experience . . . indeed, anything that sheds light on you as a learner. Try to categorise what you learn. Think about knowledge (generalisable principles or concepts as well as information). Think about skills and processes. Think about values and attitudes.

Whatever the trainees wrote would be marked in a vague and encouraging way.

The course, said this trainee, was also highly politicised.

> We were told not to give any grades for children aged from 7 to 9. I asked my tutor whether this was based on any research; she just said, read so and so. But this was something written about it from a political position. We were always referred back to the politics. But education isn't political. It's about teaching people well.
>
> We were told we had to produce 'mission statements' about how our school would enforce equal opportunities. I found this totally pointless. We spent a painful week on that. The political aspects were there all the time. The lecturers were all

anti-government. To pass the assignments, we had to refer to either gender, special education needs or racism. There was never any discussion why this was necessary. But I don't think these are the issues in schools. We were also told to represent broken families positively. In our wall displays we couldn't show mum, dad and children; we had to have just mum, or mum and her lesbian lover. We were told this was because a lot of children didn't come from families with mums and dads. But I would want to represent a normal family because although not all of us have them that's what these children would like. That's not leaving children out.

The main feelings in the group were dissatisfaction and confusion or very closed minds. We ended up with no practical help at all.

But when this trainee did her teaching practice at a school in the southern counties of England, she discovered to her dismay that the politicised approach continued in the classroom itself where the English lessons were used to correct the children's faulty attitudes.

They were doing a story called *Bill's New Frock*, all about Bill who wakes up one day as a girl and learns how horrible it is. At one point he gets wolf-whistled. The teacher asked all these 12- and 13-year-old girls to put their hands up if they had been wolf-whistled or if any of the boys had wolf-whistled. When children's sexuality is just generating, this kind of approach just isn't fair because they are out of their depth.

I've noticed that the boys have been a bit more comfortable with me, because I'm not forcing all this stuff down them. The head of department believes in alerting children to gender stereotyping. They have to look out for places in normal story books that have such stereotypes. They make their own story books when they are 14 and are told not to stereotype. This is so arrogant . . .

Child-centred learning is supposed to be the most positive way to learn. But the kids are not on task for a lot of the time. Some of the best lessons I've seen have been chalk and talk. A lot of work at the school I'm now at is project work. They don't have exercise books. The child can do any project of their choice. They do them through all five lessons, for perhaps four to five weeks. After a while it gets pretty boring. They're finding out things for themselves. But it's very diffused, there's a lot of noise and a mixture of abilities so it's not a constructive working atmosphere. The boys in particular find it very difficult. I had to threaten to send one boy to the head of the second year because he was playing up. Another boy asked me if he could be sent too because he was so desperate for some kind of structure to be imposed on him.

Politicised Education

The politicisation of teacher training is not confined to a few isolated teacher training institutions. Its centrality can be gauged from the tone of textbooks for trainee teachers, such as the Open University coursebooks for students studying for the post-graduate teacher qualification by distance learning.[12] Despite their ostensible purpose, these are not so much educational texts as ideological tracts, characterised by an emphasis on teaching 'correct' attitudes to race, sex and class and an overt hostility to the Tory government and the New Right. The authors, for example, attacked an article by Ellen Yeo, an experienced primary school teacher, apparently because it was published by a 'New Right pressure group' (sic), the Campaign for Real Education.[13] Her crime, apart from guilt by ideological association, was apparently to use 'anecdote, assertion and evaluation' to attribute 'all sorts of social and economic problems to educational practices in primary schools'. In other words, she dared to express the opinion that schools might be doing something wrong. The authors – who, it appeared, felt entitled to use assertion *instead* of evidence or evaluation – contrasted this with 'the principled comments of a classroom teacher', Jane Needham. These 'principled comments' included her concern that 'to develop the skills necessary for successful oracy, reading, writing and numeracy should not override the promotion of the children's happiness, their belief in themselves and their ability to relate to one another . . . Any competition which exists should be with oneself rather than with others . . . The most daunting fact of all is the wide gap between the co-operative environment that I as a teacher try to create and the unfair world outside.' Instead of education, here was the feel-good factor and a wholesale retreat from the 'unfair' real world to a fantasy land where having fun was set at a higher priority than learning to read and write. Of teaching in the traditional sense, there was virtually no mention.

From other texts in this series, the trainee teacher would learn that English language and literature 'empowered' people to challenge all established systems such as government and business interests; that they should be renamed as media, cultural or textual studies; that English teaching and associated texts should be carefully monitored and selected for gender, race or class bias; that great literature wasn't essential and reading lists should reflect the multicultural society; that there were no definitive definitions of grammar; and that 'home' languages should be of equal status to English.[14]

The trainee teacher would also learn that the maths curriculum was a tool used by the élite to maintain an unskilled workforce; that the New Right had been a strong influence upon government in the teaching of maths; and that multicultural maths (sic) was to be recommended along with 'cultural and social' mathematical topics, such as analysis of HIV/AIDS statistics (essential, no doubt, to every child).[15] To help with the teaching of history, the trainee teacher would learn that the curriculum concentrated too much on the teaching of dry facts by rote and not enough on investigative skills; that there was too much emphasis on British and European history and that knowledge gained from history 'empowered' pupils to question critically established institutions.[16]

It is a remarkable and distressing fact that the Open University, a noble undertaking of the 1964 Labour government designed to open the national mind, should have been diverted in this fashion by people intent instead on closing that mind. Under the guise of an educational imprint, the Open University has to some extent been reduced to pumping out crude ideological propaganda, hijacking the curriculum to 'correct' faulty thinking and brazenly enlisting the children in the classroom in the political struggle against the government. This has absolutely no relevance to teaching teachers how to teach anything at all. Instead, it instructs them in ways of indoctrinating the young. And what is particularly notable for a so-called educational project is its hostility to any point of view that challenges the authors' own position. That, of course, is an approach which is utterly inimical to liberal values. But then one of the many paradoxes of the child-centred orthodoxy is that, despite its liberal image and carefully phrased liberation rhetoric, it is anything but liberal in its practical application. It does not accept that any other approach may have any validity at all – least of all the possibility that it itself may be wrong. Debate is not entered into; it is closed down by whatever means possible – ridicule, character smears or guilt by ideological association.

This Orwellian characteristic was captured in a rare article in the *Guardian* about the way trainee teachers are taught. The writer of the article, published in 1990, sampled a teacher training course at the Roehampton Institute, one of the largest teacher training institutions in the country. He reported that no deviation was permitted from the line that children learned best through play, and that children should not be told what to do by their teachers but encouraged to learn for themselves. Trainees were taught that whole class teaching, where the teacher addressed the whole class, didn't work. Instead, teachers had to start with the children's own level of understanding and give them the right to negotiate their own curriculum.

He quoted the assistant dean of education Graham Welch as saying: 'Negotiated curriculum is an idea rooted in a concept of democracy. There is a lot of evidence to suggest that children as young as three are better motivated if they have a say in the way their day is organised.'

The idea that democracy entails equal rights for three-year-olds is, of course, a bizarre idea. Yet at Roehampton, apparently, the equalisation of rights between teacher and child was a doctrine that couldn't be questioned and was swallowed as such by the vast majority of students.

The reporter, Edward Pilkington, wrote:

In every class I attended at Roehampton I heard tutors consistently reinforce the child-centred philosophy. In classes on reading and writing, students were told: 'We want to make learning easy for children. It would be funny if we went around trying to make it more difficult.' In maths: 'We are talking about a model of education which gives children control over their own learning.' In environmental studies: 'You should not draw a distinction between work and play. Instead of saying, "You have finished work, now go and play," you should say, "You have finished the learning activity I asked you to do, now you can choose a learning activity for yourself."'

So, eleven years after Mrs Thatcher's government had come to power and two years after the Education Reform Act was passed, one of the most influential of British teacher training institutions was, it seems, entrenching itself even more deeply behind enemy lines with no-one in government appearing either to know or care. There were no longer any teachers at Roehampton who did not endorse the learning-by-play doctrines. Welch told Pilkington: 'When I came here a year ago I found a variety of opinions about how we should teach. But now we have begun to put our house in order by being up-front about our methods. We have our own ethos.'[17]

That ethos embodied something rather more deep-seated than the mere advocacy of one particular teaching technique over another, rather as if one branch of the dental profession, say, had started to advocate treatment of dental decay by laser rather than drill. The ethos of the Roehampton Institute, which was no more than a reflection of the new education orthodoxy, was nothing less than a retreat from the very idea of teaching itself.

Knowledge 'Versus' Creativity

The film *Dead Poets Society* achieved cult status in the early 1990s. It had great resonance for the current education debate. It was the story of an inspirational teacher who fell foul of the stuffy school authorities. Set in the fifties, it showed how conventional teaching methods using drills and learning by rote were soulless and mind-numbing, crushing the creativity out of pupils all so that they could be processed into 'good' professional jobs like banking or law or medicine. The message was that orthodox education was inimical to creativity.

That kind of mechanistic, formula teaching belongs to another age and cannot possibly work in today's sceptical and questioning climate. But for some reason, the idea that what they had been doing was destructive came to achieve great resonance with teachers.

Committing anything to memory became an anathema. Repetition and rote learning were viewed as a fast track to a police state. The fact that most children enjoyed such repetition because it enabled them to get lots of answers right and thus significantly enhanced the 'self-esteem' which everyone was so worried about somehow got lost for the majority of teachers. Examinations became seen as the mere 'regurgitation of facts', so much so that not merely the mode of assimilating them but the facts themselves came to be repugnant. Schooling, as George Steiner lamented, became little more than a process of 'planned amnesia'.[18]

The reason behind this deliberate atrophy of memory was the absolute pre-eminence that came to be given to a child's innate creativity and spontaneity. What mattered most was what a child was able to make of its own experience and the uses to which it was put. Other people's experiences, the mere 'facts' passed down through the generations, were in themselves devalued. To achieve any value, they had to be transmuted through the child's own experience – by discovery, investigation or the exercise of the imagination. Moreover, the imposition of such facts upon a child became viewed with intense suspicion for somehow holding back or fettering that child's creativity. Such an imposition became therefore a kind of child abuse, and anyone rash enough to advocate it was treated accordingly within educational circles as a pariah. Knowledge itself thus became redefined within such circles into a wholly subjective process.

But of course it is quite false to set knowledge and creativity in opposition to each other. Creativity is like freedom itself: it flourishes only within a clear framework of moral and intellectual boundaries. Remove the enclosing structures and all that remains is a vacuum of anarchic impulses which are deeply hostile to creativity. Many of the greatest figures of English and world culture, after all, were subjected to precisely the kind of rigid educational disciplines that so offend contemporary educationists. Conversely, we have hardly lived through a new Renaissance of creative endeavour since the elevation of creativity to its pedestal in British education; on the contrary, our culture is atrophying beneath the weight of the second- and third-rate. The idea that education is all about releasing what is already within the child has a long history, as will be discussed later. But education relies on input. The very word 'education' is commonly misconstrued to come from the third conjugation Latin verb '*educo*', meaning 'to lead out'. But it doesn't. It comes instead from an entirely different first conjugation verb: '*educo*' meaning 'to educate' or 'to put in'.[19] An elementary knowledge of Latin verb stems informs one of this fact. But then teaching Latin verb stems hardly complies with the contemporary prohibition against 'useless' facts.

The assumption that creativity is the highest good can be found on widespread display in classrooms up and down the land. It can be seen in the way the children are set to work and in the attitudes expressed by their teachers. The educationist who will justify it in these terms, however, is a most elusive creature. When one attempts to discuss this pre-eminence of the imagination, teachers become strangely shy. They deny indignantly that it is a matter of 'either/or'. They adopt a 'variety of teaching approaches' (always). Of *course* knowledge is important, they say edgily, and immediately reel off a list of 'facts' that the children have learned for themselves through discovery – the properties of light, the difference between an adjective and an adverb, what is a right angle . . . Unfortunately, when one looks at what these children are actually producing, it is all too often depressingly mediocre.

The village primary headmaster who didn't believe in the 'right answer' explained that, for him, knowledge should be acquired through creativity – without which it was worthless. 'Creativity is the highest value I would want to explore through education,' he said. 'I believe that's what life is all about. The most important aspect of education is the education of the imagination. The rest is subsidiary.'[20]

This was an astonishing statement. At a stroke, this headmaster utterly devalued what must surely be the prime duty of an education system: to teach a child to *think*. For that, children need to be equipped above all with powers of reason and analysis. Of course creativity and the imagination are important. They are the means by which human beings demonstrate their solidarity with each other and give expression to their highest aspirations. But if they haven't been taught in the first place to make sense of the world, the resulting chaos makes a mockery of creative endeavour.

One person who appeared to understand this well – surprisingly – was Professor Brian Cox. Surprisingly, since it was Cox himself who, as will be seen in detail later, played such a significant role in turning the English curriculum over to a creative agenda. But in his memoir *The Great Betrayal* published in 1992, at the height of the furore over his creative curriculum, he wrote of his own education at Nunsthorpe Elementary School in Grimsby in the 1930s in terms that gave the lie to what he himself had so recently tried to implement.

> At Nunsthorpe it would not have occurred to the teachers that accurate spelling, punctuation and grammar might inhibit a child's creativity. They took for granted that a child's imagination cannot work in a vacuum, and that he or she must come to terms with the nature of form, whether it is language, music, paint or clay. They would have been astonished if anyone had suggested to them that accuracy is a middle-class virtue, or that it is only appropriate in training courses for jobs. They believed that in a democracy everyone should be able to use words precisely, to speak clearly, to develop a lucid argument and that such attainments are essential for personal growth. They accepted the traditional view that children who cannot use language accurately are denied major forms of self-exploration and fulfilment . . .
>
> This approach to learning . . . has profound moral, psychological and cultural implications. The emphasis is on service and self-discipline, a true attention to the realities of the social world. Instead of writing disorganised personal prose, we learnt the rules of civilised discourse. The rules we learnt were a means of understanding the demands of the outside world, and of controlling them in the service of the community and our own desired forms of personal fulfilment. There was too much rote learning, and oral abilities were neglected, but the children profited in a rich variety of ways from the traditional teaching. We memorised great verse, we read extensively and learnt to write accurate prose. These were rare gifts for the working-class children who formed the majority of pupils.[21]

Unfortunately, this firm grasp of the value of rules and order in education now lies in ruins. The retreat that has taken place from the orderly teaching of form and structures, from knowledge that is 'out there' rather than 'in here', is nothing less than a retreat from the values of civilisation itself. The redefinition of knowledge as the subjective experience of the child has entailed a shift in emphasis from teaching to learning and a transfer of authority from teacher to learner. The imagination, after all, cannot be taught, merely cultivated. The most the teacher can do is to facilitate its expression. On most obvious display in the primary classroom, this orthodoxy typically involves an emphasis on learning through play, investigation and discovery. Learning, above all, has to be fun. Work must never be 'boring' or 'difficult'. It entails individualised learning 'at the child's own pace' rather than whole class teaching. Since the child's feelings must be paramount, it follows that teachers are reluctant to correct the child's work and say that an answer is wrong. So although the child will learn that certain things are right, because that leads to approval, the alternative is not 'wrong' but 'a point of view'.

The resulting moral vacuum easily lends itself to political extremism. Children's creativity is extolled as the means of continuously reasserting resistance to the established order of society. Accepting the authority of the teacher is held to be in itself a sign of 'obedience'. Unless education is a continuous project of intellectual subversion, it becomes an accomplice to tyranny. This unsettling doctrine is embedded at the heart of the educational establishment. Caroline Gipps, Reader in Education at London University's Institute of Education, has demonstrated how for her, being in receipt of information now calls into question an individual's independence of mind. After a sustained attack on the government's education reforms, she claimed in 1993 that learning 'is a process of knowledge construction, not of recording or absorbtion' and depended on the learner's interpretation. 'Self-motivated, active learners' were therefore to be set against 'obedient, compliant learners', but with no recognition on Gipps's part that to be an active learner there had to be an engagement with something that was being taught. This, however, was associated with 'predictable obedience'. 'The transmission model of teaching, in a traditional formal classroom, with strong subject and task boundaries and traditional narrow assessment, is the opposite of what we need to produce learners who can think critically, synthesise and transform, experiment and create,' she wrote.[22]

There is ample evidence that teachers have in great numbers swallowed wholesale the retreat from teaching and the redefinition of knowledge. Bernard Barker, the principal of a school called Stanground College in Peterborough, believes that:

Young minds are not baskets to be filled and understanding is not a ladder to be climbed. People forget swiftly and continuously, especially those facts which have no practical application in their lives. All those grades prove nothing about the knowledge and skills of the people. The linear, sequential acquisition of facts is, after all, a very small part of human development. Learning is unpredictable, discontinuous and often unconscious. Growth, experience and understanding are intertwined and the learner's own feelings and aspirations are as significant as the phenomena of external (*sic*) world . . . There will be less concern with authority and judgment. Children will be criticised less. Teachers will be enthusiasts, not examiners anxious to create a moral hierarchy of approved books, knowledge and behaviour.[23]

So there is to be no judgment any more, of children or of texts. Instead, there is to be equality between good and bad in the interests of value-free creativity. Knowledge is worthless unless it can be applied. What the child picks up in the street is as valid as anything the teacher can impart. Not surprisingly, this has led directly to the collapse of the authority of the teacher.

Yet once again, the proponents of this doctrine deny that this is its effect. The village headmaster we met earlier was typical of educationists who try to have it both ways:[24]

Teachers and children are all in knowledge together. Knowledge isn't simply spun out of the individual child's mind. But the knowledge of both the teacher and the child is constantly refined by the relationship between the two of them.

For example, when very young children write they use their limited skills to creative effect. They may muddle up the beginning, middle and end of a story, or move the narrative voice from the first person singular to the first person plural. These would often be seen as errors, but the child has had a reason for writing like that. And when you look at that reason, you find you reorganise your own sense of narrative structure.

In other words, the child had become the teacher of the adult. The roles had been reversed. Surely it followed, therefore, that the teacher would not tell the child it was wrong. And indeed, 'wrong' was a word that this headmaster would not accept. It caused a palpable shudder. Such 'errors' were, after all, a sign of innate creativity.

Nevertheless, said the headmaster, even he did feel obliged to 'raise a question in the child's mind' over such creative constructions. So when he asked such a child why the story had been written in that way, what did the child reply? 'The child usually says: "Because it was wrong", which isn't what I want to hear at all,' said the headmaster with engaging candour. Perhaps, then, in the strange new world of the modern classroom the child is indeed wiser than the teacher after all.

But this headmaster wouldn't accept that this approach had dealt a death blow to the authority of the teacher. 'There's no reason to be a teacher unless you think you are an authority; but that doesn't imply that the teacher has unquestionable or full authority,' he said. But he had misunderstood the nature of authority, which above all confers upon an individual the status of a repository of superior wisdom. If the nature of a teacher's wisdom can be undermined by a child's error, so much so that the child's error itself becomes the new wisdom, then it demonstrates that there is nothing superior about the teacher's wisdom at all. So it isn't just the teacher who's undermined but the authority of that wisdom itself and the fundamental rules of procedure – that the beginning comes before the end, or that in the interests of communication the first person singular is not interchangeable with the first person plural.

What is involved here is not merely a retreat from teaching but a deconstruction of knowledge itself. The proponents of this doctrine falsely characterise such anxiety as an authoritarian desire to recreate row upon row of passive, bored and repressed children as teachers talk at them non-stop for 40 minutes. This caricature of didacticism is what they understand by whole class teaching and the transmission of knowledge. But on the continent – where the term 'didactic teaching' would be regarded rightly as a tautology – whole class teaching is not only the norm but it most certainly does not turn pupils into passive receptacles. On the contrary, the teacher constantly interacts with the class, asking and answering questions, pulling pupils along, challenging them. The pupils are engaged and respond; but the key is input from the teacher with facts they do not know and questions they may raise.

English Versus Swiss Methods

The shocking differences between British and continental classroom practice were brought out in a study by the National Institute of Economic and Social Research comparing maths, science and technology teaching in England and Switzerland.[25] Despite spending less time on these subjects in Switzerland, the Swiss pupils were years ahead of their English counterparts. The 'child-centred' approach that the researchers observed in English schools, however, turned out to be anything but that:

> For most of the time pupils are left to their own resources. The teacher's role is mainly to help individual pupils when there are difficulties and to check their work. Pupils are addressed by the teacher usually only if they request it. Often several pupils need the teacher's help at the same time; they therefore put their hands up – or queue at his desk – waiting until the teacher is free to help them. Some teachers have a rule for length of queues, such as 'not more than four waiting at my desk at a time'; other teachers walk around the class with a 'crocodile' of pupils trailing behind them waiting for their questions to be answered. The pressure on teachers means that checking of pupils' work is often cursory; many pupils do not receive adequate support from the teacher to carry out their work successfully, and poor understanding by pupils frequently goes unnoticed. Average pupils, and even more so those who are below average, consequently suffer.

The dreaded didactic approach of the Swiss schools, however, paid dividends. Half to two-thirds of the lesson was devoted to continuous interaction between the teacher and the whole class. The teacher started with a problem and developed solutions and concepts through graded questions addressed to the whole class. Pupils were thus guided towards discovering the solutions themselves. Virtually the whole class was mentally engaged and the teacher could see from the responses who was weaker and needed individual help during the written exercises. 'To English teachers familiar with the long tail of under-achieving pupils in their mathematics classes who have trouble in understanding what they are expected to do, the degree of evenness amongst Swiss *Realschule* pupils in their attainments comes as a considerable revelation as to what lies within the realm of possibility,' wrote the authors.

In English science classes, they observed, the obsession with 'process' and with children finding things out for themselves wasted huge amounts of time and left the children with a very poor grasp of scientific principles.

For example, in a biology lesson concerned with the relative influence of inheritance and environment on human characteristics, most of the lesson-time (a 'double lesson' of 70 minutes) was devoted to pupils collecting data on each other's hand sizes, thumb lengths, colour of hair and eyes, and whether or not they could roll their tongue (an inherited characteristic of no known significance – as, indeed, the teacher explained). Pupils were asked to 'think about how to record, analyse and present this information', and additional time was to be spent the following lesson on pie charts, bar charts, etc.

In other words, the English pupils had spent all that time investigating and discovering information of virtually no value whatsoever. 'On our visits it seemed evident that many pupils had gained only a superficial idea of the principles underlying their individual experiments and some had developed a wrong understanding. Teachers tended to refrain from instructing pupils in the scientific principles at issue; rather, in accordance with views long commended by Her Majesty's Inspectors, they wished to encourage pupils to "think scientifically for themselves", "speculate about scientific ideas" and "pursue their own lines of inquiry".'

A footnote to the report added the following deeply alarming observation made by researchers from Leeds University: 'We have seen pupils convinced that the "froth" visible in a boiling yeast suspension meant that the enzymes present functioned more effectively at 100 degrees Centigrade than at room temperature (when,

in truth, the enzymes have been destroyed at that high temperature).
Meanwhile a harassed teacher was either unaware of the situation,
or uncertain about the . . . legitimacy (*sic*) in the currently approved
teaching context, of correcting this impression.'[26]

So English pupils were being left to flounder in their own
(un)scientific errors. By contrast, the Swiss pupils' own experiments
were often preceded by the teacher's experiment carried out in front
of the whole class, followed by discussion of the implications.
Alternatively, pupils were asked to anticipate possible results,
explain the reasons and consider the next experimental step. In
this way, the transmission of knowledge provided the essential
framework for the simultaneous development of the pupil's critical
and analytical faculties. And all the evidence was that this was far
more effective than the English approach.

The researchers also discovered that Britain was grievously failing
its least able pupils. In Switzerland, woodwork and metal work
were taught to high standards, with pupils making real artefacts
and with skills taught systematically. Similarly, cooking or sewing
were taught to the whole class who then cooked or sewed by
themselves. By comparison, what British pupils were doing in
technology under the National Curriculum was risible. Instead of
being taught anything, they were given broad themes, spent many
weeks trying to decide what projects to do and then not surprisingly
got disillusioned. One group trying to decide on 'communication'
for four fruitless weeks eventually brought in ingredients and made
a pizza instead! The actual making of objects became marginalised
almost to the point of extinction.

There was little incentive to produce good work. Among examples observed by the researchers were: 'crudely decorated cardboard
boxes which were about to fall apart; a rough wooden train which
fell to pieces when moved; a "completed" electric table lamp which
had not been wired up; and a small woven piece of cloth which
had to be done again because the pupil had not tied up the ends
of the threads'. In a spirit of hubris, the curriculum told them
to 'simulate production and assembly lines' and 'to evaluate the
quality of products and to devise modifications that would improve
their performance'. The one thing it did not do was actually teach
how to make anything properly. For all its professed preoccupation
with 'relevance' and 'experience', child-centred learning couldn't
even be relied upon to teach anyone how to sew a mail bag.

Deconstruction of Education

The deconstruction of British education has extended beyond the schools into vocational qualifications as well. Vocational education, according to Alan Smithers, professor of education and employment at Manchester University, is a 'disaster of epic proportions' and means Britain can no longer produce a decent skilled worker. 'The new courses are utterly lightweight, ridden with ideology and weak on general education,' he wrote in 1993. This failure was due to the 'philosophy of equal ability under which there must never be winners and losers, in which competition is a bad thing and testing is regarded as little better than a necessary evil'. Plumbers, untutored in elementary trigonometry, weren't able to calculate the slope for a drain. 'All rigour has been sacrificed to flexibility. The students are not stretched. Remarkably, even the basic skills of literacy and numeracy, principal concern of employers and, one might have thought, of vital importance to electricians and plumbers, are neither taught nor assessed. They are simply "inferred" from project work.'[27]

A General National Vocational Qualification intermediate level test paper sat in January 1996 showed candidates pictures of four birds. One was building a nest, another was defecating, a third was preening and the fourth was eating a worm. Candidates were asked which one was feeding itself. Another question asked whether seeds grew best in a dark cupboard, under a bench, inside a box or on a sunny window-sill. Despite the fact that such questions could be answered by infants, they were given to 16-year-olds.[28]

So the doctrine of equality, promulgated ostensibly on behalf of the people at the bottom of the heap whose feelings were never to be offended, was catastrophically failing those very people, abandoning them not only to a lack of skills but to illiteracy and innumeracy as well. The obsession with creativity was ensuring that people who were most in need of educational assistance were left incapable even of creating a wooden box or wiring a lamp. Not only did the Swiss, by contrast, manage to provide their own pupils with the skills to make artefacts with precision, but they also managed as a result to inculcate work habits such as perseverance, reliability, care, patience and precision. And because they managed to keep their lower than average pupils motivated, these children tended to persevere with academic studies. All this the Swiss managed to do because they still understood the meaning of teaching and of knowledge.

Yet British teachers and other educationists appear to be impervious to such evidence. They continue virulently to resist the notion that knowledge has to be transmitted. One secondary school teacher wrote in a letter: '. . . learning is something the *learner* must do'.[29] The teacher, it seemed, had no role any more except to facilitate this learning process.

Another letter, from a former teacher and senior lecturer in education, showed how the deconstruction of knowledge is intimately wrapped up with the belief that as knowledge might be denied to some, it must therefore be denied to all. Taking issue with the idea that education must involve a transmission of knowledge as 'simplistic', he went on:

> This is a belief which has bedevilled education for far too long. It has engendered an over-emphasis on didactic teaching at the expense of other modes of teaching and an over-valuing of academic knowledge at the expense of other forms of knowledge. It takes a simplistic view of the psychology of learning and the psychology of the learner stressing memorisation and the acquisition of facts as the main route to learning. Thus we have thousands of pupils each year 'swotting' facts for examinations, the majority of which they will forget when the examination is over . . . At a more general level it has contributed to the ridiculous and damaging division between academic and non-academic, between theory and practice, between verbal and non-verbal and to the different social status of each. This in turn has caused the majority of pupils to undervalue their own abilities and talents because they do not match up to the narrow criteria of the transmission of knowledge model, with consequent low self-esteem, lack of confidence in their own learning and low expectations of success.[30]

But the people with low expectations of pupils' success are the teachers; and the loss of self-esteem and confidence are the direct consequence of those teachers' abandonment of such pupils to their own 'creative' absence of knowledge.

than make statements or give direct instructions . . . where a simple clear statement would have saved a great deal of time . . . There was a reluctance to say openly that a particular answer to a question was wrong. Incorrect answers were sometimes ignored; more often they were praised as if they were right, and then ignored. Conversely, correct answers were sometimes treated as if they were incorrect.

The unthinking and undiscriminating use of questions, he wrote, might reflect a 'taboo on didacticism', a sense that children at all costs must not be told. 'The result . . . can be a charade of pseudo-inquiry which fools nobody, least of all the children, but which wastes a great deal of time. Similarly the indiscriminate and thus unhelpful use of praise . . . may stem from a laudable concern that children be encouraged and supported in their learning. Yet in the end too this can be counterproductive, with children becoming confused or cynical in the face of what they may begin to see as so much mere noise.'

And just to make it quite clear that the well-worn alibis wouldn't wash this time, Professor Alexander added that standards had fallen in Leeds *despite* increased spending on the schools. Any criticism of these destructive methods had been stifled through the power of local education authority advisers influencing appointments and promotions. This led to 'unthinking conformity and the loss of the professional analysis and debate which are essential to educational progress'.[36]

The following year, Alexander took part in the wider 'Three Wise Men' report on primary education. Once again, the report excoriated the primary school orthodoxy. 'Over the last few decades the progress of primary pupils has been hampered by the influence of highly questionable dogmas which have led to excessively complex classroom practices and devalued the place of subjects in the curriculum,' it said.

It went on to shred the child-centred philosophy which sought to deny the validity of a received version of knowledge.

First, to resist subjects on the grounds that they are inconsistent with children's views of the world is to confine them within their existing modes of thought and deny them access to some of the most powerful tools for making sense of the world which human beings have ever devised. Second, while it is self-evident that every individual, to an extent, constructs his/her own meanings, education is an encounter between these personal understandings and the public knowledge embodied in our cultural traditions. The teacher's key responsibility is to mediate such encounters so that the child's understanding is enriched.

Too much work consisted of 'little more than aimless and superficial copying from books'. Whole class teaching, by contrast, provided order, control, purpose and concentration because the interaction with the teacher was essential.

The problem, it acknowledged, was partly ideological. 'In some schools and local education authorities, the legitimate drive to create equal opportunities for all pupils has resulted in an obsessive fear of anything which in the jargon might be deemed "élitist". As a result, the needs of some of our most able children have quite simply not been met. There has also been a tendency to stereotype and to assume that social disadvantage leads inevitably to educational failure.'[37]

The report caused a furore in the educational world. But this time, Professor Alexander appeared remarkably shy of his own findings. Not to put too fine a point on it, he suggested that the report to which he had been a signatory had not said what it had said, blaming it all on media distortion. But there had been no media distortion. It rather appeared that Alexander might not have been able to cope with the pressure from his outraged colleagues who wouldn't forgive him for blowing the whistle in this way. At the press conference launching the report and in subsequent newspaper articles, he sought to put a gloss on the report which simply was not supported by what it actually said.[38]

Its findings are, in any event, borne out by the situation in many primary schools, which give an impression of chaos caused by the fact that the teachers appear to have set themselves goals that are simply impossible to achieve. At one such hard-pressed inner-city primary early in 1996, the teachers could be observed constantly moving from table to table in the classrooms where different groups of children were engaged on different tasks. In every classroom, there were prominently displayed lists of assignments, so that if a child was stuck and the teacher was busy the child knew what else to be getting on with. The chances were that the teacher *would* be busy, because of the impossibility of getting round to everyone who needed assistance. Wouldn't it have been more sensible, one teacher was asked, to teach the whole class at once rather than give himself such a logistical nightmare? He couldn't do that, he explained, because the children were all at different levels of ability. But that didn't appear to be a problem for all those schools on the continent, or for Sudbourne primary school in Brixton. Moreover, this way it appeared that hardly *anyone* was being taught properly. 'It certainly means that teachers have to be like jugglers,' said the headmistress. 'But if teachers are very good and very well organised, it can work.'[39] But what if they are not?

This school had a very high proportion of children from impoverished backgrounds, and a high turnover of immigrant children. It was hard not to be sympathetic to the difficulties such a situation posed for these teachers. But it was hard also not to conclude that the children's difficulties were exacerbated rather than helped by the attitudes of some of the teaching staff. 'It's up to the children to devise their own investigations in science,' said one teacher. Did she ever do experiments in front of them to demonstrate scientific principles? She looked incredulous at the very question. What did they do in history? 'They have to go round the school looking for evidence,' said another teacher. The nine-year-olds were doing a topic on mammals in which they were 'learning how to do research'. Next would come electricity. 'We encourage them to be as independent as possible and not to be spoon-fed,' said the teacher. So what would happen if they failed to find out for themselves enough information about electricity? 'Well, they wouldn't have found it out then!' came the reply.

Official reports have told a consistent story about the state of teaching in primary schools. In 1994 the education inspectorate Ofsted reported 'significant and worrying weaknesses in the teaching which left about one quarter of primary school pupils adrift'. There was a strong association between the quality of the teaching and the standards achieved. 'Poor pedagogic skills characterised 90 per cent of the lessons where pupils' standards of achievement were unsatisfactory or poor,' it said.[40]

What has been less understood, however, is the connection between those poor pedagogic skills and the doctrines of learning rather than teaching and the deconstruction of knowledge. As Chief Inspector of Schools, Chris Woodhead runs Ofsted. In his view, if standards are to be raised we must challenge certain wrong-headed beliefs which are still prevalent in too many schools. Woodhead has talked to perhaps half the primary heads in the country. He believes that while many schools are now questioning deeply held assumptions, there is still a good deal of reluctance to accept that education is above all else a matter of developing basic skills of literacy and numeracy and introducing children to a body of knowledge and understanding. There is still a view that the basic goal of primary education is to develop children as 'independent learners'. There is a continuing tendency to down-play the importance of teaching and to talk, vaguely but passionately, about how children learn.

'My basic argument is that teachers need a secure grasp of what they are teaching,' he said. 'They need high expectations of children and to be able to create an ordered environment, to explain things clearly, to question and challenge children, and so on. These seem to me to be pretty unexceptionable propositions, but they often provoke considerable controversy. The climate *is* changing, but there is still a long way to go before we have a situation where the profession as a whole has come to see that to reject didactic teaching is to reject teaching itself.'[41]

In one of the public speeches that has helped make him so unpopular in the eyes of the education establishment, Woodhead tried gamely to reassert the liberal ideal of education as the disinterested study of the best that has been thought and said, an end in itself with no further instrumental purpose. Too many teachers, he said, identified the teaching of a worthwhile cultural inheritance with Dickens's stultifying teacher Mr Gradgrind. One head had spoken to him of the irrelevance of a curriculum based on history or English Literature. '"What is the point", he argued, "in spending time on arcane knowledge which has absolutely no relevance to the needs of industry and commerce?"' Another head had then summoned up the courage to disagree and spoke of the need to know something of human experience. Woodhead commented: 'Have we really come to a situation where the received wisdom is that knowledge is that unimportant? Where we have turned our backs on any sense of initiating the next generation into the different ways in which human beings have over the years come to organise their experience of the world? Into the riches of our literary traditions and our historical heritage?'[42]

The answer, tragically, to all his questions, was 'yes'. In a subsequent speech, Woodhead reflected on the widespread retreat from teaching and from knowledge itself in favour of the cultivation of the imagination. He had been shown a document, he said, written by heads about the role of the teacher. It contained 'lots about facilitation and learning skills and empowerment but nothing, absolutely nothing, on the teacher as an authority whose job it is to teach children something which they wouldn't otherwise know'. But as the philosopher Michael Oakeshott had written, it was only when learning took place by study and not by chance, 'with the appearance of a teacher with something to impart which is not immediately connected with the current wants or "interests" of the learner, that the initiation of the newcomer into the human inheritance could become schooling at all.'[43]

Woodhead has been a lone voice in the upper reaches of the education establishment in attempting to sound the alarm about the collapse of educational values and introduce a proper debate. For his pains, he has been ridiculed and abused, written off as a Tory placeman (he is not: he doesn't even vote at general elections) or as a career opportunist who has trimmed his sails according to the prevailing political wind. In fact, he has played a brave and risky game in Whitehall: far from doing the government's bidding, he has been a consistent thorn in the side of an Education Department which has tried to pedal as fast as possible in the opposite direction from the implications of the evidence he has been producing.

But then one of the hallmarks of the new orthodoxy, in education and beyond, is the refusal to accept evidence as having any validity. In its place have come ideology, prejudice and propaganda. We are witnessing not merely a retreat from teaching and of knowledge but a denial of reason itself. Evidence of demonstrable facts lays claim not upon our imaginations but on our powers of deduction and analysis. Those powers have been devalued, written out of the script. Rooted as they are in reality, they challenge the new world of make-believe fun and therefore cannot be allowed.

The evidence of the past is transmitted through memory. Memory keeps faith with our history. Through memory we remain rooted in a culture and its institutions. We can take comfort in their immutability to provide a continuity between past and future. But the value of such fixed points has been destroyed. In its place, every child has to make its own experience afresh. Knowledge, the transmission of received wisdom from one generation to the next, has been disinherited and memory scorned.

Michael Oakeshott defined education as

> the transaction between the generations in which newcomers to the scene are initiated into the world which they are to inhabit. This is a world of understandings, imaginings, meanings, moral and religious beliefs, relationships, practices – states of mind in which the human condition is to be discerned as recognitions of and responses to the ordeal of consciousness. These states of mind can be entered into only by being themselves understood, and they can be understood only by learning to do so. To be initiated into this world is learning to become human; and to move within it freely is human, which is a 'historic' not a 'natural' condition.[44]

It is no accident that Oakeshott is deeply unfashionable in contemporary educational circles. These have repudiated the very idea of education as an initiation. Instead, it has been redefined as a rolling programme of fresh and subjective experience in which by definition nothing can be learned because there is nothing of value to hand on. Value has become a subjective concept. This repudiation has implications that stretch way beyond the school gate. Before the child gets into the classroom, its most powerful educational experiences of all are being provided by the family. Both school and family, teachers and parents, are involved in the business of handing on to the child those habits of thought and of feeling that initiate the child into the mysteries of its culture and of what it means to be human.

What has happened is nothing less than a breakdown of the accepted conventions of the transmission of our culture, which, if it is to survive, has to be handed down in an orderly and systematic manner from adults to children. It has involved a corruption of the concept of education itself and a flight from literacy and knowledge, all in the name of a rigid doctrine of equality and individual rights. This fundamentalist egalitarianism has destroyed the hierarchy of authority between the generations. The relationship between adults and children in both family and school, between parents and their offspring or between teachers and their pupils, has gone into reverse with the effect of infantilising adults and making children assume adult responsibilities well before they are ready. All external authority has been demolished.

This is not some fringe ideology which, although troublesome, has left the majority of people untouched. These attitudes now course through the bloodstream of our culture. They permeate the establishment and govern the running of our institutions. The effects upon the family, and upon the social order itself, are discussed later in the book. But the teachers' task is therefore now all but impossible, not least because their own professional culture has become subverted by the self-destructive orthodoxy of moral and cultural relativism, the doctrine that no value or activity can be held to be any better or worse than any other. Britain is in the grip of a culture war, and the most fundamental aspects of education are now in the front line.

Chapter Three

THE WAR OF THE WORDS

'Proper Literacy'

Anyone who may be troubled by the progress their child is making in learning to read would do well to study a paper delivered to a conference of English teachers held at Ruskin College, Oxford in 1991. The paper was written by Peter Traves, an English adviser with Shropshire county council. Local authority education advisers play an important role in helping influence teachers and promote desirable practice in maintained schools. One might reasonably expect them, therefore, to represent the best and most rigorous in educational thinking. One might also expect that their views illustrate the educational orthodoxy, reflecting the consensus about what should be taught and how.

In his paper, however, Traves revealed that his interpretation of literacy was very far from the ability to read words on a page which the vast majority of parents of young children would wish their offspring to acquire. For him, being able to make sense of the words on a page was the *wrong* kind of literacy, a kind actively to be discouraged. 'Proper' literacy, he wrote, quoting from a text that took the approved line, meant bringing 'your knowledge and your experience to bear on what passes before you'. 'Improper' literacy meant superficially following the words across the page.[1] 'Proper' literacy, in other words, involved tapping into what the child already knew from its own life. Teaching a child to decipher the codes of language so that the words became accessible was, it seemed, an activity to merit deep disapproval.

It remained unclear how, in the absence of explicit decoding instructions, a child's own knowledge and experience would be able to transform meaningless printed marks into intelligible words and sentences. However, 'mere reading' wasn't the point of the exercise at all. The point was political power. Through 'proper literacy', wrote Traves, people 'extended and enriched control over their life and environment'. He went on: 'Improper literacy, mere reading, on the other hand, is a reductive and destructive state of being in which the illusion of achievement is substituted for the genuine article, where the potential for power has been thwarted and channelled.'

Reading, in other words, was positively unhelpful. The population's regrettable failure to display the appropriate revolutionary zeal apparently meant that teaching children to read was not just a waste of time – it was actively oppressive. 'We need to examine our own roles as educators in this state of affairs,' he wrote. 'We need to consider what has happened in education to produce a situation where mass literacy has not resulted in a corresponding demand for greater control over the economic, political and cultural life of society. Public education has not produced an empowered people. Education has served the role of controlling and repressing the aspirations of the general population, and the teaching of reading has had a central part to play in this.'

At this point, the ordinary concerned parent might well wonder whether any child in Peter Traves's own household was taught 'mere reading' or took to the barricades instead. Helpfully, the paper enlightens us with an account of the trials and tribulations of his own young son, Richard.

'Richard's pre-school and nursery experience of reading was a very positive and fairly rich one. Although he went to school unable to 'read' in the sense of being able to decode the print of books, he behaved as a reader in almost every other respect . . . He enjoyed being read to, talked about the stories and wanted more books. He memorised stories and large chunks of the phrasing from books and then delivered them back enthusiastically in a readerly tone of voice.' Behaving *like* a reader is thus smoothly substituted for behaving *as* a reader. Memorising stories and having a 'readerly tone of voice' (whatever that may be) comprise, it seems, the 'reading experience'. As long as the child's behaviour gives the illusion of reading, he is reading – even though he isn't!

In the light of this fantasy, it was perhaps hardly surprising that Richard would encounter difficulties later on. But his father was mortified by what happened when six-year-old Richard went to another school which used a reading scheme – in other words, attempted to teach the child to read.

'Richard started at the bottom of the scheme and stayed there . . . He had seen himself as a reader. He now described himself not only as a non-reader but as generally stupid.' Which perhaps wasn't surprising considering the reaction of his parents who were clearly distraught that Richard had been unable to master 'improper reading'.

'His parents managed to make all the available mistakes,' wrote Traves. 'These included enforced reading through the books which he brought home, shouting gems such as "Look what it says!"; "Yes, you do know that word – come on, what is it?"; "You're not looking at the words, stop guessing from the picture." A great deal of tears were spilt on both sides.' And where did Richard's father place the blame for this 'painful and sobering' experience? Why, upon the pressure to demonstrate visible, measurable success and the dominance of a 'mechanistic definition' of reading that 'concentrates heavily on the capacity to read relatively decontextualised print out loud' which 'pays little or no attention to the broad and complex web of behavioural and intellectual patterns that underpin real reading'.

But reading relatively decontextualised print *is* reading. If you can't do that, you can't read. No amount of political wishful thinking can make it otherwise. The painfully obvious reason for Richard's distress was that he had never been taught to read while being told the lie that he could. Traves's paper sheds chilling light on an ideology which appears to defy reason itself, which makes words mean the opposite of what they do mean, and which produces real victims, in this case – ironically – the ideologue's own child.

It is, furthermore, an ideology which lies at the very core of the disaster that has befallen British education. Literacy is, after all, the launch-pad of the entire educational adventure. If you can't read, you are excluded not merely from participation in education but from playing a full role in society. Far from being 'empowered', people who are unable to read are comprehensively disenfranchised. If there are any rights in education, being taught to read is surely the most fundamental of all. Yet the issue of whether or how people are taught to read has become a political battleground. Peter Traves spoke not as a marginal zealot but as a proponent of an orthodoxy being promulgated by English advisers like himself and other educationists the length and breadth of the country – an orthodoxy, moreover, with implications well beyond education.

To most people, this may appear quite baffling. The fact is, however, that English teaching has become the epicentre of the culture war over identity and values which has convulsed not just British education but the whole of our wider society. English, after all, is the subject at the heart of our definition of our national cultural identity. Since English teachers are the chief custodians of that identity, we should not be surprised to find that revolutionaries intent on using the subject to transform society have gained a powerful foothold, attempting to redefine the very meaning of reading itself. But many perfectly competent teachers may well find this astonishing.

Concern over illiteracy is rising because the situation appears to be getting dramatically worse. In 1991, a survey found that 60 per cent of newly qualified teachers felt they had received too little instruction in teaching reading to beginners.[2] In 1994, the Adult Literacy and Basic Skills Unit published a report showing that 20 per cent of people fell below their 'foundation' or basic level in literacy, and only 12 per cent reached all literacy levels.[3] In 1996, the Basic Skills Agency found that more than a quarter of the people it surveyed didn't know where to put a full stop or a comma in a block of text, 60 per cent were baffled by apostrophes and four out of ten couldn't spell 'apologise' or 'unfortunately'.[4] But it is not just because so many people have these difficulties that the alarm bells are sounding. It is the perception, by employers and others, that young people who have gained notionally high grade academic qualifications can't spell or punctuate and that they have no command of grammar. They lack that elementary control over the language which they need even to write a properly constructed letter of application for a job.

The full extent of the disaster that is now unfolding was well illustrated by a remark made by Dr Peter Brown, chief examiner for the GCSE. 'It is not uncommon,' he said, 'to have a question paper criticised because the questions are not spoken on a tape-recording in addition to being printed on the paper. "What about the candidates who cannot read (or read very well)?" is the cry.'[5]

One might imagine that when levels of illiteracy have reached such a pitch that GCSE candidates have difficulty reading their exam papers, there would be unanimous agreement that something is going very badly wrong indeed with the teaching of literacy in our schools. Far from it – as the controversy over the 1996 Ofsted reading report so vividly illustrated. Among the English teachers in particular, there is either a refusal to accept that there is a problem at all or, in so far as there may be a problem, it is all put down to factors such as domestic poverty, the impact of TV, poor parenting, the high proportion of ethnic minority children for whom English is not their first language – everything, in short, except the way they are now being taught to read.

Above all, these 'scares' about reading – like all the other concerns about education standards – are laid by many educationists at the door of right-wing politicians seeking to hammer a few more nails into the coffin of an education system they wish to dismantle (or so we are told). This tension erupted in 1991 in the furore over Culloden school in East London, a school which had been acclaimed in a BBC TV series and praised by the National Union of Teachers for its good morale and high standards, especially in reading. Tests conducted, however, by the educational psychologist Martin Turner for the *Mail on Sunday* painted a rather different picture. Some 75 per cent of the 7- to 8-year-olds and 33 per cent of the 10- to 11-year-olds were non-readers. The children's performance in reading was only 3 per cent higher than average even though they were 50 per cent above the average in general ability.[6] This revelation was duly denounced as an attack by the tabloids, the school governors complained about inflammatory journalism and *The Times Educational Supplement* wrote that the school felt 'fitted up' by right-wing critics of modern teachers.[7] No-one, however, appeared unduly concerned about the children.

Three weeks later, however, the Schools Inspectorate agreed that too little attention was given to the systematic and structured teaching of reading at Culloden school. Reading standards, it found, were poor in 10 out of 14 classes, with average and less able pupils suffering most. 'The school largely follows what has come to be known as the "real books" approach, with no use of reading schemes and no systematic phonics. Many children continue to read the same few well-known texts which do not challenge them sufficiently.' Moreover, pupils 'do not always recognise the words they can read in familiar books when these occur again in other texts, nor are they able to use phonic or other word-building skills to help them decipher print'. The inspectors urgently recommended a more thorough and systematic approach including the teaching of phonics.[8]

Phonics Versus 'Real Books'

Culloden, however, was unfortunately not a one-off school run by oddball teachers. Its failure to teach reading properly through the systematic use of phonics was in line with the education orthodoxy being put into practice in primary schools up and down the land. 'Phonics' is the name given to the decoding of print through the sounding out of every letter, gradually putting those sounds together to form words. Once it has been taught, it gives children the ability to decode accurately any words they may encounter, unlike approaches based on whole word recognition which rely on memory and guesswork. However, as we shall see later, phonics fell out of favour as the political battlelines over literacy were drawn up. The belief that children had to be taught the codes before they could derive meaning from words was replaced by the belief that meaning must come first and replace the teaching of the codes. Phonic decoding was nudged aside by reading schemes such as *Look and Say* which relied on the child recognising a whole word by sight, aided by context and pictures, a flight from phonics into the 'meaning first' reading philosophy.

But it wasn't just the phonic system that became associated with 'improper reading'. It was the very concept of any systematic teaching of reading at all. A more overt political dimension had crept into the critique. Reading schemes based on methods such as *Look and Say* were also junked on the basis they were boring, middle-class and out of touch with children's lives. Since meaning was all, and since the meaning of these 'stories' didn't find favour with prevailing political prejudices, they had to go. One might say that such 'stories' were hoist with their own petard. Having laid claim to 'meaning', they were then judged on that basis and found wanting. Thus was born the principle of 'real books', in which stories written as literature for children were used in place of reading schemes. The idea was that, exposed to writing of quality, children would be so enticed by the delights of literature that they would take to reading with enthusiasm. The only problem was that no-one was now teaching them how to decode the print on the page. It was reading by osmosis. The result was that the pretence of reading, which Peter Traves had inflicted upon his own son, was now institutionalised as the best educational practice.

The outcome was as surreal as it was tragic. The teaching of reading became the un-teaching of reading, espoused with evangelical fervour by teachers who truly believed that their non-reading children were reading. If the children displayed enthusiasm for children's literature, that was considered a success – even though the children couldn't read the words on the page. Attitude replaced achievement.

A kind of madness took over. A book published in 1993 by the two Cambridge teacher-training hothouses, the University's Institute of Education and Homerton College, illustrated with devastating clarity the flight from reason that had taken place over teaching children to read.[9] One of the contributors, Helen Bromley, was deputy head of Sunnymede Infants School at Billericay in Essex. She waxed lyrical about the experience she had had in teaching a class of rising fives to read. It soon became clear, of course, that teaching them to read was the very last thing she was doing. 'Actually, I prefer to believe that I am improving and refining the reading skills that the children already have,' she wrote, without irony.

Strangely, the parents didn't seem to share her faith that their children were already programmed with innate reading skills. They would say unhelpful things such as: 'She's not looking at the words', or that there were 'too many words on the page . . . can we have something she can read herself?' So these parents had to be reprogrammed. Their expectations had to be made 'realistic'. One day a mother wrote on her child's record sheet: 'At last he is looking at the words.' Her obvious anxiety was, however, merely further cause for self-congratulation by Bromley that this mother's thought processes had been reorganised to transform evidence of failure into proof of success. 'I was particularly excited by this,' she wrote, 'because it represented such insight on the part of the parent. She had been educated to realise that looking at the text was not automatic (although most adults take it for granted) and had been astute enough to recognise the achievement of her son.'

Being realistic, it appeared, meant accepting that teaching a child to read meant not teaching it to read.

For the children were not reading. They were instead enjoying being read to. This is hardly a great achievement for a teacher, since there can scarcely be a child who does *not* enjoy being read attractively written and illustrated stories. Equally, it is hardly an achievement for a teacher of reading to allow a long delay to elapse before the child even starts looking at printed letters. Rather than acknowledging, however, that this betrayed her own low expectations of the children, Bromley gushed over their keenness to look at new books. Of course, we would all rejoice in children's enthusiasm for books, particularly if they come from backgrounds where books are a rarity. But this is hardly a substitute for being able to read them. Her pupils, Bromley enthused, knew more about books than any other children after two terms at school. 'They could discuss the relative merits of various illustrators, sort out a selection of works by a named author and were adept at reading meanings beyond the literal.' The one small problem, it appeared, was that they couldn't literally read.

'Phonics,' she claimed, 'was a continuous thread, gradually woven into the effective strategies that the children began to employ in their reading . . .' Whatever this continuous thread was, however, it wasn't phonics. That system of learning to read must be taught systematically rather than mixed up with other 'strategies' which may hopelessly confuse the child and do significant damage. Indeed, like many teachers who talk about using 'a variety of reading strategies', there was no indication that this teacher understood what phonics meant at all. This was hardly surprising. The fact that the children couldn't read appeared immaterial. 'What was most remarkable was that all the children believed that they could read the book . . .' A fairy story indeed. The children were being told a lie.

The fierce white heat of real books ideology may now have passed, but the assumptions beneath it remain firmly embedded in current education practice. Children are being labelled dyslexic and referred to psychologists when all that is wrong with them is that they haven't been taught to read properly. Despite the references to phonics in the National Curriculum, the majority of primary school teachers remain convinced that to teach children to decode the language is to separate them into successes and failures and to bore them rigid by mechanistic drills and inflexible and meaningless reading exercises. Not only do they fail to understand how phonics works, because they themselves have never been taught it; not only do they ignore the fact that phonetic reading schemes are now much more imaginative and creative; most important of all, they find it impossible to believe that virtually every child, however deprived its background, can be taught to read in this way, and *fast* – provided it is taught properly.

Good primary schools prove this over and over again. In one such, situated in a highly deprived part of London, some 98 per cent of the pupils come from Bangladeshi or other Asian homes where little or no English is spoken. When the children first arrive in the nursery class, they have to have an interpreter; some, who speak only Urdu for which there are no interpreters, have to cope with no translation at all. Nearly half of these children are on free school meals, a prime indicator of deprived backgrounds. Some turn up in the morning with faces baggy from lack of sleep; they go to bed very late because their homes are so overcrowded that they go to bed on the living-room sofa and can only sleep when their parents have gone to bed themselves.

Yet by the time they are five these children are *all* reading English story books. What's more, their reading skills help teach them to speak English as well. In one reading class, a group of five- to six-year-olds were reading together on an overhead projector from a reading scheme book, from which they deciphered new words such as 'knight', 'reached', 'beautiful' and 'brilliant'.[10] From the way they read and talked about the story collectively with the teacher, it was clear they were all enjoying it greatly. The little girls bounced up and down and hugged themselves when they were told they were going to read this story. Afterwards, the class all read together from a regular children's story book. It was hard to imagine that at home most of them would speak very little English at all.

The head teacher later remarked that many teachers in the neighbourhood just couldn't believe that such young children could learn to read so fast.[11] These teachers, she said, thought they knew what phonics was but in fact they did not. They had never been taught it, and they had only a hazy idea. They might teach children their letters, but at a very slow rate and they wouldn't teach them to put the letters together. They seemed to think that would be too difficult. This head knew the mind-set she was dealing with among her teacher colleagues. At a previous school in the 1980s, she had herself been a convert to the 'real books' approach. It had been a disaster.

> It was so patronising; I just wouldn't listen to the parents. They'd come and say, why wasn't I using the reading schemes, and in reply I'd ask them how they would feel if they were a child being marked out as being on a less advanced reading scheme than their friends. I just kept telling them that their child would learn to read and love books. I really thought this approach would make reading more exciting for the children. But our reading results took a terrible dive. I remember one little boy saying to me: 'I'm really getting the hang of reading now. I can read this book without opening it. Soon I'll be able to read with my eyes shut!' And as our results dropped further and further down, so the money the council gave for children with special needs kept going up and up. But no-one came to us and asked what was going wrong here.

Reading Crisis

This teacher radically revised her approach after she realised children at the neighbouring school who were being taught in the traditional manner were learning to read at a very young age. However, attempts to sound a national alarm by showing how school reading standards have dropped have encountered venomous resistance. In an infamous episode in 1990, the education psychologist Martin Turner lost his job and saw his reputation persistently trashed by educationists because he produced figures demonstrating a huge drop in the reading ability of hundreds of thousands of seven- and eight-year-olds in ten unnamed local authority areas south of the Wash.[12] It was, he said, the most unprecedented drop in reading standards ever recorded in peacetime. Despite the fact that his figures were subsequently upheld as accurate by a government-sponsored inquiry, Turner's research is misrepresented and rubbished to this day. This was all because he advocated teaching phonics and blamed the 'real books' approach for the decline in reading. The persecution of Martin Turner and the disinformation that has been generated in the attempt to destroy the authority of his evidence provide graphic illustration of the desperate struggle that is still being waged to redefine literacy on politicised grounds. They also illustrate a surprising degree of fanaticism among his opponents.

It was in 1984, when Turner joined Croydon education authority, that he first came across the 'real books' phenomenon. 'There was a tremendous polarisation between the education psychologists and the advisers who were evangelical about these new methods of teaching reading,' he said. 'It was very born again; these people had a fervour. I couldn't understand it. They were getting terribly excited about books like *Mr Gumpy's Outing* which I had been reading with my own four-year-old. I went back to the classic texts about how children learned to read and thought, this is just a superstition.'[13]

In 1990 Turner noted that Croydon's reading test scores had been going down for five consecutive years. 'Many more children were now in the bottom category for reading. And they were all coming knocking on our door. The psychology service was being asked all the time to help these children. Were they dyslexic?'

In one school which had gone over to 'real books', Turner was asked to provide psychological help for no fewer than 18 children from two cohorts in one year alone. 'I tested their reading and language ability. These were all capable children but they had no reading skills.' In the staff room there was war, with traditionalists and revolutionaries no longer on speaking terms.

'In 1990, I heard there were falling standards in Surrey. They had been doing intelligence as well as reading tests and so could show that the average intelligence of those who were failing in reading was very little different from those who could read. When I realised it was a national problem I lost interest in Croydon. I advertised on the electronic mail and got a number of replies from other psychologists. I organised a conference in Croydon that June. I knew it was important.' Eight authorities were measured and all showed the same sharp downwards trend. 'The decline in standards was huge,' he said. 'It was the equivalent of losing seven months over a four year period.'

The Times Educational Supplement got wind of Turner's evidence and the education world then exploded. In particular Croydon education authority, which was highly embarrassed by Turner's figures (even though he had been careful not to single it out) denied there was any drop in standards, forbade him to speak to the press and moved him abruptly from his post to work with colleagues who were unsympathetic. 'At no time,' Turner records, 'did any of them appear to express concern for the children.'

Turner's research was written up and published by Stuart Sexton, a former education adviser to the Conservative government. His card was then well and truly marked, even though previously he had had nothing to do with the political right. 'I was very naïve,' he said. In October 1990 he was asked to see the Education Secretary, John MacGregor. 'He was surrounded by his junior ministers, Eric Bolton (HM Chief Inspector of Schools) and his civil servants,' Turner claims. 'Bolton leant across the table and said: "This is only happening in Croydon" (referring to whole [real] books teaching). This became a constant refrain in the damage limitation strategy.'

Bolton's reaction to this account, however, illustrated the ambiguities at the heart of the debate about reading. Bolton, whose role in curriculum politics is explored more fully later, has denied Turner's version of the MacGregor meeting. 'I did not say it was only happening in Croydon,' he said.[14] 'I said HMI's evidence was that the proportion of primary schools that taught only through real books or phonics was very small.' The view that most schools were teaching a variety of methods, he added, had been borne out by the 'Three Wise Men' report (mentioned earlier) and by the 1994/95 annual report from Chris Woodhead, the Chief Inspector of Schools.

But far from endorsing Bolton's sanguine view, Woodhead's report indicated that the 'variety of methods' approach wasn't working. 'Approaches to the teaching of reading generally include some phonic work,' Woodhead wrote. 'The place and purpose of teaching phonics, however, rarely feature strongly in school reading policies. Consequently, the teaching of phonic skills is not as thorough as it should be and is often used mainly for those who are showing signs of reading failure rather than as an established part of a well-structured reading programme for all pupils.'[15]

In other words, too many teachers are paying lip-service to phonics. They claim it is one of the methods they use but in fact they do not apply it properly. And while it is of course true that children learn to read better if given access to different types of books, it is clearly disingenuous to use that argument to mask the fact that many teachers are not teaching children to read in the first place.

In the event, research carried out by the HMI (the school inspectorate) under Eric Bolton helped marginalise Turner's work. John MacGregor, the Education Secretary, told both the HMI and the then education standards authority, the School Examination and Assessment Council, to check out Turner's claims. The HMI, whose 'research' simply consisted of listening to schoolchildren reading rather than conducting any formal tests, said that although the reading standards of one fifth of the children were unsatisfactory, there was no evidence of a decline in standards since the late seventies.[16] The House of Commons Select Committee on Education, taking its cue from the HMI, was similarly equivocal. There was no crisis, it said, or any need for a 'magic solution'.[17]

However, the study then commissioned by the School Examinations and Assessment Council from the National Foundation for Educational Research found that Turner had been right.[18] Out of 26 local authorities tested, it found evidence of a statistically significant decline in reading standards among 19, with the decline most clearly visible after 1985, particularly in 1988–90. Most schools, it said, used a combination of reading schemes and real books and all but one said their staff used phonics. Nevertheless, the report went on: 'To say that there has been no radical change in teaching methods in the last decade is not to discount the fact that the real books or apprenticeship approach to teaching reading has influenced teaching and that, in some instances, the principles underlying the approach – all of which are open to question and misinterpretation – have been misapplied. On a separate issue there is no doubt that the current dearth of reading schemes with a systematic phonics component has made it difficult for teachers to provide structured instruction in phonics for those pupils who require it.'

The National Foundation for Educational Research went beyond its brief and suggested its own explanation for the decline in reading scores: the rise in one-parent families, substitute care and social disadvantage. Of 15 schools where reading scores fell, it said, seven were in large centres of population or industrial areas. So far from making the point, however, it appeared to disprove it since by its own account the majority of schools were *not* in such socially pressured areas.

The effect of the National Foundation for Educational Research study on the curriculum body was electric. Its members understood only too well that regardless of the spurious social deprivation 'explanation', the important point was that Turner's research had been vindicated. There *was* a reading crisis. Something was going very wrong in the schools at the most basic level. Much of their discussion of the document was devoted to how its findings could be hushed up.[19] Eventually, it tossed this hot potato to the then Education Secretary Kenneth Clarke who did agree to publish it.

Turner has now left the main education battleground of schools and education authorities to run the Dyslexia Institute instead. The main use of systematic phonics, he said, now lay in teaching children who had failed to learn to read – because they were never taught phonics in the first place. 'Many of these teachers never come face to face with their victims. The children and their families retreat. I see the depth of these children's misery. And it's not just the child. The family is isolated in the street, the dad tries to teach the child to read and fails and gets angry, the child feels a failure and is possibly labelled dyslexic. There's no real test of conscience in this matter because these teachers are never brought face to face with what they've done. I'd like to take a coach load of these children and dump them in front of them and say, look at what you did to them.'

Turner's research administered a shock to the system. A number of teachers began to focus on reading and altered their practices to some extent. Cynics might say this has merely made teachers much more careful to pay lip-service to phonics even though they may not be using the method properly. Even so, its impact has been limited. Educationists have lost no opportunity to attempt to marginalise Turner and dismiss his findings. The intellectual gymnastics involved in this counter-attack indicate how high the stakes are here. This argument is about something much more significant than an otherwise arcane dispute over a classroom technique. The prize which Turner's opponents feared he would snatch from them was nothing less than the intellectual soul of society itself.

The weapons used against Martin Turner in this metaphysical

battle have involved smears and distortions. He has been dismissed as a right-wing ideologue in books written for teachers and students of education. These books themselves tell us much about the way in which education has been hijacked for other ends. Ostensibly part of a scholarly canon of educational texts aimed at classroom teachers, they are themselves often little more than vehicles for ideology and political propaganda. A sustained attack on Turner was included in one such book, *Language, Literacy and Learning in Educational Practice*,[20] published for an Open University MA course. The preface sowed the first seeds of suspicion that perhaps this book had a hidden agenda. Although it was about language and literacy its contributors dealt, it said, 'with many of the "big issues" relating to life in the postmodern world – 'identity, social relations, social control, ideology, freedom, democracy, power, aesthetics, pleasure . . .' In other words, this was a highly politicised book in which the idea that literacy might possibly be about teaching children to read was hardly central.

The book contained an edited version of Martin Turner's *Sponsored Reading Failure* followed by a 'critique' from Dr Barry Stierer, senior lecturer in the school of education at the Open University and one of the book's editors. The two chapters purported to offer students a 'debate' on the subject. But Stierer's chapter, which dismissed Turner's work as an 'ideological' report which had a 'disproportionate impact' on public opinion, comprised a stream of opinionated and highly questionable assertions to back up this claim. For example, he used the National Foundation for Educational Research study to support some of his arguments, while omitting to mention the crucial fact that it had vindicated Turner's findings. He quoted Professor Asher Cashdan, director of the Learning and Teaching Institute at Sheffield Hallam University, as saying Turner's research 'would not have been allowed to appear in its present form in any publication of standing'.[21] But Stierer chose not to add that in the light of the National Foundation for Educational Research study Cashdan subsequently wrote: 'As one of those who has consistently refused to accept that there has been a demonstrated drop in children's reading attainment in the latter part of the 1980s I now have to revise this opinion and admit that the Foundation's systematic study does seem to confirm such a fall.'[22]

Britain's universities once enjoyed the reputation of being the centres of truth-seeking, temples of disinterested knowledge and objective evidence. Books like this, however – and unfortunately they are now a commonplace – betray that tradition utterly. Not only do they substitute flimsy assertions and propaganda for the objective and balanced information they pretend to offer, but tellingly they seek to deny the validity of evidence itself. This renunciation of that fundamental belief in the workings of reason, which has characterised thinking since the Enlightenment, is a remarkable and widespread phenomenon of our times.

In the course of his onslaught upon Turner, Stierer also turned his fire upon the present author. The particular accusation here was to have been bamboozled (in an article in the *Guardian*[23]) into a credulous belief that Turner was right because of his pseudo-scientific mystique as a psychologist. But of course, the reason why Turner was believed was that he relied not upon opinionated assertions but upon *evidence*, factually and dispassionately collected for everyone to see. But evidence which does not support the approved political position is apparently not evidence at all. Stierer claimed that the reading tests upon which Turner's evidence was based had been 'widely discredited'. But they have not been discredited. The people who reject them do so, as Stierer himself explained, because 'they only provide a crude measure of children's ability to decipher decontextualised print, or to comprehend unseen text which is read for no real purpose'. In other words, they are discredited because they only reveal *whether or not a child can read*. And as we have seen from Peter Traves, reading no longer counts as 'proper' literacy for the *sans-culottes* of the English teachers' revolution.

The perversity of this counter-evidential culture is plain from any study of the main authorities on the teaching of reading which show overwhelmingly that phonetic teaching is far more successful than any other approach. These writers are authorities precisely because they use demonstrable evidence, both from analysed test results and from personal, direct observation, of how real children actually get on with these different approaches. The vast majority of so-called experts who dismiss such evidence as bunk do not write from direct experience of teaching children to read.

Yet the evidence is all around and points in the same direction. Experience in British schools has demonstrated time and again a systematic improvement in children's reading ability where phonics has been reintroduced. At Woods Loke primary school in Lowestoft and Raglan school at Bromley, phonic teaching was introduced after desperate parents organised a petition for change. As a result, seven-year-olds then achieved a reading age of nine-year-olds.[24] St Clare's school in Handsworth, Birmingham made a similar change but the head teacher Kevin Cassidy had to write about it under a pseudonym because of the hostility being expressed to his views.[25]

One of the most authoritative experts in this field is the American Jeanne Chall. In a classic work published in 1967 she produced an analysis of hundreds of studies, both by other authorities and by herself, which showed that children taught to read by cracking the codes were better at both reading and spelling – and with no lessening of interest in books – than children who had been taught by the meaning-first approach.[26] When Chall updated her research in 1983, she wrote that the evidence in support of code-cracking had got even stronger. Yet curiously, despite the fact that research into reading has been pointing overwhelmingly in the same direction for so long, very few teachers or teacher educators in Britain have been taught about it. It certainly does not feature on the reading lists given to student primary school teachers by teacher training institutions, as we shall see later.

As Martin Turner comments, it is not only the ignorance which is so striking; it is the fact that there is a widespread lack of understanding that these are empirical issues *at all*, capable of being investigated experimentally and proved. As people like Barry Stierer demonstrate, there appears to be a resistance not just to unpalatable evidence but to *any* evidence. Facts seem to be an affront. This shows not merely how social science tends to crumble when it has to argue its corner on properly scientific grounds. In Turner's words, 'ideas with a durable emotional appeal enable ideological managers in the state sector to secure and enhance their hold on power'.[27] Truth, in other words, is an obstacle to political progress.

The reason why learning to read through whole word recognition or picture clues is so profoundly unsatisfactory is perfectly simple. The English language is organised on an alphabetic system rather than one based on pictures, symbols or logograms – like Chinese, for example. An alphabetic script is a transcription of speech sounds into written symbols. But the meaning-first approach, which involves either memorising whole words or working out the meaning from any accompanying pictures, grafts onto an alphabetic system an approach geared instead to the translation of symbols. As the American educator Samuel Blumenfeld has written:

> It is known that by imposing look/say (or whole word/whole language) teaching techniques on an alphabetic writing system, one can artificially induce dyslexia, thereby creating a learning block or reading neurosis. Reading disability is a form of behaviour disorganisation induced by the look/say method, because look/say sets up two mutually exclusive tendencies: the tendency to look at written English as an ideographic system, like Chinese, and the tendency to look at written English as a phonetic system because it is alphabetic. The alphabetic system is in harmony with the spoken language because it is based on it. But the ideographic look/say system is in opposition to the spoken language because it is an entirely separate system of graphic symbols with no direct relation to any specific spoken language.[28]

But another crucial point about the alphabetic system suggests a further, profoundly unsettling reason for this dissonance. The alphabet is a primary tool of democracy. As the psychologist and reading expert Marilyn Jager Adams has written, the alphabetic system makes language potentially accessible to everyone.[29] 'Writing systems that were comprised of large numbers of symbols were naturally the possession of the élite. They were passed on only to those few whom it was felt should and, among those, could and would invest the amount of time and study required for their memorisation.' The Chinese writing system, for example, contains as many as 80,000 basic logograms but most Chinese adults are said to have a working familiarity with only about 4,000 to 5,000. By contrast, alphabetic systems generally comprise between 20 and 35 characters, few enough to be memorised by almost anyone.

The 'real books' or 'whole language' movement, as it is sometimes called, while purporting to be about a transfer of power actually serves to concentrate power in its proponents. After all, there are few more effective ways of stripping people of their democratic rights to inclusion in a society than by depriving them of the ability to read and write adequately.

The hidden agenda of the key advocates of 'real books' and whole language would appear to be political power. Far from promoting democracy, they are notable for their displays of prejudice and intolerance. To illustrate the point, Martin Turner quotes the experience of a specialist reading teacher at a conference on reading dominated by gurus of the whole language approach.[30]

> The lecturer, when discussing evaluation, made the following statements. We need to protect ourselves. We need to be able to say we know what we are doing. They – politicians, parents, the business community, ex-teachers – don't need to know. She made the analogy of the doctor using specialised rhetoric to prevent patients from arguing with them. Teachers, she said, need to do the same. She advised them to gain control of the rhetoric so that they would have power over their opponents! . . . The conference finished with the reading of three letters to the press, one from a concerned businessman, one from a concerned parent and the last from a concerned teacher: all were held up to ridicule (I agreed with every word in them) and they had a good laugh. I was appalled.

New Literacy

The battle over reading is but one element in a far wider collision of values with major implications for society as a whole. A radical individualism is being used quite explicitly in an attempt to subvert the established political order of liberal democracy. Knowledge itself has been privatised, with the source of authority being transferred from the teacher to the child in a so-called attempt to relocate power away from 'oppressive' adults – but which, in depriving the child of knowledge, transfers that power to none other than the revolutionary teachers themselves. Subjects that have been enlisted in this struggle have been renamed ('New History', 'New Maths') to signal their rebirth in heroic colours. And at the epicentre of this struggle sits the New Literacy, of whose proponents Martin Turner fell so foul.

The New Literacy is a movement which seeks to redefine literacy into a social and political programme. It is against systematic instruction, phonics, reading schemes, standardised tests and the teaching of the literary canon. It favours 'real books' for teaching reading and a 'whole language' approach which doesn't believe in paying attention to spelling, punctuation or grammar in case these might stifle a child's spontaneity. Creative writing is emphasised at the expense of formal structures; the idea that it is necessary to teach those formal structures in order to unlock a child's creativity, and that not to do so merely stifles children's ability to express themselves, is simply not recognised. The New Literacy encourages non-standard forms of English, emphasises race and gender issues, regards familiarity with the media as a valid form of literacy and advocates the use of English to 'empower' children and correct social inequalities.

In a key text which both explained and promulgated this movement John Willinsky, an associate professor in the Department of Curriculum and Instruction at the University of Calgary, USA, unfurled the banner.[31] Quoting a British acolyte, Margaret Mathieson, he endorsed her definition of the new politics of language as 'a moral and ideological effort to "remedy social injustice" perpetuated against working class students'.[32] What the movement was *against* was 'a literacy which is defined as the ability to perform at a certain level on a standardised test and which asks education for preparation and practice in that ability. It is to resist treating literacy simply as a competence that people have or do not have at some arbitrary level.'

Instead of being all about whether or not children could read, Willinsky told us: 'The New Literacy represents a faith in the social agency of public education as a forum for mobilising and empowering the young.' Stripped of its jargon, what this meant was that children's own experience in interpreting the world was to be considered as more valid than anything that could be taught or tested. Discussed on its own terms, of course, it all made very little sense, because what this amounted to was not education at all but nothing less than abandoning children to their own ignorance. At the same time, it was dressed up with an expectation that the children would achieve a highly sophisticated level of competence.

This last paradox is a phenomenon that characterises the New Curriculum. While parents may scratch their heads over the fact that their children can't multiply eight by six or have never heard of the Battle of Trafalgar, the New Curriculum acolytes riposte that since they are able instead to do set theory in maths or discuss the validity of different historical sources, standards are actually rising rather than falling. As the teacher and education writer Jennifer Chew has noted, the New Literacy shares the belief with New Maths and New History that children can tackle advanced work without first having mastered preliminary instruction and practice.[33] That is why the 'real books' approach works downwards from meaning to the words, rather than upwards from the mastery of letters and sounds to the words' meaning.

According to Willinsky, using an analogy with cycling, children must know where they are going from the start; they must learn to read by reading just as they learn to ride by riding. But as Chew has observed, young children don't know where they want to go, and want and need plenty of practice first. The outcome is that these children's apparently sophisticated knowledge is built on sand. They may know about set theory without understanding the notion of mathematical exactness; they may be able to discourse upon an author's *oeuvre* without being able to read the words on the page. Even Willinsky had to admit that, according to his own research, children taught to write by traditional methods demonstrated a *more* positive approach to writing than those who have been through the New Literacy 'let-it-all-hang-out-so-creativity-may-flourish-unhindered' approach.

The cycling analogy is, of course, ridiculous. One might as well say that a barrister should accumulate knowledge of statute or case law merely by virtue of appearing in court; or that a surgeon should get to know the workings of the human body by operating on a few. So why should the New Literates foist such a manifest absurdity upon a child?

The answer is that the New Literacy involves a transfer of function from teacher to child in an attempt to do nothing less than redefine the very concept of childhood itself out of existence. Children are expected to display adult competence by reading in an adult way because they are expected to behave and respond as adults without having been allowed to be children first. As we shall see later in the book, this is a crucial development that links education with discipline and family life in a comprehensive reversal of roles between adult and child. It has led to a breakdown in the understanding of the adult's primary duty to care for a child, with utterly disastrous consequences.

Willinsky himself made the revolutionary nature of this project very clear. 'This division over how best to go about reading is part of a larger philosophical struggle over the nature of the child. How are we, as educators, to take and fashion this creature? Is the child a blank slate, the empiricist's *tabula rasa*, that must be carefully inscribed with the needed information and procedures to operate in this world? Or is the child an autonomous meaning-maker in search of increasingly more sophisticated ways of developing a sense of the world?'

For the New Literates, the child is indeed an 'autonomous meaning-maker'. They reject entirely the fact that children's immaturity and lack of experience of the world's meaning makes them by definition dependent upon adults to guide them and instruct them to make sense of life. They believe that knowledge resides innately inside the child. Precisely how and when this lethal superstition took such tenacious hold will be discussed later. But the belief that children are able to make their own meaning has become the orthodoxy. And it follows from that, of course, that the role of the teacher has had to be completely redefined. As Willinsky explained, the New Literacy envisaged the 'withering away of the teacher'. Since 'students are sources of experience and meaning', the teacher turned responsibility over to the children effectively to teach themselves: hence the emphasis on 'learning' rather then 'teaching' which has become such a notable feature of current education discourse.

The idea of instruction has thus become a taboo. Barry Stierer has claimed: 'Instruction is a term that is very rarely used in the British context. Not only is it rarely used, but I think that for most British teachers it has quite explicit American overtones of programmed learning, of highly mechanistic conceptions of curriculum, teaching and learning. In fact, I don't think it's a term that would be tolerated by many British teachers – certainly not in relation to reading, or even any kind of language or English work in British schools.'[34] Equally intolerable, it appeared, was the notion of accuracy. 'Any conception of accuracy in the teaching of reading,' he wrote, 'will be based on unequal power relations between the reader and the arbiter of accuracy.'

In other words, correcting a child's mistakes in reading was not merely tactless. The very authority of the teacher to say that a word was correct or not was suspect. Power, after all, had drained away from the teacher. So it followed that if a child 'read' a completely different word from the one in the text, no-one must tell the child this was wrong. This resulted in a crazy, *Alice in Wonderland* world in which the ideologues of whole language came to glory in children's reading *failure* as evidence of *success*.

In a letter in the *Guardian*, Stierer wrote that traditional reading tests

> assess children's ability to read words and sentences removed from a meaningful context . . . Children taught by more modern methods, by which they are encouraged to 'read for meaning', will find such tests difficult until they have begun to recognise and take a detailed interest in the spelling patterns of words. Such children may not display an ability to decipher decontextualised print until the age of eight or nine but will be significantly more motivated to read for enjoyment later in life than children who are taught that reading is a mechanical chore. It is therefore not surprising that a growing proportion of seven-year-olds are performing badly on traditional tests. Indeed, in an important respect this can be interpreted as an index of the success of the nation's infant teachers, who are increasingly inviting young children to approach reading from the beginning as a meaningful and enjoyable activity.[35]

It would be wrong to imagine that the explicit revolutionary tenor of the New Literacy was merely another legacy of the sixties. Its philosophical antecedents, which go back at least to the last century and the early years of this, were always explicitly political in intent. Nurtured chiefly in recent decades by American educators such as Frank Smith and Kenneth and Yetta Goodman, it owes much to the writings of John Dewey, the American philosopher whose theories of education were the single most important influence in transforming not just the American educational scene but education theory and practice in Britain. We shall return to Dewey at greater length later in the book. But the significance of Dewey to reading and writing was that he believed that high literacy was an obstacle to socialism. Any division between the learned and the unlearned cordoned off one social class from another. It was the task of education, as he saw it, to break down any such stratifications. Reading and writing were not so much irrelevant as inimical to the political project of breaking down class distinctions and bringing forth the egalitarian society.

The education historian John Brubacher has noted that when Dewey founded his own school in 1894, he said it was an 'archaic practice' for elementary schools to spend 75 to 80 per cent of their time on verbal studies.[36] It amounted, he argued, to forcing a middle- and upper-class education on the mass of the population. Instead, education should be centred in the household occupations of home and community. Reading and writing began, he thought, when children kept their own records of the humdrum of their daily lives – without, of course, ever being formally taught to do so. In these thoughts, we hear the unmistakeable echoes of our contemporary educationists and their obsession with children finding meaning within their own experience, with breaking down the boundaries between classroom and everyday life and with learning to read through osmosis.

But in fact, the 'meaning-first' philosophy pre-dated Dewey. Around the time Dewey was born in the middle of the nineteenth century, the Secretary of the Massachusetts Board of Education, Horace Mann, was evolving his influential theories about reading. As Marilyn Jager Adams has chronicled,[37] the blossoming of American literature and the broadening social value of reading gave rise, as that century wore on, to anxieties about the way reading was being taught and about whether the phonetic drills prevented children from gaining proper access to meaning and ideas. In an uncanny premonition of today's controversies, phonics and comprehension in mid-century America not only came to be seen as mutually incompatible but the issue aroused the most intense emotions. In his annual reports Horace Mann decried the 'odour and funguousness of spelling book paper' from which 'a soporific effluvium seems to emanate . . . steeping (children's) faculties in lethargy'. He described the letters of the alphabet as 'skeleton-shaped, bloodless, ghostly apparitions' and said children should be taught to read whole, meaningful words first.

It wasn't until the 1930s, however, that the meaning-first curriculum became dominant. Education was then being promoted as the way to produce a new, productive and creative America. From the thirties and forties, meaning-first reading programmes were introduced, backed up by pictures, and phonetics became little more than an ancillary tool.

Whole language may have been originally an American system but it recruited some of its most evangelical disciples from the universities of England. In 1992 the American whole language guru Kenneth Goodman, Professor of Language, Reading and Culture at the University of Arizona, wrote: 'The basic concepts of whole language (largely without the term) have become institutionalised in British schools and parents like what their kids are doing in school.'[38] The vast majority of English teachers probably knew little or nothing at the time of these seismic changes in thought. Teachers in England have always been practical types, suspicious (rightly) of theories that appear to bear little relation to the day-to-day reality they experience in the classroom. Nevertheless, without their knowing it, theory does play a highly significant role in the shaping of the education culture, in promoting a way of thinking that seeps down from the universities through the teacher training institutions, specialist books and the activities of advisers, inspectors and others of influence in the education world.

So it was and still remains with the teaching of reading. It might be thought that this madness was a short-lived affair that has now run its course. This is not so. This movement has put down very deep roots indeed among the English teachers, many of whom – now in middle age – would themselves have been taught English along these lines when they were at school. Willinsky himself paid tribute to the 'progressive consensus' that emerged among English teachers during the 1970s. Classes turned into 'workshops'; teachers became 'entrepreneurs' or 'consultants' and pupils became 'participants' and 'researchers'. The *Language in Use* programme of teaching, that became available in Britain in 1971, was designed to 'broaden the democratic mandate of equal opportunity in education'. This meant that instead of lessons about literature, pupils dealt with the language of street and home; instead of being taught the correct forms of language, they were told there was no such thing as correctness.

The first chapter of Willinsky's book has been used as the first chapter of the Open University reader for the MA in Education which was published in 1994.[39] Links between the New Literacy and teacher training have been forged not merely through the Open University but also through London University's prestigious Institute of Education. Margaret Meek, repeatedly cited by Willinsky, taught at the Institute until 1990. In 1992 the Institute published

New Readings: Contributions to an Understanding of Literacy to mark Meek's retirement. This book, out of whose 22 contributors 16 were teacher trainers, was devoted entirely to New Literacy thinking. As Willinsky himself wrote, the London Institute was hugely influential in promoting the New Literacy because of the large number of acolytes who were based there. 'Through the 1970s, their influence spread as the London School became the dominant force in the National Association for the Teaching of English, eclipsing the "Cambridge School" under the leadership of literary critic F. R. Leavis and those who carried his sense of English's moral mission to the schools . . .'

The result was that instead of pursuing a moral mission, English teachers in London turned their attention to 'the lives, culture and language of the working class'. In other words, what they taught these pupils was not, as the great Victorian thinker Matthew Arnold had it, 'the best that is known and thought in the world'; they taught them merely to look at their own reflection in the mirror. In doing so, of course, they did not 'enfranchise' the working class at all; on the contrary, they impoverished it by deliberately depriving it of its rightful inheritance: access to the common culture.

The vast majority of teachers are not political ideologues. Most English teachers who are not teaching their pupils to read properly are failing simply because they themselves were never taught how to teach children to read. The evidence suggests that this is still largely the case with new teachers. Judging from the reading lists available to students in teacher training institutions, the bias is overwhelmingly towards whole language or real books, regardless of the endlessly repeated claim that teachers use 'a variety of methods' to teach reading – that giveaway phrase that suggests by definition a failure to appreciate the systematic nature of proper language teaching.

In 1989, a government sponsored inquiry undertaken by the education researcher Tom Gorman into what books trainee teachers used to learn how to teach reading found that most of them reflected the now orthodox New Literacy approach.[40] Gorman commented that this approach had led to greater involvement of parents, more access to interesting books and a turning away from reading schemes. 'However,' he went on, 'such a change in direction stems in some cases from a misunderstanding of practice and theory.'

The approach didn't require teachers to think analytically about the language of pupils or the texts, and tended to underplay the amount of knowledge teachers had to have about the sound system and the written system to help children understand the relationship between the two. It was easy for teachers in training to misunderstand and misapply the principles underlying this approach, all of which were open to question and misinterpretation. 'I concluded from this brief inquiry, therefore, that it is likely that many teachers in training are not being provided with the information that they need to provide information to beginning readers,' he wrote.

In 1991, a subsequent study by the National Foundation for Educational Research reiterated Gorman's findings.[41] Some 60 per cent of newly qualified teachers, it reported, said they had been taught little or nothing about phonics. 'It appeared that, in general, lecturers do not make, and do not encourage students to make, a division between teaching children the skills and strategies needed to decipher text and developing the attitudes and the skills characteristic of fluent and enthusiastic readers. These two objectives of the teaching of reading are regarded neither as divisible, nor as chronologically sequenced (i.e. a first stage in which children learn to decode text; and a second in which they become interested and self-motivated readers).' The result was that no fewer than three-fifths of teaching graduates said they did not feel confident about teaching children to read.

Considering the scale of the disaster implied by these findings, the comments by Gorman and the National Foundation for Educational Research might have been thought relatively mild. However, the book which Gorman had found was topping the teacher training charts, *Read With Me* by Liz Waterland, was subjected to a devastating critique by two education psychologists, Roger Beard and Jane Oakhill, who claimed that its fundamental assumptions about the way children learned language were arrant nonsense.

Read With Me was published in 1985 and remains a classic of the genre. Gorman quoted some of its assertions: 'In many ways the acquisition of written language is comparable to that of spoken language'; 'Essentially, reading cannot be taught in a formal, sequenced way any more than speech can be'; 'Reading is not a series of small skills fluently used; it is a process of getting meaning'; 'The text offered to the child is crucially important';

and the teacher's position was reduced to 'the role of the adult as a guiding friend'.[42] The first three of these assertions are factually wrong and the last two are misguided and dangerous.

By 1992, wrote Beard and Oakhill, this book had become the most recommended book on initial teacher training reading lists.[43] It had never been properly evaluated, despite the fact that this approach had been the only one tacitly endorsed by the non-statutory guidance for the teaching of reading in the first version of the National Curriculum in 1990. Yet it was at odds with a number of major studies on early reading. *Read With Me* assumed that if children used 'real books', they would recognise words as wholes by a series of clues including phonics. This was, however, nonsense because phonics couldn't be used to recognise whole words.

It proposed to teach reading not in the generally understood manner but to provide conditions in which children would 'catch it, like a cough'. It offered 'a misguided view of what reading should be about' and drew overwhelmingly on 'controversial' exponents of whole language such as the American educationists Frank Smith and Kenneth and Yetta Goodman. Phonics was treated almost as an anathema – 'the dreaded subject' – ignoring the fact, wrote Beard and Oakhill, that the basis of the English writing system was the representation of its speech sounds by alphabet letters. Liz Waterland had rejected reading schemes as 'worthless, dreary, impoverished, meaningless' without any evidence to support such value judgments. How extraordinary, observed Beard and Oakhill, to reject a book *just because* it had been written to teach someone to read!

Extraordinary indeed. Extraordinary, also, that a book seemingly based on ignorance and prejudice should have been elevated to such influence in teacher training institutions. And most extraordinary of all, despite this savage critique by two highly qualified experts, and despite the findings of two inquiries which revealed how trainee teachers were being sent out into the classrooms wholly unequipped to teach children to read, no-one saw fit to address this situation. It was not until June 1996 that the government signalled any concern about teacher training when it proposed the introduction of a core curriculum for teacher training institutions to ensure that new teachers could teach reading, writing and arithmetic.

However, this announcement appeared to be merely a political panic measure in the face of combined pressure from the Tory right-wing and tougher rhetoric by the Labour Party. There was no carefully considered approach to tackling the ingrained culture of teacher training colleges which continue to pump out their counter-evidence about the teaching of reading.

Reading University, for example, has produced a pamphlet to help teachers to teach reading.[44] This document, written by two language teachers, solemnly observed that public dismay at the failure by 28 per cent of seven-year-olds to reach the required level in the first National Curriculum Tests was an overreaction; the books were too difficult, and it was likely that some of those children were able to read but fell down on specific requirements like using a dictionary. Briefly intoning the mantra: 'Teachers use many approaches', the authors then revealed their true colours. Reading, they opined, involved making 'intelligent guesses'.

To illustrate this insight, they used a story book picture with the legend: 'The terrible, terrible tiger he roared . . . and leapt at me.' 'An inexperienced reader,' they wrote, 'might well read, "The terrible, terrible tiger he roared and jumped on me."' Although this is not accurate, the child would be showing that she knows the story must make sense and she would be using the picture and her own knowledge of language to do just that. This is an intelligent reading strategy. As long as her mistake or "miscue" does not change the meaning, she should not be corrected. As she becomes more experienced, this kind of miscue will gradually disappear.'

But of course this is not an intelligent reading strategy, nor indeed any kind of reading strategy. It is simply guesswork, which isn't surprising since such children will not have been taught to read. It also does *not* accurately convey the meaning; a leap is more forceful than a jump. Such a child would not be reading the print but speaking to the picture, which may be an enjoyable activity but is not reading.

One of the most disturbing features of such books is that they dress up propaganda and ideology in the borrowed clothes of dispassionate academic inquiry. *The Politics of Reading*, for example, from Cambridge University, proclaimed: 'A dangerous preference for assertion, at the expense of argument, is rife . . . Reading is too important, and our pupils too vulnerable, to let ill-conceived and politically motivated contributions to the debate carry the day.'[45] But its own contribution was nothing other than a set of politically motivated assertions which perpetuated certain myths about reading which have now passed into the cultural folk memory. 'Certain ideological viewpoints have come to prominence of late, with one particular interest group in reading dominating powerful positions on bodies such as the National Curriculum Council and the School Examinations and Assessment Council,' it said. As we shall see, those bodies were consumed by fighting over these matters but at no stage were they dominated, as the Cambridge book and so many others have implied, by the phonics lobby: quite the contrary. 'Many of those loudly expressing their points of view do so out of ignorance,' it stated. Yet its authors appeared to be quite ignorant of all those studies reported by Jeanne Chall which have revealed that phonics is a far more successful method than any other. 'It is almost impossible to provide lively texts out of the simple words children can master phonetically in the early stages of reading,' it said. Yet the class of five-year-olds referred to earlier in this chapter demonstrated considerable enthusiasm for their reading scheme books, as do many other children.

For these Cambridge authors, phonic teaching committed the crime of being orderly and logical by teaching children one step at a time. One might think that this was a desirable state of affairs. Not so. The authors appeared to prefer learning that was 'uneven and untidy, individual and unpredictable'. Yet such disorder is profoundly unsettling for a child and does damage to its educational progress. The authors confused a child's ability to speak with the ability to read, even though these are quite different activities and are learned in quite different ways. They displayed ignorance of and prejudice towards phonics, and appeared to admire the whole language approach because it treated children 'with a proper respect for their emotional and intellectual powers'. It was yet another example of the romantic fallacy of the child as the autonomous meaning-maker.

A key contributor to this book was Margaret Meek, the immensely influential educationist from the London Institute. Her chapter illustrated the kind of thinking that made her a household name in the world of early years education. Literacy, she wrote, was embedded in everyday life. Children didn't need to be taught it. 'Instead of having confirmed their ability to read the print in the world around them, children are still expected to bring home words on slips of paper in a tin in order to practise "word recognition".

Comparative studies show that literacy is locally learned, yet claims are made for teaching methods as universals.' On the contrary, however, the studies show that literacy has to be taught. The precise process by which Meek imagined children learned to relate sounds to letters and letters to words remained here, as in every book of this type, profoundly mysterious.

But maybe it wasn't surprising that Meek denied the evidence. For she appeared to have given up on rationality and progress altogether.

> At the end of this century, historians and anthropologists have turned us away from 'autonomous' literacy, the notion that there is a state of literacy which guarantees 'progress', 'civilisation', individual liberty and social mobility as the result of schooling in rational, cognitive 'disciplines'. We can no longer regard literacy as an absolute. We know, yet continue to act as if we don't, that no single test can reveal the range of literate competencies possessed, or required, by children and adults . . . If, however, our literate practices and texts continue to become more diverse, who will say what *counts* as literacy?

Who indeed. For advocates of the New Literacy such as Meek, watching TV or playing on the computer might well be classified as a type of literacy that is just as valid, if not more so, than reading.

Meek finally revealed the real agenda behind her otherwise bizarre argument. 'The powerful literacy of the great literate tradition in English is still exclusive. Those who complain about standards sit within it and know that their education keeps them, *hors concours*, in the top stream.' In other words, this was all about perceptions of power and privilege and the need to impose the egalitarian nirvana. Literacy divided sheep from goats; such division was unacceptable; therefore literacy must go. Let them watch videos instead. But again, this argument stood logic on its head. Literacy does *not* divide the population. It enfranchises everyone. Almost every child can be taught to read, the overwhelming majority with relatively few problems. The divisions are caused by teachers failing to deliver.

According to Styles and Drummond: 'Most children who do not

make progress in reading are suffering, not from specific learning difficulties, but from a lack of belief in themselves as readers.' But most children who can't read are suffering from nothing other than the fact they have never been taught properly; and with books like these pumping out this kind of ideology with no apparent attempt whatsoever at balance, published under the aegis of prestigious teacher training institutions, there is little chance they ever will be.

It is difficult to overstate the comprehensive reach of these attempts to rubbish the evidence of Martin Turner and other experts on the teaching of reading. The attacks feed off each other and have created an orthodoxy out of a mutually reinforced mythology. Wherever one looks, one finds educationists involved in teacher training arguing that anxiety about reading standards is spurious, but in terms which merely reinforce the fundamental errors involved. Professor Ted Wragg of Exeter University is a prominent teacher-trainer who has consistently rubbished these concerns. In 1990 he ridiculed the psychologists for their findings;[46] in 1991 he ridiculed phonics as 'phobics'[47]; but 11 days after that an article in the Daily Mail quoted his own students as saying: 'We receive no training in the formal instruction of reading. We feel we are not being adequately prepared to teach young children.'[48]

Some educationists resort to the internally inconsistent argument that there is no evidence of any problem with the teaching of reading and in so far as there is a problem, it's not the teachers' fault. One such critique by Henrietta Dombey was published in 1992 by the National Association of Advisers in English,[49] and was introduced by David Allen, the chairman of the Association, as a 'sane and balanced' antidote to 'the present flurry of short-term panic'. Referring to Martin Turner's 'alarmist publication', Dombey conceded that Buckinghamshire had shown a similar decline to the National Foundation for Educational Research's figures among its own seven- and eight-year-olds since 1985. But the increasing numbers of children falling behind, she said, was the result 'not of faulty teaching but of social factors outside the teachers' control'. She produced no evidence to back up this assertion, only further unproved assertions: that more and more children were starting school less well-equipped to learn to read because they were poor; that the National Curriculum was requiring children to read more complicated things; and that there was more pressure on children because of greater competition in society. In short, it wasn't teaching that was going wrong, but learning; the problem lay not with the teacher but with the child.

The question remains, however, why so many teachers have resisted the large amount of very sound evidence. Very few teachers, after all, are politically motivated and the vast majority have no time for theory. In 1972 Brian Cox, whose position was later to become bewilderingly contradictory, told the Bullock committee:

> The problem for our children today is that these research findings are resisted emotionally by many teachers, administrators and college lecturers. Some of these people fear that the new emphasis on phonic instruction means a return to the dull content, the drill and mechanical word by word reading characteristic of some schools pre-1930. This is not necessary . . . Most systematic phonics programmes depend on didactic teaching, with the teacher explaining and the children practising the sound and letter relations. This is the main reason why their successes are so vigorously opposed. Advocates of child-centred learning find such methods repugnant to their ideology.[50]

In other words, the teaching of reading was resisted primarily because it involved *teaching* which, as we saw in the previous chapter, increasingly became an activity to be frowned on. But as Keith Gardner, an expert in the teaching of reading, wrote in 1968, the shift from teaching to learning, the passing of responsibility from adult to child, started as long ago as the immediate post-war years. 'In the post-war infant school it has been considered slightly old-fashioned to teach reading at all. The belief is that children will learn to read in their own way and in their own good time. Anxious parents have been fobbed off with such pious statements as "He will learn to read when he is ready". Inspectors have actually criticised schools that try to teach reading. In the modern craze for child-centred education, reading has become something that is acquired – not taught.'[51]

When one considers that those children who were so inadequately taught to read in the fifties and sixties are now themselves middle-aged parents and teachers, the dismaying dimensions of this problem immediately become apparent. The reason why so many of those parents and teachers have such low expectations of their children and pupils, and suffer from such ignorance about how children learn to read, is surely at least partly explained by the fact that they themselves were never taught properly and so never flourished as readers. We are grappling here not with some brief passing fad of recent provenance but with a deeply entrenched problem that has now passed down through two or even three generations.

The reason why conventional teaching became a taboo was bound up with the idea that a child's innate creativity and spontaneity would be damaged and stunted by the imposition of rules or codes or influences brought to bear by the adult world. The history of this idea, as is discussed later in the book, goes back to Rousseau and even earlier thinkers. But it meant that the child's imagination became the quality that was prized more highly than anything else. The new emphasis on creativity lay behind a more informal approach to spelling and punctuation. It was considered not only more important for a child to be free to express itself without being constrained by rules of spelling or punctuation, but also that such rules were inimical to creativity. The understanding that they were expressly designed to help communication and understanding of a language capable of a very high level of subtlety and differentiation simply collapsed in the face of undiluted hostility to the learning of absolute rules.

The results were as widely noted by the general public as they were fiercely denied by much of the education establishment. Employers in particular repeatedly complained about sloppy writing with lamentable standards of punctuation and spelling. Jennifer Chew, an English teacher at a sixth form college in Surrey, charted the decline among her own students. At the beginning of their A-level course, she gave them all the same spelling test. With the exception of one year, 1988, when the scores went up slightly, she found that in every year from 1984 to 1993 the standard of spelling dropped. The most disturbing trend was the average spelling score of pupils with GCSE grades A to C, which dropped sharply after the GCSE exam was introduced – and kept dropping. (In 1994 and 1995 the scores rose slightly, a fact she has ascribed to the diminution of coursework in GCSE English.)

Yet the original proponents of creative writing never dreamed that their approach would produce such an effect. As Chew herself recalled: 'We were originally told in teacher training that if we allowed children to write spontaneously they would become more accurate in spelling, punctuation and grammar. It was inconceivable that those things were to be abandoned.'[52]

Not only were they abandoned, however, but their disappearance was hailed as evidence that children's innate creativity was now being allowed full expression at long last. When, after the introduction of the National Curriculum, the government started to insist that more attention should be paid to spelling and punctuation, there was heavy resistance from within the teaching profession. One example of this was an article published in 1992 by Pat D'Arcy, a former English adviser for Wiltshire education authority.[53] Ms D'Arcy reported that the Wiltshire language curriculum support team believed that there were three aspects of language: grasping the code, handling the medium and shaping meaning, and that the first should not be over-emphasised at the expense of the other two. In other words, the structure of the language – spelling, punctuation and grammar – should not take precedence over constructing a good narrative line or constructing a poem. The idea that structure was essential to produce a narrative or a poem was, it seemed, a foreign concept to Wiltshire's language experts.

To illustrate her point, Ms D'Arcy reproduced a story written by ten-year-old Paul. This is what Paul wrote:

One sunny day a Ant went for a walk. His mother said to watch out for that hunggre Anteater. The ant ceeped walking until he got a bit tierd. He stopped and looked around to see if there where aney hunggre anteaters ariund but he could not see aney. So he lead down and fal a sleep. As he was a sleep the anteater was spieing on Him from a tree. The anteater thourght about ant on toest with a bit of ralish. He got so exsitted he fall of the tree with a bag. The ant wocke up and as qiuck of a flash the anteater pounced on him. The anteater said 'I am going to eat you up'. but the ant thourght a mint and he came up with an I dear? Eat me them but down throu me in that hole. The anteater siad I am going to throught you in that hole. So he did but the anteater did not now that it was the army ants base. When ant got to the bottem he preasted a red bottom and all these army ants came out and attacked the anteater. and the ant and his mother lived happly ever after. The End.

One might think it regrettable that at the age of ten a child had such a poor grasp of literacy. Ms D'Arcy, however, saw things rather differently. The story, she said, should not be judged on its errors, since of the 38 misspellings, 21 comprised *only one error*. 'In other words, over half of Paul's 38 misspellings are 99 per cent correct.' Astonishingly, this seemed to negate the importance of the mistakes altogether. What was much more important to Ms D'Arcy was Paul's mastery of fable, his clear beginning and ending and sequence of events which all made this 'a highly competent narrative'. What, one wonders, would have been her definition of incompetence? But to this former education adviser, exactness in language appeared to have little function except to stunt a child's spontaneity. 'If Paul's teacher had indeed conveyed the message to the children that above all else correctness was of prime importance, stories such as this would never have been composed. Instead, the children would have checked every word against a word bank or dictionary before daring to write it down.' This would have prevented the flow of thinking essential for 'the construction and communication of meaning'.

But children who have been taught properly write imaginatively and creatively *and* can spell and punctuate; indeed, spelling and punctuation are essential aids to their creativity because they help them express complexity and subtlety. The idea that accuracy entails a child rushing to the dictionary before being able to write a word is risible. It is simply a matter of being taught the skills of spelling and punctuation so that the eyes and ears become trained and the skills become automatic. What stifles creativity is the kind of un-teaching that fails to provide a child with mastery of its own language.

Any doubts that Ms D'Arcy spoke for very many English teachers would have been dispelled by the letters from teachers received by the author after a critical article was published on the issues involved in Paul's story.[54] These teachers dismissed his mistakes as unimportant and repeated the misapprehension that mastery of the language was a bar to creativity. There seemed to be no longer any grasp of their duty to teach a child to be literate, or even, for some of them, their duty to teach at all. One English teacher wrote: 'The classroom teaching of spelling is a profligate waste of time since it serves no purpose whatever for the significant minority of "natural" spellers and is a source of frustration for the "dyslexics" at the other end of the spectrum.'[55] For this teacher, then, there was nothing in between innate spelling ability (picked up, no doubt, by the same mysterious process of osmosis that enabled children to learn to read without being taught) and dyslexia. The children who needed to be taught to spell had simply been written out of the script.

At the same time, however, others wrote with a very different perspective. Parents wrote with hair-raising tales, not merely of teachers' refusals to teach correct English but their inability to do so. One Wiltshire father wrote that he and another parent had been worried about the fact that a probationary English teacher was not only leaving their children's errors of punctuation and spelling untouched, but that she was 'correcting' words that were spelt correctly. To his surprise, however, the headmaster and the school's head of English had disagreed with them. The probationer, they said, was the best young English teacher they had had for years. Surprise turned to horror, however, when this father read notes prepared by the head of English on how to structure a lesson, using as an example Shakespeare's play *Julius Caesar*. The notes were badly punctuated and full of spelling errors, and the eponymous hero was referred to throughout as 'Ceasar'. The father wrote: 'When I taxed him with this problem, the head of English cheerfully admitted I was right. He knew he couldn't spell and couldn't punctuate either. He was, he said, a victim of circumstance; he happened to have gone to school in a time when punctuation and spelling were not considered important; self-expression was all.'

The result, commented this father, was that his daughter – with three A grades at A-level – was given to writing things like: '"I would of gone but I missed the bus."'[56]

In April 1996, researchers produced evidence which reinforced the impression that the command of language was slipping. A study by the University of Cambridge Local Examinations Syndicate found that English GCSE students in 1994 were up to three times worse at spelling than O-level pupils in 1980, had a narrower range of vocabulary and were six times as likely to use non-standard English.[57] The disastrous effects of the New Literacy, and the extent of its penetration into the education system, could no longer be avoided.

Chapter Four

THE EDUCATED PHILISTINE
The Revolt Against Grammar

It is perhaps in the revolt against the teaching of grammar that one can see the most chilling effects of ideology. It has led to the systematic destruction of our children's ability to become the masters of language and thus of communication, in both their native and in foreign tongues. Grammar has become such a dirty word that, as we saw earlier, some teachers who want to teach it to their pupils are driven to do so in secret. What these teachers are trying to do is to teach children something that previous generations had accepted without question as the right of every child: the codes to crack the mysteries of language. Yet, like the reading codes, teaching the codes of sentence structure has also became a taboo. It has been identified with tedious drills and rote learning, founded upon fundamental misapprehensions about the way children learn and yielding instead to the new orthodoxies of free expression and 'relevance'.

As we saw earlier, the results have been utterly disastrous. Children and young people are increasingly unable to manipulate English and speak, write or read foreign languages beyond a pidgin level of ability. This is hardly surprising, since depriving them of grammatical competence deprives them of the ability to decode language at all.

Educationists revolted against teaching the codes of grammar for a number of reasons. In English, the distaste for drills and instruction and the emphasis we saw earlier on creativity stretched from the non-teaching of reading, spelling and punctuation into the non-teaching of grammar. Among those who taught foreign languages, there was a perception – no doubt fostered by the increasing availability of foreign travel – that pupils might be able to read literary texts in foreign languages but were unable to hold their own in conversations abroad. There was undoubtedly a measure of truth in that perception. But this small window of reality was slammed shut by the gales of ideology blowing through the entire education culture. Foreign language teachers, just like their counterparts in the English departments, fell victim to the mind-set that children could somehow pick up knowledge by their own efforts, and that what they did manage to pick up through their everyday lives was of greater value than anything that might be taught to them formally in the classroom.

The conviction grew in education circles that young children couldn't cope with abstractions. And certainly, it's not easy to grasp abstract concepts such as subjects or objects in a sentence. Nevertheless, generation after generation of children has previously done so when properly taught, and it is perfectly possible to teach grammar in a lively and imaginative way. But the memory of what was undoubtedly bad and unimaginative teaching of grammar became enmeshed with the broader educational imperative that nothing should be taught which might cause any child to fail or to feel that others were progressing any faster. Rules of grammar, like all other rules, became a taboo, to be replaced by learning from life. Just as with reading, spelling or punctuation, what the child couldn't pick up from its surrounding environment somehow became an improper use of language.

Conveniently, the ideas of the linguistics guru Professor Noam Chomsky appeared to endorse this position. Chomsky held that every child inherited a piece of complex and fixed mental machinery, programmed into its brain at birth, that gave all children the principles of universal grammar, and that certain aspects of grammar and syntax were common to all languages. Crucially, Chomsky was writing about spoken rather than written language. Nevertheless, his theories were used to imply that grammar didn't have to be taught at all; the child, that 'autonomous meaning machine', would make its own grammar. This in turn fitted neatly with the egalitarian political position which held that all languages and dialects were equally good. There could be no correct way of handling language because no forms of language were inferior. In 1976, Professor Michael Stubbs of London University's Institute of Education wrote: 'It is accepted by linguists that no language or dialect is inherently superior or inferior to any other, and that all languages and dialects are suited to the communities they serve.'[1] This was, of course, an absurd non-sequitur. Languages comprising merely a few thousand words might well be adequate for simple, primitive societies; but that hardly meant that all varieties of language were therefore adequate for *any* speakers' needs in *any* society. But logic hardly came into the picture. The revolt against teaching the rules of grammar became part of the wider repudiation of all external forms of authority and any discrimination that might flow from them. The power of ideology turned the original impatience with the unimaginative teaching of grammar into a visceral hatred. No teacher could afford not to be consumed by it, to avoid being written off as an antediluvian fossil.

So it became the orthodoxy that a child could learn language without being taught concepts such as prepositions or subordinate clauses. There was a growing confusion between the way children learn to speak their native tongue and the way they learn to speak a foreign language. Because very young children pick up the spoken language by being immersed in it, the belief grew that older children could pick up foreign languages in exactly the same way. Both in English and in foreign languages, the theory went, children would assimilate the rules of grammar without anyone needing to teach them.

But this was absurd. For a start, people use language without necessarily being able to analyse the way they are putting sentences together. If they have never been taught any insight into their own use of language, they certainly cannot apply those rules to make another language work because they have no idea of what rules they are unconsciously using. Moreover, very young children pick up language easily first because they *are* very young, and second because they really are 'immersed' in the language they are learning. It is all round them all day: at home, in the streets, in the shops, wherever they go. There is nothing to compete with it as their minds develop. (Bilingual babies appear to be able to handle two languages simultaneously precisely because the assimilation *is* simultaneous.) But when they are older and learning a second language, that has to compete with the codes of the first language that by now have been deeply assimilated into the child's thought processes. The structure of that second language must therefore be taught. The child will never be 'immersed' in that second language if it is merely taught for a few hours at school; the language in which the child is actually immersed is, hardly surprisingly, its native tongue.

Like so much of education ideology, therefore, the immersion theory of foreign language teaching and the revolt against the teaching of English grammar are counter-evidential, irrational and destructive. Nevertheless, they have been implemented with evangelical zeal. An entire edifice of false deductions has been erected on wholly untrue foundations, leading to ever more fantastic justifications which defy reason itself. It has taken on the character of an obsession.

The communicative obsession has meant that we have moved from the original perception that pupils weren't sufficiently fluent in idiomatic French or German and arrived at the point where examination boards are awarding high A-level grades to illiterate foreign language students on the grounds that all they need to do is to show they can make themselves understood abroad. So if the word endings are all wrong or the adjectives don't agree, who cares?

This attitude is on brazen display among those educationists who write guidance for foreign language schoolteachers. One such coursebook writer, Barry Jones, writes in *Grammar!*, a book of principles to help French teachers, that the prime goal of teaching French is the communication of meaning, and therefore teachers should focus on particular language structures *only* if the task in hand requires them to do so.[2]

In other words, teachers should go to some lengths to avoid teaching their pupils any 'unnecessary' grammar – unnecessary being defined, it would appear, as anything that might require systematic learning. So if there should be a number of different grammatical structures within a text, teachers should avoid explaining them all. Instead, Jones helpfully suggests, some should be taught 'as vocabulary'. In other words, phrases such as 'je voudrais', or 'je suis allée' should be taught as phrases whose meaning – 'I would like' or 'I have gone' – can then be slotted into a sentence ready-made.

What teachers should not bother to do until much later, suggests Jones, is teach pupils the conditional or perfect verb forms along with the correct word endings. The result is that such pupils are quite unable to work out for themselves what '*we* would like' or '*they* have gone' might be in French. They do not understand, for they have not been taught, how a French or German sentence is constructed, why the words are in a particular order, why endings change or any of the many otherwise baffling conventions of the language. They are instead expected to commit to memory a stock of ready-made expressions, which leaves them unable to crack the codes of French or German just as their infant counterparts are expected merely to commit whole words to memory and so left unable to read at all. Instead of mastering a foreign language, they become its servants instead. They end up speaking pidgin French or German. Not surprisingly, this does not go down too well in France or Germany where such students are held to be illiterate; as indeed they are.

Yet such is the strength of the prejudice among teachers that even when they are forced to accept that their teaching methods have failed lamentably, they cannot bring themselves to agree that the teaching of grammar should be restored. Indeed, they go to almost comical lengths to avoid doing so while at the same time trying to work out ways to remedy the problem they themselves have created.

As Lid King, one of the editors of *Grammar!* put it, grammar was now on the agenda not just because of changes required through the National Curriculum. There was 'a more generalised feeling expressed in different ways by many teachers that somehow the communicative revolution had undervalued the importance of structural understanding'.[3]

Or, in simple English, these pupils couldn't speak, read or write the language with any measure of precision or accuracy. They certainly couldn't manage it at A-level, and to the irritation of native foreign language speakers, they were talking no more than pidgin language. The teachers had managed to throw out the baby with the bath water, a fact they appeared only dimly to understand. Lid King ironically summed up the corner into which language teachers had managed to box themselves: 'We have seemed to offer a choice: on the one hand, learners can correctly translate (the) mad slaves (are baptising the camels) into Greek and conjugate the verb, to baptise; on the other, they can successfully buy a chocolate ice-cream (but do little else).'

But the teaching of grammar was never designed, despite the prejudices of its detractors, to engage pupils in such pointless exercises. Undoubtedly, it was sometimes taught badly; but it was taught out of necessity, because without it children could not gain control over the use of language. As Richard Johnstone, Professor of Education at Stirling University observed in *Grammar!*: 'Without grammar control, whether based on implicit or explicit knowledge, there can be little if any systematic creativity or accuracy in language use.'

Nevertheless, it seems these teachers simply could not bring themselves to advocate that grammar should be taught. Instead, they danced around the subject with ever more convoluted formulations to suggest how children might learn grammar without teachers having to teach it. So, for example, according to James Burch, a teacher trainer from Lancashire, 'language structure' should arise naturally in classroom talk but not (perish the thought) as part of a taught syllabus. No attempt was to be made to explain that structure, to tell pupils why they must use words in a certain way. Instead, correct usage was to be 'emphasised' to 'raise the learners' awareness of the structure of the language they are using'. How their awareness was to be raised without any explanation being offered was a puzzle that would surely tax the prestidigitators of the Magic Circle. 'Learners will use language correctly; but this correctness will depend more on carefully nurtured instinct and feel than conscious understanding,' wrote Burch. This would appear to advocate a grasp of language no more substantial than conjuring up information through an ouija board. Yet Burch was a teacher trainer, a fount of wisdom for brand-new teachers who were expected to put his theories into practice in the classroom.

The reason for Burch's intellectual gymnastics appeared nothing less than the taboo on teaching itself. Far from being the source of knowledge, the teacher had become instead a nuisance to be nullified. 'If the learners' own suggestions and ideas become centre-stage, then it will be their view of reality and not the teacher's that is shaping the area of classroom interaction where activities are undertaken to help with retention and fluency. There is probably no better route into the realm of the subconscious and long-term retention.'

'Communicative Competence'

The double-thinking of the language experts scaled ever more fantastic heights. Brian Page, past president of the Association of Language Learning, a prime mover of this way of thinking and, according to *Grammar!*, 'one of the most influential people in modern language teaching', claimed that it wasn't true that the communicative approach meant not bothering about grammar: it was just that different values were being employed. This meant that the old way of assessing an error as wrong was now considered to be – wrong. 'Nowadays we are more enlightened and have marking criteria which reward efficient communication.' In other words, approximations and errors would be considered quite acceptable as long as the general meaning was more or less intelligible. The concept of right and wrong had been abolished.

This is precisely the philosophy behind the collapse of the marking system in foreign language A-levels which so distressed David Horrocks, the former A-level chief examiner we met earlier, and caused him to resign in 1990. Indeed, Page went on to criticise one board's marking criteria for French because it held that top marks should only be given where there were very few grammatical inaccuracies. Now most people might imagine that this is exactly what should happen. Correct use of the language should get marked up; incorrect use of the language should get marked down. Not so, according to Page, for whom this was now a heresy to be stamped out.

'This implicitly tells the learner, "You must get it right because you must get it right" and that, to my mind, is not enough.' In other words, the very notion of having absolute rules *at all* is under attack. They have to be justified afresh every single time they are used; and since, in language use, this is patently impossible, they should not be used at all but discarded. One might as well say – to extend this approach to other areas of life – that drivers shouldn't stop at red traffic lights unless obeying such a rule is justified afresh at every single road junction. The idea that the rules of language are essential precisely *because* they are universally applicable does not seem to have occurred to these leading lights of British modern language teaching.

On the contrary, wrote Page: 'We do not know what correct grammar is for.' But despite such perplexity, he grudgingly acknowledged, there was a problem: ungrammatical language made people think the user was stupid. Something therefore had to be done, he conceded, to teach foreign languages in a way that was acceptable to foreign countries. But what? Grammar – that 'endless series of meaningless exercises to learn correct forms' – was out of the question. The self-evident solution was thus rejected out of hand.

Two other language education experts, John Thorogood and Betty Hunt, echoed this surrealistic approach in *Grammar!*. They rejected 'inappropriate rigour' in teaching 'decontextualised grammar points' because in their view this would undermine the whole communicative philosophy of introducing language on the basis of the pupils' need to know particular words at a particular time, rather than according to an 'arbitrarily preordained system'. But it seemed not to have occurred to them that there was nothing arbitrary about the fundamental concept that a pupil's needs cannot be predicted. No-one, after all, can predict what words they might need in the next sixty seconds, let alone next week or next year; the whole point of rules is precisely that they are universals which can be applied to any unpredictable situation.

That is the point of having a code. It is the key to unlock the use of language. Without it, the foreign language student cannot unlock that language, any more than the infant deprived of phonetic teaching can unlock the mysteries of print. And just like the teaching of early readers, the communicative obsession in foreign languages is based on an utter failure to recognise the function and necessity of universally applicable systems. They have become instead a taboo.

But not, it appears, universally. For despite their purported egalitarian intentions in transferring skills from teacher to learner, the brutal truth was that these expert educationists considered some pupils simply too stupid to be taught grammar. 'The low attainer entering at basic level,' wrote Thorogood and Hunt, 'is mercifully spared the futility of trying to learn a full and unnecessary range of grammatical forms. But for the ablest learner, opportunities should be provided as a matter of course for mastering the full morphological system . . . Priority should be determined by observed usefulness.' So only clever children, it appeared, should be taught grammar; the less able, however they might be defined, should by contrast be denied the opportunity to master the language.

According to another contributor to *Grammar!*, in a chapter entitled 'Grammar and the "Less Able Learner"', some teachers might consider it a waste of time teaching grammar in a foreign language to less able pupils in the traditional way. 'We may feel that these pupils will have difficulties in grasping grammatical concepts and therefore tend to teach vocabulary and phrases in topic or interest chunks without a lot of deliberate grammatical sequencing and with little structural consolidation,' said Patricia McLagan of the Centre for Information on Language Teaching and Research. 'With these pupils, we tend to avoid abstract concepts, explicit grammar teaching and analysis, as rules appear to be internalised only very slowly . . . Few will end up able to manipulate language, to read for their own interest or able to create new sentences.'

Mrs McLagan proposes a number of methods of teaching less able pupils by sensitising them to the structure of that language through a variety of listening and reading stimuli – for example, songs and TV – in order to make it fun. But children don't learn grammar as they go along, they have to be taught it explicitly. And there's no reason why slower learners can't learn it as a structure of rules that must be taught systematically as a code-breaking exercise to enable such pupils to build sentences from scratch.

European foreign-language teachers would no doubt regard any suggestions that they shouldn't teach explicit grammar to less able pupils as ridiculous, since without formal grammar they couldn't learn any languages – and many European children display an impressive command of English. If British schoolchildren are unable to learn grammar it is less likely this is because they are incapable than because they are being wrongly taught.

Indeed, where schools have resorted to grammar teaching the improvement in their results has been spectacular. In one comprehensive school in Wigan, for example, after grammar teaching was reintroduced the GCSE pass rate doubled.[4]

Nevertheless, it is a commonplace among teachers that grammar should not be taught because it is too hard or an archaic irrelevance. One English teacher in a comprehensive school said children learned grammar when it cropped up naturally through their creative writing instead of being taught rules which they could then apply to what they wrote.[5] Formal grammar, she thought, was too hard because it wasn't expressed in ways children could understand. 'Intelligent children don't relate to these concepts,' she said. 'Even my own son, who is very bright, couldn't understand these things in abstract terms. And where children come from backgrounds where they don't read, then to introduce a lot of jargon like "this is the perfect tense" or "don't put this word in front of that word" won't help them at all.' But how can they be helped if they *aren't* taught such things, and end up as a result unable to control their own language, let alone anyone else's?

However, the revolt against grammar has now begun to erode the ability of teachers to teach it even if they want to do so. In a survey, researchers from Newcastle University found more than half of university graduates training to be primary teachers had such a poor grasp of English grammar they couldn't recognise that 'and' was a conjunction and 'in' was a preposition; nor could they explain the difference between a clause and a phrase; and several struggled with the terms 'adverb' and 'pronoun'.[6]

At the heart of this whole series of revolutionary changes in attitude towards the teaching of language lies a paradox. The cultural critic George Steiner has observed that these trends have involved a catastrophic 'retreat from the word', a collapse into sub-literacy. 'We empty of their humanity those to whom we deny speech,' he said. 'Deliberate violence is being done to those primary ties of identity and social cohesion produced by a common language.'[7]

At the same time, one of the dominant themes of this transformation is the emphasis on the creativity and spontaneity of the child. The whole shift from teacher to pupil, from adult to child, the revolt against external rules and the distrust of hierarchies and order implicit in the rejection of formal grammar, punctuation, spelling and phonetics are all bound up with the rhetoric of creativity and the primacy of the imagination. The fact that the rules and constraints of language actually promote creativity, as the classic writers who were schooled in these testing disciplines so eloquently demonstrate, is yet another self-evident fact that has bitten the ideological dust. It's now argued that the child has to be liberated from those pointless and harmful constraints previously imposed by the adult world so that innate creativity, now elevated to the highest order of achievement, can flourish.

Decline of the Essay

It is now the orthodoxy among English schoolteachers that children's writing should principally consist of creative writing: in other words, stories, poems and works of fiction. All the rules and disciplines of the language are subordinated to the exercise of the imagination. What the child learns about language is learned through creative expression. Of course, creative writing should have a place. But this contemporary mind-set has had the effect – quite apart from the neglect of language structures – of marginalising the essay. It is now common to find a look of utter incomprehension on the faces of English teachers if one asks whether their pupils write essays as well as stories. Well of course, they say, thrusting forward the pile of stories, here are their essays. If one presses the point further and, by way of explanation, asks whether the children perhaps write essays on books they may have read, more examples are thrust forward of creative stories inspired by those books. It takes a considerable amount of time, with an alarming number of teachers, for them to grasp that there are significant and important differences between the essay and the story.

Asked whether her pupils subjected the books they read to any kind of critical analysis, one comprehensive school English teacher replied that this was surely far too advanced for pupils aged between eleven and sixteen.[8] Yet the essay teaches children from a young age different skills from those learned through creative writing. The essay teaches them about objectivity, how to handle evidence, the difference between fact and fiction and how to structure an argument. Above all, it teaches them about order and judgment – which is why it has fallen so heavily out of favour. Order and judgment are now modern taboos, not just in education theory but way beyond the schoolroom.

A key – albeit late – disciple of this doctrine of creativity is Brian Cox, professor of English at Manchester University. The seminal role played by Cox in the development of the English curriculum, and his many bewildering shifts of position, are dealt with in more detail later. However, in outlining his contribution to the English curriculum he explains why the craft of creative writing should actually supplant the pre-eminence of the essay in the syllabus.[9] Too many students and pupils, he observed in 1991, wrote 'dull and discursive prose'. Instead of mainly writing essays, he thought they should be encouraged to write in a wide range of forms including diaries, formal letters, chronological accounts, reports, pamphlets, reviews, essays, newspaper articles, biography, poems, stories, plays and TV scripts. His reasoning was highly revealing.

> For many years the essay has dominated the syllabus, from school certificate for 16-year-olds to university degree examinations. I am not decrying its great importance as a means of evaluating evidence, developing critical arguments and organising ideas in rational form. But the essay is usually a closed form, not allowing students to admit to uncertainty in their thinking, not allowing them to explore their ideas in imaginative and open-ended ways. The essay form has been dominated by scientific models of objectivity, and, as I have already said, students have relied heavily on the repetition of the views of their teachers or the critics, views often not in accord with their own personal response to the texts being studied. Images, ambiguity, dramatic tensions, all central features of twentieth-century modern literature, are not usually allowed as the student is marked for coherence, order and objectivity. Teaching of the craft of writing releases pupils from imprisonment in the essay form and from passive repetition of other people's ideas.

In this passage can be found many of the central fantasies embedded in the flight from formal language teaching. There is the sidelining of objectivity, the retreat from facts, order and evidence into the world of make-believe as a higher order of experience. There is the distrust of teachers, whose influence on their pupils is apparently wholly malign, whose every utterance fetters children's capacity for independent thought and from whom children therefore need to be protected. And above all there is the primacy of the imagination. As has already been said, the development of the imagination is certainly desirable and there should indeed be a place in the syllabus for creative expression. But the essay once enjoyed pre-eminence because what it does above all is to teach children how to think. And that should be at the very heart of a child's education.

The tremendous emphasis that has been placed on creativity, however, has greatly diminished the role of the essay in the classroom. The result is that one of the prime tools for teaching children how to think has been devalued. Instead of being taught how to think, they are now mainly allowed to imagine. It is another example of the shift that has taken place from teacher to learner. Imagination, after all, can scarcely be taught. In the circumstances, therefore, it is hardly surprising that the University of East Anglia (as we saw earlier) has had to resort to giving its new undergraduates an idiot's guide to essay writing; nor is it surprising that college and university teachers report in despair that the one thing their students can't seem to do any more is to think for themselves.

Above all, the decline of the essay indicates the deep disfavour into which the exercise of critical judgment has now fallen. Education, as the primary site for the transmission of a culture, can no longer easily transmit a hierarchy of cultural values because there is no longer general agreement that any mode of expression is better or worse than any other. This is cultural relativism. In the schools, it flows inevitably from the destruction of all external authority and the relocation of that authority within the child. And as will be suggested later, it is first cousin to that moral relativism by which moral values have also been privatised, so that everyone has become their own individual arbiter of conduct with no-one else permitted to pass judgment. Judgment itself is taboo for an intelligentsia which has declared an egalitarianism of values. Aesthetics have become politicised, and English language and literature contentious.

Standard English

The growth in popularity of creative writing, Brian Cox told the Bullock Committee in 1972, had to be seen in the context of this debate.[10] Children who cracked the language codes, he observed, were more successful than those who did not; they were the ones who became leaders in society and participants in high culture.

Left-wing teachers now incline to oppose this process. They want a popular culture, and they think that cultures are to be seen as different, not as inherently superior or inferior. In education this 'value-free' concept of culture leads to the belief that 'correct' English, the old grammar school tradition of clarity, is not superior to the vitality of working-class speech. This assumption is one cause of the dilution of English teaching today and the reaction against spelling and grammar. In these matters a strong vein of primitivism is at work, together with a dislike for rational forms . . . An explicit grammar is an acceptance of order; the new revolutionaries deliberately reject such structuring of experience.

Far from promoting true egalitarian values, however, these revolutionary 'primitives' – who happen to command the heights of the education culture – betray over and over again a deep contempt for their pupils' abilities, and particularly for those pupils who come from economically impoverished backgrounds. For example, in 1995 Professor Asher Cashdan, a teacher trainer, wrote in an educational magazine about the National Curriculum English Order:

> Standard English is, in my view, still greatly over-emphasised . . . This push for Standard English still smacks of a right-wing witch-hunt against so-called 'slovenly' speech, spelling and punctuation. I am continually amazed at teachers meekly accepting the imposition of Shakespeare plays on all pupils. Although there is talk in the Order of smoothing 'access' for the less advantaged pupils, it remains a fact that most of them (and plenty of us too) will make very little of a Shakespeare play and may be turned off all drama and literature by having the Bard stuffed down their throats at an early age. When will teachers insist that high quality literature is only a part of what should be studied in school, and that much contemporary work outside the canon of English literature can be rewarding, motivating and educationally worthwhile?[11]

Precisely what of value Professor Cashdan believes pupils obtain from the study of inadequate or trashy material is unclear. What seems all too obvious, however, is his assumption that children from poor backgrounds are stupid, which is not only deeply offensive but demonstrably untrue. Such an attitude illustrates precisely how the most needy pupils of all have been the most bitterly betrayed. Children from homes where there are no books and where theatre visits are an activity that might as well take place on Mars rely absolutely and desperately upon the schools to provide an education that will lift them out of such disadvantage. Taught properly, Shakespeare is often a liberation for such disadvantaged pupils, as many good and dedicated teachers can eloquently testify. Shakespeare is not an 'imposition' upon pupils but their entitlement, whatever background they may come from.

In a subsequent edition of this magazine Cashdan was rebuked. In a letter to the editor, Nicholas Lee pointed out that according to Professor Randolph Quirk, the foremost grammarian of this century, standard English was the form of language which was the most widely accepted and understood within the English-speaking world. Cashdan, he observed, had of course written his article in standard English and spelled it correctly, a facility he would apparently wish to deny to others.

There are, naturally, many teachers and educationists who would similarly reject Cashdan's approach. Nevertheless, cultural relativism has penetrated deeply into British society, and a significant body of teachers has assimilated its assumptions. With such powerful figures in the education world advocating the destruction of literacy and sounding the 'retreat from the word', some parents can now be disconcerted to find themselves talking a different language from their children's teachers. One Labour MP, who was concerned that his young son wasn't making good enough progress in reading, approached his teacher only to be told: 'No need to worry. None of these kids needs to read books any more. They can watch TV instead.'[12]

To the education radicals, TV is an example of what are sometimes called 'the emerging literacies' – in other words, one of a number of media which they promote as of equal value to literary texts. Media education has an explicitly political and ideological purpose. As David Buckingham, a lecturer at the London Institute, has written, it has aimed (too ambitiously, he judiciously conceded) to 'subvert dominant ideologies', 'empower the oppressed' and 'revolutionise the school curriculum'.[13] 'Throughout its history,' he wrote, 'media education has been regarded by its advocates as a movement which has sought to bring about radical political changes, both in the consciousness of students and in the education system itself.'

It would achieve these ends, apparently, by making available to students what had previously been hidden from them about the ways in which the political system did them down. 'Thus, telling students about the ways in which media institutions operate – for example, about how patterns of ownership and control serve to marginalise or exclude oppositional perspectives – is seen as a means of "opening their eyes" to the covert operations of capitalism', not to mention gender or race stereotypes. In addition to putting out such propaganda, it appeared, media education also promised to subvert the whole notion of transmitted authority. 'Primarily by virtue of its content, media education has the potential to challenge traditional notions of what counts as valid knowledge and culture. In the process, it is argued, it makes for much more egalitarian relationships between teachers and students: the students are now the "experts", while the teacher is no longer the main source of authority.'

Such nostrums are not the preserve of some far-out clique. On the contrary, they have captured the citadels of culture, in the universities and elsewhere. At Oxford University, the Rupert Murdoch Chair of Language and Communication is held by Jean Aitchison, who believes that there is no such thing as correct English. 'I don't like the word "correct" at all because I don't think there is a correct English,' she said. 'I actually deplore the use of the word "correct" and "incorrect" in relation to language. I use the word "appropriate" or "acceptable" much more readily . . . I really hate people being given inferiority complexes over the way they talk or write.'[14] It would appear, therefore, that the main reason for Professor Aitchison's aesthetic distaste is her concern to spare the feelings of people who might think worse of themselves if people tell them they have used the language wrongly. The very concept of wrongness, therefore, has to be written out of the script altogether. Sensibility alone rules. Modes of speech are not better or worse, only different – a repudiation of value judgments which, as we shall see later, has extended beyond education into many other areas of conduct and behaviour, and with disastrous results.

So impressed, however, is the British establishment by such a philosophy, so profound and important are deemed to be the thought processes of Professor Aitchison, that the BBC invited her to deliver the 1996 Reith Lectures, the prestigious link with that tradition of high-minded culture which once distinguished public service broadcasting. Her broadcasts were devoted to promulgating the simple message that the English language was not deteriorating, only changing. The same could not be said, unfortunately, of the Reith Lectures themselves.

Nor could it be said of the Oxford BA course in English Language and Literature in the hands of Professor Aitchison and her colleague, Dr Margarita Stocker, who in 1995 launched between them a course in 'Language, media and film' as part of the degree. Among the essay topics facing undergraduates taking this paper were women's film, gangster movies and horror. Said Dr Stocker: 'Students are more visually literate than literate-literate nowadays and we are working with that. We need to face the fact that we are living in a multi-media 20th century. For most students, the important thing is not so much the content of what undergraduates study but that they have a range of skills. You can do that with film as much as with a degree in English or history.'[15]

Relevance, therefore, has become the essential precondition of learning. Skills and process are all-important, ends in themselves rather than means. Knowledge is old hat. Culture is privatised. Literacy is subjective. The idea that it might be possible to encourage students at Oxford to become 'literate-literate' by reading books, and that being 'literate-literate' has an intrinsic merit over and above watching films or TV and is the rationale for studying English, is clearly far too unsophisticated an analysis, especially for an initiative associated with the holder of the Rupert Murdoch Chair in Language and Communication.

Education and culture have passed into the hands of philistines. From the failure to teach infants to read, through the repudiation of grammar, punctuation and spelling, all the way to teaching Oxford English undergraduates about gangster movies, the new illiteracy was blessed by a cowed and enfeebled establishment that no longer believed it had the right to engage its critical faculties except to support the notion that anything goes.

Cultural Relativism

This linguistic scepticism was intimately related to the rise of philosophical scepticism which we shall explore later. The result was a deconstruction of reality which undermined not just what was taught in schools but in the universities as well. Under the impact of a relentless cultural relativism, in which no absolute values could any longer be asserted, the whole aesthetic culture started to disintegrate. It wasn't just the validity of language or teaching that came under fatal attack but the validity of the texts themselves. Intellectuals were no longer content to draw from classic texts the lessons of antiquity. In the modern spirit of individualistic hubris, they believed they could improve upon them. So the literary canon was recreated, in accordance with the doctrine that creativity was the highest form of human activity, and to promote the ideologies of gender, race and class.

A landmark event in this steady relativisation of English literature was the month-long international seminar for English teachers held at Dartmouth College in America in 1966. Although there were only some fifty participants, the influence of that meeting was to persist, in the words of the New Literacy advocate John Dixon, for nearly two decades, and affected some of the key research groups which would influence the thinking of the early curriculum body, the Schools Council.[16] Not only did this meeting equate responses to literature with reaction to films, TV plays and the students' own personal writings or spoken narratives, it also invited the reader to redefine the value of a text in the light of his or her own experience. The process was seen as an explicit means of challenging the forces of 'powerful institutional centres'. Relative values were turning into a political weapon.

Meanwhile, individual texts were being reduced through the application of modern theories into higher orders of meaninglessness. The only meaning derived from this arrogant and nihilistic exercise appeared to attach to the reputations and careers of the academic deconstructionists themselves. In the 1970s the French critic Jacques Derrida helped set off the boom in literary theory, defined as the sceptical interrogation of artistic forms. In the words of the American literary academic Mark Edmundson, it tried to prove the kinship of art with propaganda, pornography and imperial oppression. Seen in this light, great poets no longer appeared as the acme of wisdom or human triumph but as shrewd deceivers, apologists for race and gender oppression and 'bagmen for the bourgeoisie'.[17]

The study of English literature became riven by every passing fad: psychoanalysis, deconstruction, Marxism, feminism, cultural materialism and the new historicism. It provided the opportunity for cultural revolutionaries to claim that great literature was a means of political domination. The radicals took up linguistics and psychology and used them to fashion the study of English into a weapon to deconstruct personal and social identity. The study of English literature became fused with the new identity politics revolving around race and gender. Shakespeare in particular was repeatedly 'reclaimed' for key minorities including homosexuals and ethnic groups. There was a fundamental failure to understand that the reason for studying great works was precisely because they transcended particulars of politics or place and put us in touch with universal values.

But by then the emphasis had shifted away from knowledge of universal values – or indeed, knowledge of anything much at all – to creativity. If intellectuals weren't actually able to create works of literature themselves, then the next best thing was to 'recreate' them through critical theory. It no longer mattered what the text said to you as much as what you might say about the text. And in the process, these texts weren't of course recreated but effectively deconstructed out of existence. It was an exercise not of creativity at all but of intellectual vandalism.

As the American historian Gertrude Himmelfarb observed in 1995, whereas there used to be arguments about the validity of certain truths in academic lectures, now there was a disregard of the very idea of truth itself. Scholars had become instead advocates for a radical scepticism and relativism that rejected truth, knowledge, reason or objectivity and refused even to aspire to such ideas on the grounds that they were authoritarian and repressive. This marked a complete break with normal academic scepticism. It meant that no truth was to be derived from history, no objectivity, no events; only texts to be interpreted in accord with the historian's own interests and disposition. The humanities, she wrote in despair, had been relativised, subjectified, problematised and politicised. The scholar's own feelings and experiences were exalted and the first personal pronoun was dominant. Scholarship itself was being redefined, and Nietzsche's famous prediction was now a reality: 'Nothing is true; everything is permitted.'[18]

This postmodern crisis in scholarship was rooted in trends in philosophy. From the 1930s, a number of philosophers had repudiated the idea that the individual attaches words to things and derives valid meaning from them. It was rather, they thought, that the social group shared the same language; words and meanings therefore existed only as social constructs. Meaning therefore became meaningless; anyone's meaning was as valid as anyone else's. Jacques Derrida had argued that language referred only to itself but not to anything else. 'Il n'y a pas de "hors texte",' he wrote: there was nothing of validity outside the written text. Another French philosopher, Jean-François Lyotard, wrote that postmodernism had destroyed the consensus around knowledge, a consensus traditionally encapsulated in narratives of one kind or another.[19] Narratives of any kind now no longer had any authority. In other words, there was no longer any way of understanding reality, since reality had been deconstructed out of existence. The only texts that didn't appear to qualify for this passport to meaninglessness were the writings of these philosophers themselves, who retained their university employment even while they appeared to be making the very concept of a university redundant.

As the educational philosopher Nigel Blake has written, this approach lends itself to factions fighting each other over their own definitions of reality. Such sectional interests eventually 'undermine traditional academic values of open-mindedness and self-criticism, and even to militate against sharing values'. The university itself is thus threatened by such an approach. 'Its separatist implications undermine any guarantee of the growth of knowledge. Without commitment to some conception of that growth which transcends their own concerns, subject areas may cease to be fully critical enterprises which guarantee progress. They may cease, in effect, to be disciplines. Academic authority may edge closer to the authoritarianism of dogma.'[20]

The impulse that has driven this otherwise bizarre and inexplicable collapse of critical judgment is a fundamentalist egalitarianism of values in which anyone's criteria are as good as anyone else's. The effect of such a deconstruction of the text has also been a deconstruction of the culture. In 1971 the cultural critic George Steiner anticipated postmodernism when he wrote in his book *In Bluebeard's Castle* of 'our present feeling of disarray, of a regress into violence, into moral obtuseness; our ready impression of a central failure of values in the arts, in the comeliness of personal and social modes; our fears of a new "dark age" in which civilisation itself, as

we have known it, may disappear'. The democratisation of culture
had been brought on by a failure of nerve within culture itself; the
retreat from the word and the arrival of the counter-culture signalled
an end to any agreed hierarchy of values. Steiner wrote: 'It is the
collapse, more or less complete, more or less conscious, of these
hierarchised, definitional value-gradients (and can there be value
without hierarchy?) which is now the major fact of our intellectual
and social circumstance.'[21]

But how could it have been otherwise? Culture itself was simply
no match against the radical egalitarian temper of the age. It had
come to be seen as the preserve of an élite, a citadel provocatively
inviting capture. Élites were, after all, responsible for preserving
the higher values of a society through its culture. But if values
themselves were now taboo, the role of that élite was reprehensible
and the culture itself an unacceptable anachronism. In a mass
democracy whose defining value was sameness, such a definition
of culture set itself against the defining political ideal.

As the cultural critic Robert Hewison has recorded, these pressures were visible as long ago as 1957 when the BBC reorganised
the Third Programme, in response to a combination of financial
pressures and the rise in TV viewing, to achieve a wider audience
for the minority channel. Half its output had been speech; now an
increase in music diluted its missionary character as an all-round
high culture medium.[22]

The revolt against high culture and the corresponding rise of the
demotic had started even before the Angry Young Men burst upon
the stage in 1956 with John Osborne's play Look Back in Anger. As
Hewison has chronicled, an article in Encounter in 1953 remarked
upon the rebellion of lower middlebrows, led by redbrick university
teachers, against the intellectual trends of Oxford, Cambridge and
London universities. Out of this 'popular' revolt, and nurtured by
the Workers' Education Association classes and the growing public
perception of the many injustices in the lives of working people,
came the new discipline of cultural studies, whose seminal texts
were Richard Hoggart's The Uses of Literacy in 1957, and Raymond
Williams's Culture and Society in 1958. But at this stage, there was
no talk of rewriting the canon. On the contrary, cultural studies
were rooted in respect for the values of the past as embodied in
great literature; the desire was rather to make such glories more

democratically available. Williams didn't question the canon until the seventies and Hoggart, the champion of working-class culture, always accepted it.[23]

In view of this early history it is an ironic, if not a tragic commentary on our times that in 1995 Richard Hoggart published a lament in which he painfully charted the manifold ways in which culture had by now become thoroughly demeaned, trivialised, vulgarised and relativised.[24] In his view, attitudes displayed by the Arts Council revealed a paralysed inability to make judgments about artistic value. 'One hugely devoted member of staff was embarrassed when asked to consider that there might be differences of quality in human creativeness. "Well," he at last said, "some people like Mahler and some like driving motorbikes. We shouldn't judge between them."'[25]

As Hoggart wrote in despair, many people now almost entirely rejected the notion of 'great' works of art in any form. 'They avoid vertical judgments, in favour of the endlessly horizontal. By the Nineties a senior official with Radio 3 could announce: "There is no art; only culture." A spiritual mate adds: "Each man is his own culture."' Yet although the impulse behind such a flight from judgment was ostensibly egalitarian and hostile to élitism, the people who were its most poignant victims were those whom the doctrine purported to champion. 'The thought that such a creed is being offered to people who still live in a bookless culture, as a justification for being satisfied with the popular press, the shoddier television programmes and other such barbarisms, is yet another instance of the "stay as sweet as you are" syndrome – all too often used by "good democrats" who seem not to see they are consigning other people to the worst aspects of consumer capitalism.'[26]

The way in which this philistine cultural relativism has been abetted by the operation of the free-market promoted so assiduously by the Thatcher/Major government is discussed in detail later in this book. But Hoggart himself had played his own part in the process. He had, after all, launched the first centre for cultural studies at Birmingham University in 1964. Since he was committed to absolute cultural standards, there were tensions right from the start over the collision with popular culture. And although Hoggart held the line, by definition such a centre relativised aesthetic values. As Hoggart himself wrote with understated candour: 'The invasion [of relativism] into Cultural Studies began with the founding of the subject itself; which is often markedly non-traditional, and concerns itself with the phenomena of modern life at all levels, and especially with mass culture and popular culture in all their forms. It is therefore in the thick of the relativist society and is itself affected by that relativism, sometimes disablingly.'[27]

Hoggart no doubt wrote with feeling. In 1972, he was replaced as director of Birmingham University's Centre for Contemporary Cultural Studies by the radical-chic critic Stuart Hall. The centre promptly broke with the literary traditions of its founders in favour of a Marxist perspective which saw the literary tradition as a form of ideological domination. Under Hall, the Birmingham Centre became the catalyst for every passing fad in critical theory and significantly helped the culture to implode under the pressures of relativism. But what was once avant-garde soon came to redefine the traditional centre ground. Gradually during the 1980s, cultural studies were absorbed into the academic mainstream.

As the critic Angela McRobbie put it: 'Take a tradition of textual analysis long established in the English departments of a hundred US campuses, update it with the language of structuralism, post-structuralism and deconstruction, extend this mode to texts of popular culture (film, TV, video, popular fiction), add the blessings of cultural relativism which make it "legit" to take the popular seriously, canonise the new curriculum by appealing to feminism, black or popular politics (with the final of these acting as a kind of substitute for class) and we are all in business.'[28]

In 1992 the Marxist Terry Eagleton – the self-appointed 'barbarian in the citadel'– as Thomas Wharton Professor of English Literature at Oxford illustrated how thoroughly the defences had been penetrated. As the American professor Harold Bloom mordantly predicted: 'What are now called "Departments of English" will be renamed departments of "Cultural Studies" where *Batman* comics, Mormon theme parks, television, movies, and rock will replace Chaucer, Shakespeare, Milton, Wordsworth and Wallace Stevens. Major, once-élitist universities and colleges will still offer a few courses in Shakespeare, Milton and their peers, but these will be taught by departments of three or four scholars, equivalent to teachers of ancient Greek or Latin.'[29]

Bloom's elegiac note was sounded in the knowledge that his

voice was in the minority. The way in which cultural relativism was shoe-horned into the Conservative government's National Curriculum will be explored in some detail in the next chapter. But one of the key players in that process, Professor Brian Cox, illustrated in his own book how cultural relativism was explicitly used as a weapon of political protest. Despite repeated protestations that he occupied a besieged middle ground, Cox seemed to endorse the deconstruction of literary worth. He accepted the need to teach the destructive critique of the canon, not just in the usual terms that it comprised 'dead white males' but also by stressing the importance of the spoken language in the media. 'Many teachers are now persuaded that English lessons should not be solely confined to "great" literature but should include all kinds of "texts", spoken and written, together with popular forms such as television commercials, bestsellers and comics as well as TV and video.' Thus literature, just another set of 'texts', was insouciantly stripped of any inherent value. All became of equal value since what was seen as valuable was not the content but the process.

Moreover, although Cox appeared to take satisfaction in the fact that this was fundamentally a political rather than a cultural programme, he didn't seem to grasp that it would create indoctrination and propaganda rather than education. He wrote: 'Traditional teaching of English literature has been attacked because it creates in the pupil a mood of passivity, an instinctive submission to authority. Throughout their secondary school careers, pupils are often persuaded to repeat their teachers' opinions rather than to develop their own points of view. In the classroom, cultural analysis encourages students to examine for themselves the underlying assumptions in the texts they are studying.'[30]

But none of this was true. Properly taught, the study of English literature transmits a questioning and sceptical habit of mind, helped by the often radical or subversive literature being studied. Moreover, the very last thing cultural studies did was to allow students to think for themselves. On the contrary, it told them there was one analysis only that was permitted, that these texts had an agenda of class, gender or ethnic domination. Any students who were rash enough, in this pedagogic climate, to confront the new orthodoxy by writing that these 'hidden agendas' were no more than politically correct chimeras would find that their grades soon took a dive. The teaching of culture had become a political and ideological battleground. The attempt to retrieve it before it fragmented completely was to prove a more difficult task than the politicians had expected.

Chapter Five

THE NATIONAL CURRICULUM DEBACLE
Pressure for Reform

One of the puzzles about education in Britain is that the destructive ideologies that so tenaciously grip it reached their high point during the 1980s. Yet that was the very decade when Britain was governed by Mrs Thatcher, as she then was, the most ideological Prime Minister in modern memory, and a leader who was ostensibly committed to root out precisely such attitudes, in education as elsewhere. Indeed, in 1988 the National Curriculum was brought in precisely to remedy the gross deficiencies in teaching. The theory behind this unprecedented extension of state control over education was that by laying down prescribed targets for study, teachers would be forced to bring their pupils up to this basic minimum standard.

It might have been assumed that this policy would work. Mrs Thatcher, after all, had altered an entire political culture, tackling each of the professions one by one. Opposition was marginalised, 'wet' ministers were famously dispatched, and civil servants were politicised, with the national interest they had once served now identified with the interests of the Conservative Party. It seemed unlikely that such a single-minded and ruthless Prime Minister would fail to bend education to her will in the same way she had established her dominance over local government, the health service and the legal profession.

But it didn't happen like that. The education establishment fought back with every weapon at its command. The Thatcher government found itself embroiled in a tenaciously sustained and debilitating guerrilla war in which it was outgunned and outmanoeuvred at every turn. The National Curriculum became a battleground in which the attempt to bring basic educational concepts back into schools was ferociously and to a large extent successfully resisted. Civil servants elsewhere may have been cowed or convinced by the Thatcherite ethos, but the Education Department was a ministry apart. Whitehall civil servants forged an astonishing alliance with educationists to frustrate or dilute ministers' aims and to substitute their own agenda wherever possible. Political will squared up to an entrenched culture and lost. The reasons were ministerial hubris, incompetence and – despite all the rhetoric and the millions of pounds spent on the policy – a deeply rooted indifference by politicians to the plight of children in state education. Ministers may have become angry and irritated, but all the passion belonged to the other side. The result was that, despite bringing about some improvements, the National Curriculum actually made matters worse in some important ways, by institutionalising some of the worst attitudes and giving them the force of law.

The culture that ministers found themselves up against was not confined to a few extremists working for far-left Labour-run local authorities. It was, rather, a mind-set that characterised virtually the entire education establishment. The doctrines of cultural relativism and child-centred, progressive teaching methods had been absorbed into the professional bloodstream. To argue against them was to encounter not merely incomprehension or repudiation but a moral rage. There was an unshakeable faith that these theories were in the best interests of children and therefore that those who denied them were not merely in error but intent on doing children harm. There was consequently an absolute denial of the harm these theories were themselves doing to children. Where there was clear and demonstrable evidence of such harm, it was either ignored or denied or blamed on every surrounding social factor that could be thought of – parents, television, poverty, unemployment – on anything but the way children were being taught.

In the universities, the dominance of these ideas was near-total among educationists, the professors and lecturers in education, whose views were often significantly at odds with their colleagues in individual subject specialties. Those very few educationists who held out against the orthodoxy were stigmatised, ostracised and generally made miserable. Like everyone else who tried to reverse the trend, they were denounced and written off as 'right-wing', regardless of their political preferences.

Many if not most of the university-based educationists, including those who taught in the teacher training colleges, hadn't taught schoolchildren for years, if at all. But among the teachers at the chalk-face, the picture was less consistent. Very few teachers have much time for theories of any kind; they are far too preoccupied with the demanding job of daily teaching in the classroom. And a significant number of them did resist the assumption that they were no longer teachers but facilitators, that learning was more important than teaching and that knowledge and authority were to be found inside the child rather than outside. But although most of them were not political activists and may never have heard of the more arcane theorists, they absorbed these assumptions and never thought to challenge them. How could it have been otherwise? There was no challenge to these views in the culture of education. The articles the teachers read in the educational and mainstream press, the books they were directed towards on their training courses and the shared premises of their colleagues all created a closed world of thinking that was muddle-headed to the point of menace.

Most crucially of all, this thinking was enforced by those who were employed to reinforce and police the education system. Local authority education advisers and inspectors played a vital role, not only disseminating these beliefs but also in dragooning into line those teachers who thought such views were wrong. Traditional teachers found their teaching methods were held against them and blocked them from gaining promotions or new jobs. They became too frightened to teach in the manner they believed to be best for their pupils. When the education inspectors came round, such teachers had to hide their reading schemes and rearrange their orderly and attentive classrooms into approved disorder for fear they would be criticised for stifling the children's creativity.

One Wiltshire teacher recorded that, to his personal knowledge, good teachers had been intimidated by local authority advisers who blighted their careers if they dared to teach children to write correct English. This meant, he said, that a number of schools tended to be stuffed in their upper echelons by placemen and favourites; as a result, rank-and-file teachers had little option but to obey such diktats.[1]

An education psychologist painted a chilling picture of the damage being done by this enclosed world of closed minds.

I greet with relief the beginning of debate about modern teaching practices. In my job I see small children whose listening skills and ability to stay focused on a task are chronic – yet they are put to learn in an environment which no undergraduate would have to suffer . . . Sometimes a brave teacher does arrange the tables so that each child has a space of his or her own; the children love it, but the teacher has to move the tables back again.

Children do not learn through play . . . but through instruction, explanation, guidance, motivation from an adult. Children need to be taught to make connections, to look for meanings. They do not learn from Wendy houses or computers, they learn from people. And whenever I say this to a group of teachers, the older and wiser members of the group come to me afterwards and thank me for saying it; they have been waiting for years for someone to make this point. But for some reason they cannot say this in public. And neither can I; which is why I do not want my name published. My job is important to me and public condemnation of teaching methods will not be approved. But in an odd way I cannot really pinpoint *who* will disapprove.[2]

The tightness of the grip exerted by the body of inspectors and advisers cannot be exaggerated. It was through them that the ideologies espoused by university-based educationists were most effectively transmitted into the classroom. And a curious political symmetry appeared to be at work. The power of this culture seemed to increase in direct proportion to its loss of power in the national political arena. The successive General Election victories of Mrs Thatcher, and more particularly the way her administration appeared to squeeze all dissent or serious challenge out of the system, made some people even more determined to exercise a contrary power in the social arena. While Thatcherism was busy consolidating its hold at Westminster, a quite different war was being waged on the battlefield of the mind. 'Best practice' became the rubric deployed by the front line of ideology. 'When I hear that phrase "best practice",' said one war-weary political insider, 'I reach for my gun.'[3]

The ruthless way in which this orthodoxy was applied has produced a small but significant group of prominent dissidents. The way in which the reading specialist Martin Turner was sidelined has already been described. There were other examples. Events at the Lewes Priory comprehensive school in Sussex, for example, were so disturbing they led to a debate in the House of Lords.

In 1987, the extremely well-qualified history teachers at Lewes Priory decided that the GCSE history exam wasn't up to scratch. It was stopping them from teaching history as they understood it to be. Its slant towards subjectivity, through course-work and empathy (exercises in which pupils were asked to imagine themselves in the position of individuals from the past), was preventing their pupils from being examined properly in their historical knowledge. One member of the department kept his head down and so avoided the ensuing storm. The other teachers – Anthony Freeman, who held a doctorate, Chris McGovern, who had a first-class history degree and was head of the department, and Arthur Franklin, who had a master's degree – decided to make a stand. In addition to teaching GCSE, they decided to offer their pupils an alternative syllabus, the Scottish O-grade, which they believed was a more rigorous exam.

The authorities strongly disagreed. In a series of explosive rows, the teachers were refused permission even to discuss the Scottish option with parents on the school premises. So the teachers met interested parents in their own time and away from the school. The result of their persistence was that they were accused of insubordination and mutiny and lost their jobs through 'redeployment'. But that wasn't the worst of it. Although Franklin was eventually re-employed in a tertiary college, McGovern and Freeman were effectively blackballed from teaching in the state sector. It was local education authority policy to redeploy teachers in comparable posts. But Freeman was frozen out and rejected by job after job. The only person in post eligible for the position of head of history in a new 11 to 16 school which was being formed from Lewes Priory, he was rejected unanimously by the school's board of governors. The education authority then formally overrode that decision – but only, according to Freeman, on condition he turned down the job that was then to be formally offered to him and left teaching altogether. He then applied for a job at another school, where he was told he wasn't sufficiently committed to the 'gender awareness' policies of the history department. A further school doubted whether he was able to teach across the ability range; yet another rejected him on the grounds that he would be teaching the sixth form in a way that would be too complicated. Eventually, this highly qualified history teacher whose competence had never been in doubt was retrained by the local authority as a schools tour operator.

McGovern continued to make waves and managed to get the affair raised in the House of Lords in 1989, only to get the disingenuous reply from government spokesmen that the teachers had done nothing wrong and had merely failed to be reappointed following the reorganisation of local schools. In 1990, McGovern reported on a test he had carried out with permission at a local primary school in which, he claimed, most 10- and 11-year-olds had scored the equivalent of reasonable GCSE grades on one of the previous year's GCSE papers, upholding his contention that GCSE marks depended not on historical knowledge but on the mere application of commonsense skills. The result of that was that he was told by East Sussex education authority that it was unlikely that he would get a permanent job. He retrained as a primary school teacher and now teaches in a prep school. 'The fact is that here are two teachers with four degrees and thirty years' experience between us; yet all that expertise is now being denied to state school pupils,' said McGovern. 'My skills are now being used on children whose parents are paying a lot of money for the privilege, whereas state school children could have had them for free.'[4]

In the House of Lords debate, the Conservative peer Lord Beloff said that during the affair political control had changed on East Sussex council from a Labour/Alliance coalition to Conservative. 'But as far as I can see no difference whatever was made. The councillors of my party were as supine in the hands of their bureaucratic officials as their predecessors – at any rate, most of them. With hardly any exceptions, the parent governors of the school who had sealed the teachers' fate were connected either with the University of Sussex or Brighton Polytechnic. They are all persons professionally engaged either directly in education or are officials of bodies concerned with advancing certain educational theories and practices,' he said.[5]

Ironically, considering that the 'crime' of McGovern and Freeman was to offer the Scottish O-grade at Lewes Priory, an increasing number of English schools is now turning in desperation to Scottish examinations to provide an intellectually more rigorous exam at age 16. The Scottish Standard grade and the Highers are being taken by more than 1200 English pupils every year, even though the results are officially ignored in compiling the English school league tables.[6]

A third incident not only demonstrated the systematic manner by which the education establishment managed to marginalise its dissidents but also provided an early indication of the less than disinterested role being played by civil servants in the Education Department. For more than 20 years, the education activists John Marks and Caroline Cox have been campaigning against the prevailing orthodoxy and calling for a return to traditional standards in education. The education establishment has consistently written them off as right-wingers pursuing a political programme which could therefore be safely ignored. The problem was that the facts they produced were correct.

In the early 1980s, before the days of school league tables, Cox and Marks decided to research the performance of different types of schools, since the Education Department had refused to assemble figures about exam results which schools had only recently started to publish as required by the 1980 Education Act. They obtained figures relating to 350,000 pupils in more than half the country's education authorities, and found evidence of wide differences. In particular, they found that children from comprehensive schools performed relatively badly compared with groups of children from similar social classes in selective schools. Secondary modern schools were sometimes getting better results than comprehensives.[7]

Their research was published in June 1983 and given minimal coverage in the educational press. But immediately a massive campaign was launched to discredit their findings. There was a barrage of hostile comment. *The Teacher*, the journal of the National Union of Teachers, refused to publish any reply from Cox and Marks answering their critics. *The Times Educational Supplement* ran an article claiming that neither Cox nor Marks had any direct knowledge of comprehensive schools.[8] In fact, they both sent their own children to comprehensive schools – which in large measure explained their anxiety and concern.

But the real campaign against them was based in the Education Department, which was institutionally wedded to the philosophy of comprehensive education. In a rare and revealing public comment, Nick Stuart, a senior civil servant in the Department, talked about 'a fundamental and philosophical divide with research just one of the instruments of battle'.[9] Cox and Marks were on one side of that divide and the Education Department – regardless of the opinions of ministers – was firmly entrenched on the other.

In September 1983, the Department prepared its own confidential commentary on the Cox and Marks research. Its main criticisms were that they had misrepresented the proportion of grammar schools in their study, and that they had used out-of-date census data. The research, it said, was seriously flawed. According to Cox and Marks, this commentary was withheld from them on grounds of confidentiality, but leaked to the press over a period of several weeks, despite the fact that its criticisms were based on incorrect interpretation of the grammar school figures and census data.[10] At a meeting with the Department's statisticians and ministers on 23 November 1983, they wrote, the statisticians admitted these errors and apologised. At a further meeting, the head of statistics said he had never said their work was seriously flawed and dissociated himself from such a statement. A few days later the then Education Secretary Sir Keith Joseph made a statement to the House of Commons vindicating their research. But by then the damage from the constant stream of leaks had been done. On October 23, Cox and Marks were surprised to be telephoned by the *Guardian* asking them to comment on a letter sent to them by Joseph which they hadn't yet received. On October 19, the National Union of Teachers had published a critical analysis of their research making exactly the same error as the Department's statisticians had made in the commentary which had not seen the light of day. And despite Joseph's vindication of the authors, the *Guardian* insisted that the research remained discredited by his statisticians[11] as did the *Teacher, The Times Educational Supplement* and the *Observer*.[12]

Such was the culture that the Thatcher government took on when it decided to launch its education reforms. But anxiety about Britain's educational performance had certainly not originated with Mrs Thatcher. It was in 1976 that the Labour Prime Minister, James Callaghan, launched what was called the Great Debate on education. In a speech at Ruskin College, Oxford, he expressed concern about the teaching of the three Rs, the relevance of the curriculum to the needs of the economy, the adequacy of the exam system and popular unease about informal methods of teaching.

Callaghan spoke against the background of the oil crisis, the new fragility of public expenditure and the new mantra of getting value for money. The concern was therefore not so much about whether children were being properly educated to become responsible members of society and well equipped for life itself. It was more about whether education was adequately meeting the needs of the economy. But this concern had predated even Callaghan's speech. In 1972, the government set up a committee chaired by Alan Bullock in response to alarm about high rates of illiteracy.

Indeed, anxiety about education had been rumbling virtually ever since the post-war education settlement was finally put in place, some 20 years after the Education Act was passed in 1944. According to the former Chief Inspector of Schools, Professor Eric Bolton, Callaghan's speech was a watershed. Until 1976, the government had expressed no interest in the curriculum and people would have considered it quite improper if it had done so. But Bolton argued that by the time Labour left office in 1979, the climate had shifted and it was accepted that the government *did* have a legitimate interest in the curriculum.[13]

Even though some problems were being acknowledged, the early years of Mrs Thatcher's government were marked by a distinct lack of interest in education. It just wasn't seen as a priority. At the political level, there was certainly no great ideological division. By and large, there was a consensus in favour of comprehensive schools among both Labour and Tory politicians. 'There were plenty of Tories in local government who felt this was the thing to do. It was a common culture,' said Stuart Sexton, an adviser to successive Conservative education ministers during the 1970s and 1980s.[14] It was Mrs Thatcher after all, during her time as Education Secretary in the 1970s, who had shut down more grammar schools than any other minister. When Sir Keith Joseph was appointed Education Secretary in 1981, he was troubled by low standards. But even in those early years, it was clear that unholy alliances were being forged which meant that virtually from the beginning government ministers unwittingly connived at their loss of control over their own agenda.

Joseph's preoccupation was with the low standards of achievement of the bottom 40 per-cent of the pupil population who he believed were being failed by the system. This meant he was receptive to the idea of the GCSE, an exam which combined the more academic O-level with the less academic Certificate of

Secondary Education. The aim was to create a public examination that the majority of children could pass and that would not divide children into sheep and goats. But according to Stuart Sexton, the impetus for the GCSE had come from the National Union of Teachers which had wanted to get rid of O-levels. The GCSE had originally been approved by Mark Carlisle, a minister with little knowledge of education. 'Carlisle was totally unaware that the average level of achievement was so low, at CSE grade four,' said Sexton.[15] In other words, Carlisle had little or no idea how low the standard of the new GCSE exam would have to be to establish that middle ground of achievement.

According to Professor Bolton, by Joseph's time the primary concerns among ministers were the large numbers of pupils who left school at 16 and the fact that very few stayed in touch with a broad curriculum. Joseph wanted to make the curriculum less subject to variations, more broadly based and more relevant; he wanted to end an education system divided between a successful élite at one end and a group of failures at the other. And he thought he could achieve all this through the new GCSE exam. In the event, it was his successor Kenneth Baker who finally approved the start of GCSE. 'There was pressure from Number 10 and the political right not to approve it,' said Sexton. 'But in the end Mrs Thatcher couldn't stop him. After all, it had been her own mentor, Sir Keith Joseph, who had said we could no longer continue to have a very well-educated élite and the lumpen remainder.'

So it came about that it was none other than this ideological Conservative government, ostensibly committed to traditional standards in education, that introduced the GCSE exam. It came about through bumbling serendipity: a combination of ideological pressure from the main teachers' union, pushed along by a combination of ministerial ignorance, idiosyncratic unworldliness and a dash of showmanship, and smoothed into being by a civil service which was quietly signed up to its underlying egalitarian agenda. Yet this Tory exam, which falsely promised to do the impossible in squaring a circle of achievement, embodied the very philosophy which struck at the heart of the meritocracy the government purported to promote, and which was to undermine education from top to bottom. It was the philosophy that no child could be allowed to fail. All must have prizes instead.

*   *   *

Until Kenneth Baker became Education Secretary, however, no minister had thought of creating a national curriculum. Despite Bolton's view that people had been moving in that direction ever since Callaghan's speech, Sexton insists that this was not so. 'We thought it wasn't the job of ministers to interfere with the content of education,' he said. 'We were concerned about structure, and we were aware that teacher training colleges were putting out some strange ideas.' Like several education insiders, Sexton maintains that the idea for a national curriculum came originally from the civil service. 'The civil servants tried out the idea of a national curriculum on both Shirley Williams and on Mark Carlisle, but they wouldn't hear of it,' he said.

The civil servants wanted a national curriculum because *they* wanted direct control over education. They were, at that stage, singularly powerless. The schools were run by local education authorities; the teachers were trained independently; the exam boards set the exams and effectively laid down what schools should teach. People were given to asking what precisely the Education Department was for. Officials were desperately looking for a role to justify their existence. It appears that the civil servants eventually managed to bounce ministers into setting up a national curriculum that they had always intended to control.

Bolton, one of those civil servants thought to have favoured precisely such a move, denies that ministers were bounced.

It was true that the Department was running out of things to do. The agenda was moving into the quality of teaching. If the department didn't change what it was doing, it would have had no role left. But everything in the 1988 White Paper (that preceded the Education Reform Act establishing the National Curriculum) had been in Keith Joseph's White Paper. He had said this was what should happen. Baker asked whether the teachers would do it. The answer was that they would not do it. So work started on the Bill, not knowing there would be an election the following year.

The fact was, however, that Joseph was firmly opposed to a national curriculum, and that no minister before Baker had seriously contemplated state control over what children were taught in school. 'Baker was a centralist and believed in corporate control from the centre in education, which was a reversal of all we had worked for,' said Sexton. 'And so the same civil servants who had failed to sell their idea to previous Secretaries of State then sold Baker their plan for a national curriculum to solve the problem of poor standards.'

When the Education Reform Bill was published in 1987, Tory activists – including those who had helped prepare the manifesto for the general election a few months previously – were aghast. The activists had wanted to remove schools from local authority control to give them more autonomy and the parents more information and choice. Instead, these activists were confronted by a ten-subject curriculum which would not only enmesh ministers in a web of detailed prescriptions about what children should be taught but would also entail a fiendishly complicated system of attainment targets and tests to ensure the curriculum was actually delivered in the classroom. It was a logistical nightmare; but it was also incoherent. At the behest of Tory right-wingers desperate to reclaim the initiative, unprecedented state control over the curriculum was yoked together with proposals for schools to opt out of local authority control. Local control was eroded, only to be replaced by unprecedented central control.

The Prime Minister had wanted some central direction over the basics, but nothing like this. 'Thatcher's original idea was to raise standards in primary and secondary schools through tests in basic subjects: maths, English and science,' said one insider. 'This was simple, even simplistic, and certainly naïve. She never understood the depth of the problem. Certainly, there was never any intention in that 1987 manifesto for a prescriptive curriculum. But almost immediately Baker was sold the idea of a highly prescriptive curriculum by the very civil servants who had failed to sell the idea to Shirley Williams. Baker was very easily flattered into thinking that this was what the country wanted. But none of this had been discussed before the election. Where did such a substantial Bill come from so quickly?'[16]

The answer, as Eric Bolton has revealed, is that it was in the process of preparation before the 1987 election was called. Mrs Thatcher was apparently dismayed. 'When she was presented with a draft of the National Curriculum,' said Sexton, ' she is alleged to have said: "All I asked for was the Three Rs!"' The great question, however, remains: how was it possible for such a strong-minded, ideological Prime Minister to have gone along with a policy of such magnitude which she didn't want?

Answers to this question vary. 'Neither Mrs Thatcher nor Brian Griffiths [then head of the Number Ten Policy Unit] *was* opposed to it,' said Bolton. Stuart Sexton, however, had a different view. 'Mrs Thatcher was simply outflanked by Baker who had a lot of support in Cabinet.' Despite her image, on this and other issues she was outnumbered. And another government insider commented: 'People just don't realise quite how difficult it is for government to get what it wants through. There's a terrific mismatch between government and machine. It really doesn't all work together.'[17]

It certainly didn't. Right from the start, Baker's plan to use the National Curriculum to put a rocket under bad teaching and raise educational standards started to disintegrate.

Baker acknowledged what he was up against. He recognised the near universal spread within the education world of attitudes he thought were inimical to education. It followed, therefore, that he couldn't trust that world to put its own house in order. He therefore decided that the market was the only mechanism that would work. League tables and more consumer choice, along with the information about schools provided by the new curriculum tests, would so expose teachers to the chill winds of competition and public transparency that they would have no choice but to raise their standards forthwith. What they were taught in the teacher training colleges and the philosophy embedded in the educational texts they may have been reading would accordingly be dragged into line. The strategy essentially placed parents in the front line of the battle that was now to take place between arcane educational theories and the day-to-day reality of children's performance in school.[18]

This approach, however, had within it two fundamental contradictions. The first was that the weight of responsibility placed on parents' shoulders could not adequately be discharged. Many parents didn't have the confidence or authority to challenge what teachers were doing. Many of these parents, after all, had themselves been so inadequately educated that their own expectations were very low. And even if they did exercise their consumer rights and moved their children to 'better' schools, the criteria by which they were judging these schools were to a large extent based on their relative successes in the public examination system. But the parents had no power at all to question the worth of that examination system or to challenge the assumptions beneath the whole framework of achievement by which schools were being measured.

In response to this concern, Baker wheeled out the National Curriculum to ensure that those educational values would indeed be upheld. But this was the second great error of thinking. He didn't appear to understand that the people he would be using to devise the curriculum would inevitably be drawn from those very people he had decided were so intellectually corrupted they could not be trusted to remedy the problems in education without the force of law. He would be turning poachers into gamekeepers, despite his appreciation of their incorrigibility. He was handing unprecedented power over the curriculum to people who by his own admission couldn't be trusted with it. So it was entirely predictable from the beginning that his plans would go hopelessly awry. And they did.

Curriculum Chaos

Ministers may have viewed the National Curriculum as a means of turning back the tide of progressive teaching. But the education world was determined to prevent the restoration of 'didactic' teaching and the imposition of tests which would label children as failures. Yet membership of the two quangos established to implement the curriculum and its associated tests, the National Curriculum Council (NCC) and the School Examinations and Assessment Council (SEAC), was heavily weighted towards pre-cisely such educationists. Aware of the risk of being accused of rigging these bodies, Baker took advice from the education inspectors and others to draw upon what were considered the best brains in the country. But several of these 'best brains' who were recommended to him by his own civil servants just happened to be progressive revolutionaries.

The result was havoc. Not only did the emerging curriculum bear all the hallmarks of the cultural relativism and retreat from teaching it was set up to combat, it also became submerged, almost immediately, under the weight of a ridiculously complex testing scheme whose insupportable intricacy was the direct result of a conflict of aims between government and quango. Ministers wanted simple 'pencil and paper' tests because they wanted the *public* to know whether children were by and large succeeding or failing. This way, they would discover whether the *teachers* were succeeding or failing. But the teachers had very different ideas. They abhorred the idea that any children should be seen to fail (and although they didn't say so, one might reasonably surmise they equally abhorred the idea that any of *them* should be seen to fail). If there were to be tests, they insisted that they should be 'diagnostic' – in other words, that their use was primarily for the *teacher* to discover what each individual child needed; and that teachers should mark these tests themselves. The attempt to create a testing system that satisfied both aims created a monster. The 'ten-level scale', as it became known, was so complicated and unwieldy that it was hardly surprising that the teachers eventually buckled under the strain.

Yet it was a strain imposed by the demands of their own side. The reform appeared to be primed with an inbuilt self-destruct mechanism. Considering the attitudes expressed by some of the quango members, this was hardly surprising.[19] In June 1988, Tony Edwards, head of the education department at Newcastle University, said of the curriculum reforms that there was 'ample scope for creative subversion . . . there is almost everything still to play for'.[20] Later, he was appointed to the NCC's curriculum review team.

In 1989 Martin Davies was appointed director of the National Curriculum Council. As director of education in Newcastle, he had presided over appalling GCSE results. He had previously attacked the education reforms in the *Newcastle Evening Chronicle* as having 'the potential for catastrophe as far as the public education system is concerned'.[21] A few months later, he wrote: 'The major criticisms of these proposals is that they are subject-based many years after the HMI [school inspectors] had identified areas of experience as a more valid way of designing the curriculum and in spite of the growing practice in many schools of planning their curriculum in a more progressive way.'[22] When Davies was appointed to the NCC, a Newcastle City councillor was moved to write to the paper. G. F. Keating wrote of Davies: 'In September 1987 he wrote a report which made six major criticisms of the Government's proposals for a National Curriculum and contained no favourable comment. It therefore looks as if he is taking a job which involves implementing proposals with which he is in disagreement.'[23] When Davies was appointed to the NCC, the paper quoted him as saying: 'Taking this job is a chance to influence from within having criticised certain aspects from outside.'[24]

The ten-level scale which did so much damage was devised by a mathematician, Professor Paul Black, deputy chairman of the NCC from 1988–91. He appeared to be hostile to much more than simple and objective tests. He wrote: 'A return to traditional didactic teaching, a return to traditional testing, a return to the O-level, are on the agenda again. Such regression cannot of course be dismissed simply because it is regression. It can be dismissed if it is clear that the old methods never served our purposes, or that the mores and needs of society are now very different. In education, both of these arguments hold.'[25] This was nothing less than the un-teaching agenda – espoused by the man at the centre of the government's testing programme. He made it clear that he favoured coursework over written tests, even though coursework offers opportunities for cheating, favours middle-class children who receive parental support and involves work set by the teachers themselves. Written tests, he said, 'will be unreliable, so they might harm individual pupils unless parents and schools decide to take little notice of them'.[26] Yet ministers had said specifically they wanted such tests to take place. Black's comments were little less than a brazen announcement of sabotage.

The first chairman of the NCC was similarly unsuited to the job in hand. Duncan Graham, who chaired the body from 1988–91, clearly imagined he was attempting to strike a balance between warring extremes. But his remarks indicated that he too had largely swallowed the failure-free, let-it-all-hang-out agenda. In his own book *A Lesson for Us All* he wrote that the education culture of the 1960s and 1970s had consisted of 'a number of teachers working in teacher centres all over the country linked with equally well-intentioned local authority advisers and teacher trainers and concerned together in a benign conspiracy, reinventing a thousand wheels a day'.[27] But as an ideologue Graham appeared not even to have the courage of his own convictions. He appeared to think progressive education was a good thing which had just gone too far. He wrote that the education establishment had got itself hooked on perfectly respectable philosophies: child-centred education, learning by doing, projects rather than subjects. As a teacher trainer in Scotland, he recalled, he himself had moved children out of serried rows in the classroom into small discovery groups. But these methods had to be part of a balanced curriculum, not ends in themselves, he observed. Yet the curriculum he created seemed hell-bent on institutionalising the very extremes he appeared to criticise. Indeed, what he utterly failed to grasp was that what he thought 'moderate' – the transfer of authority from teacher to child, the retreat from knowledge to creativity – was itself an extreme.

The result, under his chairmanship, was a display of muddle and confusion. The NCC appeared deliberately to be causing chaos by loading complexity upon complexity. Graham wrote with apparent pride that the council had knowingly crammed too much into the system. 'The council accepted that it was better to over-prescribe and then draw back than to under-prescribe and try to tighten up and believed that the best time for change would be in two or three years' time. None of us had any doubts that the master-plan could only work if subjected to drastic pruning.'[28] In other words, the classrooms were to be used as a huge experimental laboratory with the foreknowledge that the experiment wouldn't work very well. But there were real children being educated in those classrooms, who would have only one educational chance, and real teachers who were going to have to bear the brunt of unrealistic expectations. It is hard to credit such cavalier indifference towards a matter of such desperate importance.

But there appeared to be a measure of method in the madness. The incoherence and confusion were also being used to serve an ideological end. In 1990, the NCC prophesied: 'In due course it is likely that schools will throw all the attainment targets in a heap on the floor and reassemble them in a way which provides for them the very basis of a whole curriculum.'[29] The 'whole curriculum' was the jargon term for the generalised approach to education which rejected subject-based teaching. Yet the National Curriculum was being brought in to beef up individual subjects and squeeze out the unrigorous, generalised approach. So the NCC was quite explicitly subverting that intention.

Given what so many of these key players had said about the education reforms, which suggested that they were not only hostile to their intentions but intent on undermining them, it seems almost beyond belief that ministers made these appointments in the first place and then sat back and did nothing as the project predictably sank deeper and deeper into the mire. For sink it did. Duncan Graham got at least one thing right when he observed: 'Many of the later troubles which dogged the introduction of the curriculum stemmed from the fact that nobody realised the enormity of the task being undertaken.'[30] No-one realised – least of all Graham – that since almost every curriculum subject was the site of a culture war between subjective relativism – the equalisation of all values – on the one hand, and objective, externally applied facts and knowledge on the other, all attempts to draft the curriculum would be subject to intolerable internal pressures. The fight was particularly intense over English, history, maths and science. But then it wasn't surprising Graham appeared bemused by the passions that were unleashed. He broadly endorsed the relativist philosophy embraced by the majority of his committee. He therefore couldn't understand why a minority of 'extremists' were making such a fuss.

The fact that Graham was a significant part of the problem rather than the solution was illustrated by the woolliness of his own thinking about the subjects in the curriculum. On maths, for example, he enthusiastically endorsed the conclusions of the 1982 Cockcroft committee which he said had 'broken exciting new ground in the teaching of mathematics'. This exciting new ground had actually helped undermine the entire foundation of mathematics itself. 'Cockcroft took people into areas such as estimation and the concept of having to be exact in some things but not in others, and the idea that mathematics could be pursued from a practical basis,' said Graham. He didn't appear to grasp that the 'solving of real-life problems' by which he set such store couldn't actually be done without understanding the importance of being exact. He himself didn't appear to realise the importance of exactness, or why mathematics couldn't exist without it.

He therefore couldn't understand why one dissident member of the maths group, Professor Sig Prais, was so uncompromising and 'ratty'. Graham thought Prais was an extremist. Yet Prais knew, from his research in Britain and in Europe, not only how far behind British schoolchildren were in maths but that this was because of the kind of thinking espoused by people like Cockcroft and Graham. Instead of making use of Prais's unrivalled data and analysis, however, Graham and his council marginalised him so that before long Prais despaired and resigned. In the note of dissent he had previously written to the maths group's interim report, Prais observed that the group's failure to understand that German or Japanese children were performing better because they had been properly taught multiplication or division or the areas of circles had helped him understand 'why so many of you believe that "having fun" is one of the prime objectives of mathematical teaching'.

Behind that sardonic comment lay the bitter frustration of someone who had experienced at first hand the apparent impossibility of breaking through the culture of wishful thinking. But it was by no means only maths that suffered. The whole of the early National Curriculum was suffused with these de-educational attitudes, given the force of statutory application after battles in which the hopelessly outnumbered and outmanoeuvred traditionalists were comprehensively routed. Ministers' objections were circumnavigated and finessed.

History

In the teaching of history, the argument was exceptionally highly charged. This was hardly surprising, since whoever controlled the curriculum would effectively control the story the nation's children were to be taught about themselves. There was accordingly considerable fear that Kenneth Baker would hijack history and turn it into political propaganda. But those who expressed these very proper fears never acknowledged that they wished to gain control of the history curriculum for precisely the same ends.

Yet that was indeed the case. The struggle between cultural relativism and objectivity was perhaps set out most starkly in the battles over the teaching of history. It was an argument between those who believed that the teaching of history had to involve transmitting a body of factual knowledge about the past, and those who believed that what the primary lesson children had to learn from history was that *no* facts were true and that everything was a matter of subjective interpretation – from which flowed the emphasis on imaginative 'empathy' and source work. Empathy and source work meant children had to find things out for themselves instead of being taught – which meant local history took precedence over national or international events.

This subjective view of history won hands down. And once again, ministers brought this about by conniving at their own manipulation. Baker's choice of chairman for the NCC's history working group was, in the words of Duncan Graham, 'extraordinary'. Michael Saunders Watson was the Old Etonian owner of Rockingham Castle. He had no connection with history education except for the fact that his stately home was used by many schools. He turned out to be Baker's spectacular own goal. As Graham put it: 'Everybody feared the worst and characterised him as a right-wing amateur who would follow the party line. He went on to be a great disappointment to Mrs Thatcher and the right wing and a great but welcome surprise to history teachers.'

In other words, Saunders Watson's apparent amateurish ignorance of the world of political machination meant he was no match for those who were well schooled in its ways. Very soon, the history working group set itself on a collision course with the government. Ministers wanted facts put back into history; the group decided, in Graham's words, 'that it would not be possible to have attainment targets based on facts in any rational way. Like it or not the attainment targets would have to be based on skills of interpretation and critical awareness.' Since the attainment targets were what children would actually be tested on, it was hardly surprising that ministers insisted that these consisted of facts rather than creative or subjective skills.

Baker's successor, John MacGregor, made it plain he was exceedingly unhappy with the proposed history curriculum. Despite his unease, however, nothing substantial changed. Graham wrote that 'when it came to the basic principle, nothing had been altered, indeed little could be altered as the working group's conclusions were unarguable. A knowledge of facts does not equal a knowledge and understanding of history.' An impasse was reached. MacGregor delayed publishing the group's report because Mrs Thatcher would not approve it. Nick Stuart, the civil servant responsible for the curriculum, realised the report was in danger of being rejected. So he persuaded MacGregor to set up an investigation. 'By some inspired rearrangement, we persuaded MacGregor that while facts could not be included in the history attainment targets they were none the less an integral part of the course,' wrote Graham. But they weren't, since children weren't to be tested on them. The civil servants had effectively perpetrated a sleight of hand to get the project out of trouble. Factual knowledge in history teaching had been marginalised.

Not that this was recognised. On the contrary, the involvement of the Prime Minister and the subsequent interference by the new Education Secretary Kenneth Clarke with the definition of modern history created the understandable fear that history teaching was now the subject of political interference. This was undoubtedly true. Yet the result was to achieve the worst of both worlds. Clarke's insistence at first that history teaching should end at 1945 and then as a compromise that it should end 20 years before the date of teaching was indeed a brazen political intrusion into a field which no politician should properly enter. Yet at the same time children were still not being taught properly in schools. Pupils would be awarded marks in the history curriculum tests if their answers were 'plausible', even if they were incorrect. 'Skills' – the use of imagination and the application of doubt – were still to the fore, as critics in the Campaign for Real Education noticed in the history tests for seven-year-olds published in 1993. 'Pupils were to be "examined" on what Guy Fawkes might have said when he tried to blow up Parliament; whether Florence Nightingale, Jemima Nicholas and Cinderella were real people; and why cooks have abandoned coal ranges for gas hobs.'[31] The testing quangos SEAC defended these tests as reflecting the national curriculum for history. 'The standard assessment tasks do what the National Curriculum expects teachers to have been teaching.'[32]

Modern Languages

The fight over the history curriculum created huge ructions because of the fear of grossly improper political interference with the national story. The subversion of other subjects with less political resonance occurred without anyone noticing. The curriculum institutionalised, for example, the slack, unstructured and correction-free communicative ideology of modern language teaching. The premise behind the curriculum's approach to modern languages was set out in a report written by a specialist in Spanish and French linguistics, Professor Martin Harris. Harris broadly endorsed the idea that modern languages should be taught in the target language – otherwise known as the 'Berlitz' method.[33]

'Harris says you have to teach syntax in the target language: I know for a fact you can't do it, especially when the children don't have these concepts in their own language,' said one very angry university tutor. 'But he gave this idea legitimacy by institutionalising it through the National Curriculum.'³⁴

In his report, Harris acknowledged that the importance of communication had led to 'some neglect of awareness of structure as an aid to language learning' and that structure was important. However, he then went on to echo the refusal at all costs to teach grammar that characterised the attitude of his colleagues in the modern languages world, illustrated earlier. 'To be effective, the necessary training in the recognition and use of structures needs to be an integral part of the way in which the target language is taught and practised rather than the subject of theoretical exposition. There is certainly a place for short explanations, using such grammatical terms as prove helpful, but it is easy to over-estimate the value of either for many pupils, including the more able.'

Later, Harris wrote of the importance of teaching tense, gender, case or word order, only to approve of the 'communicative competence' which 'learners' achieved not by learning universal rules but by continually extending their repertoire of chunks of language. For example, pupils could learn a limited number of verbs in the past tense before learning the past tense as a structure. And he urged teachers not to correct their pupils' mistakes. 'When pupils are trying to communicate in the target language, they need to do so as spontaneously as possible. Attempts to correct mistakes or bring out underlying principles can easily interrupt the flow and inhibit further production.' Careful monitoring, he hastily added, was needed so that mistakes could be corrected later. When? And after such a delay, to what effect?

Above all, pupils should not be taught explicitly. They were to be expected instead, it appeared, not only to assimilate language structures but to understand spontaneously what they had assimilated and what innate rules were being applied. 'Once they have thoroughly absorbed a set of related chunks of language, learners need to explore and if necessary be shown how the underlying model works, not be told about it. Exploration and demonstration have much in common . . . Both are a matter of working with the language *in the language* in such a way that the underlying pattern becomes clear to the learner.'

How this was to become clear to them if they weren't taught remained obscure. But then the underlying anti-teaching philosophy emerged. Pupils should effectively teach themselves. 'Learners need to feel that they have some control over their own learning. They can help the teacher to identify the expressions which they need in order to say what they want to say. They should be encouraged to gather and learn useful words and phrases from classwork and homework or their own reading; they need to be taught how to use dictionaries to help them in this; they can be asked to produce their own lists of what they claim to have learnt and to test each other's knowledge.'

Graciously, he conceded that knowledge of grammar as the skeleton of any language 'can be an important ingredient in learners' progress towards a truly independent use of language'. But this was still very tentative. 'If learners can be helped to see the common features of the chunks of language which they have learnt, they will be better able to adapt them to the demands of different situations and increasingly to check their own production.' *Can* be helped? Only by being taught that these structures are universals can pupils put words together to make language. Yet Harris ruled out 'the attempt to describe these structures in more or less formal terms, rarely using the target language'. In other words, grammar, in so far as it should be learnt at all, must be absorbed osmotically and by immersion in the target language. Yet it is perfectly obvious to anyone who spends ten minutes with pupils that expecting them to learn the language first and the codes that unlock that language second is an absurd doctrine which bears no relation to reality whatsoever. Pupils having to bear the burden of this obsession with immersion merely drown in a deluge of incomprehension and bewilderment.

The Harris report was higher-order nonsense. But this was not a contribution to some introverted colloquium for like-minded ideologues. This was nothing less than the government's prescription for the foreign languages curriculum which had been introduced to get rid of such nonsense from the schools once and for all. Instead, the Harris report had given it the weight of statutory authority. Instead of providing the solution, it had itself become an important part of the problem.

English

The way ministers were wrongfooted over the English curriculum was even more startling. Kenneth Baker, well aware of the grip that New Literacy ideas had achieved over such matters as the teaching of grammar, punctuation and spelling and the devaluation of classic texts, appointed as chairman of the English curriculum working group Brian Cox, professor of English at Manchester University. Unlike the stately Saunders Watson, Cox appeared to be the safest of safe bets for the government's purpose. He had been one of the authors of the *Black Papers* on education back in the 1970s, which had excoriated progressive education and stigmatised all its authors henceforth as hate figures belonging to the political right. The restoration of the basics of English teaching appeared therefore to be a foregone conclusion.

But it didn't turn out like that at all. As soon as the first part of the Cox committee report was published on the teaching of English in primary schools, it was clear that the committee had swallowed the orthodoxy that rejected formal grammar teaching and devalued Standard English. Their report paid lip-service to their desirability, but in a way that effectively neutralised them. The committee clearly couldn't bear to use the word grammar at all, replacing it with a new euphemism: 'linguistic terminology'. Cox was later to claim that this was grammar and his critics were wrong to say he had rejected grammar teaching. But this was not so. His committee had redefined grammar into meaninglessness. Its report advocated that children should be taught *about* language, which is very different indeed from saying they should be taught to obey its rules. But then, notions of rules or correctness found no place in the cultural relativists' universe.

Cox advocated that Standard English should be taught, but only because it had 'social prestige'. It was merely 'a technical term to refer to a dialect which has particular uses' and it was not to be confused with 'proper, good or correct English'. Since its only claim to usefulness was apparently social prestige, it followed that its non-use was 'rarely more than a social irritant'.[35] Why then should any teacher feel obliged to teach it in any more than a cursory or grudging manner? The underlying claim was that Standard English was élitist. Yet Standard English is the most democratic form of the language available. Because it *is* standard, it is common to everyone rather than being peculiar to any region or class. It therefore puts everyone on an equal footing – provided, of course, that it is actually taught.

Baker told them to go back to the drawing board. Their proposals ran so much counter to what was plainly needed that even Duncan Graham opposed them. Yet such were the fissures over English teaching that Graham discovered problems among his own NCC staff, a handful of whom threatened to resign 'after being told that the council would have to rewrite at least some parts of the Cox report and in particular to insist that some grammar – even if it was called a knowledge of language – was taught in primary schools'.[36] Despite all the pressure, however, including disapproval by Downing Street, neither the primary schools report nor the secondary schools report which followed, and which provided the basis for the 1989 English curriculum, was substantially modified. The English curriculum enshrined in statutory form the very corrosion of literacy it had been set up to combat.

It is worth looking in a little more detail at the positions taken by Brian Cox himself. It was hardly surprising that Baker had thought he would deliver, considering what Cox had previously written about education in general and the teaching of English in particular. For he had been a passionate critic of the very attitudes he seemed now to be with equal passion espousing, even though he now maintained that he was simply carving out a compromise.

In his personal memoir *The Great Betrayal*, Cox wrote that opposition to the traditionalist *Black Papers* which he had co-written had come from members of middle- or upper-class families like Anthony Crosland, Shirley Williams or the English Professor Brian Simon. 'Their reforming zeal thrived on misunderstandings of the real needs of the working-class child,' he observed.[37] The *Black Papers*, by contrast, expressed the 'hidden feelings and thoughts of thousands of ordinary teachers and parents, breaking the grip of the taboos of progressive education'. And twenty years later, Cox said he had in no way resiled from the position he had taken then. 'In the 1960s,' he wrote, 'a great wave of utopian ideology swept across British schools, undermining the authority of the teacher, damaging to the real needs of children.'

He vividly described the madness in education attitudes that had so shocked him into writing the *Black Papers*: children inflicting discipline upon teachers, chronic misbehaviour, schools where children wandered around aimlessly all day, constant uproar and chaos in the classroom, the priority given to freedom of expression, the revolt against marking and correcting for fear of imparting a sense of failure. 'I hold just as strongly today that . . . the abdication of authority by teachers has fundamentally damaged our society,' he wrote. He was shocked by the utopian simplicities of progressive ideology.

And yet by the time he wrote this memoir in 1992, Cox had helped institutionalise some of the very features he still so powerfully deplored. In *Cox on Cox*, his book explaining the background to his curriculum report, he wrote that those who claimed his report didn't advocate the teaching of grammar were being malicious. But his formula for such teaching was so vague as to be meaningless. He said Latinate grammar didn't suit English, when it happens to be the basis for it. And he said that teaching about language should not be a discrete activity but something to be discussed during the course of other work. The chances were, therefore, that it wouldn't come up at all. His justification for refusing to give language teaching its own position in the timetable amounted to a curious mixture of fashionable fallacy and realistic despair. There was the overloading of the curriculum: 'There are constant pressures from relatively new areas of study, such as information technology, film, television and so on.' So, in accordance with the doctrines of the New Literacy, mastering the language was now less important than being taught about film or TV. But then he added, 'substantial programmes of teacher training are required if teachers are themselves to know enough to enable them to design with confidence programmes of study about language'.[38] In other words, teachers themselves didn't know enough grammar to be able to teach it.

His apparent reversals of opinion have been quite bewildering. In 1982, Cox wrote that left-wingers had been trying to bring about revolution by transforming institutions from within. 'Many teachers have not understood this. They fail to realise how teaching styles, classroom routines or the lack of them, the existence of certain standards, reflect political attitudes.'[39] Yet in 1992 he wrote: 'The teaching methods in the schools in the 1930s depended heavily on rote learning and on boring exercises such as parsing and clause analysis . . . In the 1950s and 1960s, good teachers started rejecting these old fashioned, boring and inaccurate ways of teaching English.'[40]

In 1982 he warned that 'Shakespeare and Dickens will be

replaced by contemporary texts supposedly relevant to modern social problems'.[41] Yet in 1993 he said: 'If you are trying to get children interested in the classics you can often lead them from TV soaps to Dickens.'[42] And in 1990 he wrote: 'This shift of emphasis (from reading and writing to speaking and listening) is revolutionary in its consequences . . . It could be said, without too much exaggeration, that this new initiative could change the English national character.'[43]

He also went into reverse over the teaching of reading, rehearsing in his latter writings the familiar fallacy that teachers were using a mixture of methods with phonics well represented. Twenty years previously, however, he had told the Bullock committee about the 'fashionable assumptions' that were prevalent, which were 'quite often in direct contradiction to the findings of proper research'. He wrote then of 'advocates of child-centred learning' who found didactic teaching 'repugnant to the ideology' and said that research, by contrast, showed children learned to read best when they were taught to learn the code. 'The consensus of recent research is that reading programmes for young children should emphasise phonic methods. The committee should publicise these findings.'[44] But when Turner said much the same thing in 1990, Cox attacked him in the educational press, claiming that his misgivings were 'absurd' and that 'it is generally agreed among teachers that the real books approach is relatively uncommon in schools. Many schools use a mix of approaches, including phonic methods, which Mr Turner himself advocates.'[45]

In *Cox on Cox*, he wrote of the need for a variety of 'reading cues' – which means children guess the meaning of words through clues such as the context of the story or the pictures – and said of children: 'They need to be able to recognise on sight a large proportion of the words they encounter and to be able to predict meaning on the basis of phonic, idiomatic and grammatical regularities and of what makes sense in context; children should be encouraged to make informed guesses.' But to advocate that children should 'learn' to read by guesswork is not only nonsensical in itself; it makes a mockery of his apparent complacency over the use of phonics in schools, since phonics properly taught involves a systematic approach in which guesswork plays no part.

In 1992, however, he appeared to be retreating from his line that phonics was adequately represented. In a personal letter to a critic who believed his curriculum had devalued phonics, he wrote:

> We did not want to commit ourselves to saying how much emphasis should be given to phonics in the teaching of reading, for that is still very controversial, but I insisted that the word must appear in the attainment targets and the programmes of study. If we had known about the incredible debate on reading which took place after our report was finished (over Martin Turner's research), we would have repeated our view that phonics are an essential part of reading instruction. I sat on a CATE [Council for the Accreditation of Teacher Education] committee last year which discovered that there are still teacher trainees (*sic*) who give their students almost no information about how to use phonics in the teaching of reading. We *did* emphasise that all children must speak and write Standard English but I've seen programmes of study for the National Curriculum which ignore this . . . Educational battles are never finally won.[46]

Nevertheless in yet another book published in 1995, he appeared to have abandoned this claim to the middle ground and to have thrown in his lot uncompromisingly with the child-centred orthodoxy. He wrote disparagingly of traditionalists who 'wanted an increased use of phonics in the teaching of reading . . .' and returned to the attack on Martin Turner: 'The statistics Turner provided did not stand up to careful scrutiny'.[47] He extolled the very doctrines he had once excoriated, and damned what he had once so passionately believed. He endorsed the way in which teachers 'who now see themselves as guiding and intervening in individual development are being required to convert to a transmission style of teaching, seeing themselves much more as custodians of, and inductors into, established knowledge . . .' And in criticising changes made to 'his' curriculum, he objected to the frequency in the new document of the use of the word 'taught'. 'The authoritarian "taught" left over from the past sits uneasily with the more enlightened "enjoyment",' he wrote. The representation of teaching as authoritarian, and the superior claim of enjoyment to which teaching was, in his view, apparently inimical, surely represented the final betrayal of the stand Cox had once made on behalf of children and against those detached and supercilious ideologues whose measure he had once so devastatingly taken.

The impact of Cox's thinking on the national curriculum has been highly significant. His proposals for English, though watered down, remain the bedrock of the English curriculum and continue to be used as a justification for the New Literacy culture. The damage that has been done to children's education in the most fundamental curricular area of all is therefore not to be underestimated. The question remains, though, why he performed such an astounding apparent volte-face. As his writings make clear, he was extremely wounded by the vilification to which he was subjected in his *Black Papers* days and was obviously anxious not to remain a pariah in education circles. But there was more to it than that. As chairman of the Arvon Foundation which teaches creative writing, Cox was now placing creativity at the centre of the educational project, a view which brought him squarely into line with the basic precepts of child-centred thinking. But no-one had told Kenneth Baker or his junior minister, Angela Rumbold, that their gamekeeper seemed to have turned poacher.

Cox wrote:

> Neither Mr Baker nor Mrs Rumbold knew very much about the complex debate that has been going on at least since Rousseau about progressive education, and they did not realise that many groups would be strongly opposed to Mrs Thatcher's views about grammar and rote learning. The politicians were amateurs, instinctively confident that common sense was sufficient to guide them in making judgments about the professional standing of the interviewees. I suspect they did not realise that words such as 'grammar' or 'progressive' reflect very different meanings according to context, or that the language of educational discussion had changed radically since they were at school ... I was well known as the chief editor of the *Black Papers* on education (1969–77), supposedly traditional and right-wing in their views of education. I presume that neither Mr Baker nor Mrs Rumbold was aware that for over ten years I had been conducting a campaign to make creative writing a central feature of the English curriculum.[48]

But Cox had been a member of the Kingman committee on English, whose report in 1988 had similarly failed to advocate a return to structured English teaching. Cox had not dissented from its conclusions. The politicians should have realised from that alone that Cox was not the safe bet he seemed. Politicians are often careless of detail. But if the politicians were ignorant of this development, their civil servants would surely have known about it.

Chapter Six

THE SELF-DESTRUCTIVE ESTABLISHMENT
The Education Department

According to British constitutional conventions, the civil service maintains a dispassionate, apolitical neutrality. Whatever the complexion of the government of the day, the same civil servants remain in post to do its bidding. They are supposed to have no agenda of their own. Answerable above all to 'the public interest', the definition of which becomes ever more elusive, they traditionally discharge the duty of pointing out inconvenient facts. They are supposed to warn ministers of the likely consequences of their actions, for example, or alert them to contemporary research that does not support the thrust of government policy. But what officials should not do is to further their own ideological beliefs.

Evidence abounds, however, that some officials at the Education Department, motivated both by ideology and by a bureaucratic desire for control, played a crucial role in sabotaging the government's education reforms. 'I have never known any other civil servants quite like these in the Education Department with such a view that they were as important as the policy-makers,' said one political insider.[1] 'There was a real climate of fear; more junior civil servants were intimidated by their superiors,' said another.[2]

This is strong language, open to the obvious counter-charge that it is itself distorted by political prejudice in a policy arena characterised by the most intense and vicious ideological battles. It is also complicated by the fact that serving civil servants are prevented by Whitehall rules from giving their own account of events.[3] Those who have retired tend to dismiss such claims with contumely. Asked to respond to them, Sir Geoffrey Holland, who was the Permanent Secretary at the Department from 1993–94, wrote 'my short response is that they are beneath contempt. There is no truth whatsoever in them.'[4] But there is simply too much evidence, from too many sources, to ignore. The Education Department officials were an important element of what was called 'the secret garden', the closed world of education that resisted all comers. They had their own agenda.

Sir Geoffrey himself turned out to have strong views on education which appeared entirely in line with the anti-teaching orthodoxy. Indeed, in a speech he delivered in 1995, he outlined a revolutionary change to the premise that knowledge should be transmitted from adult to child. The priority for schools, he said, should be 'learning, not teaching . . . The world these young people are going to own and live in is not a world in which people sit neatly and tidily in separate rooms in rows, and change every forty minutes from one subject to another. It is not a world of bits and pieces, it is not a world of didacticism, it is not a world of certain answers and it is not a world of prearranged sets of circumstances. It is a world where the individual has to learn and make her or his way for herself or himself, and therefore where the teacher has to become the supporter of the learner which requires a fundamental reversal of the traditional role of the teacher.'[5] It was therefore the fact that – for a short while at least – the most senior civil servant at the Education Department, one who by all accounts was more openly involved and proactive than his predecessors, was a keen subscriber to the very philosophy that has brought education in Britain to its knees.

Before Sir Geoffrey had even got his feet under the table at the Department, Kenneth Baker himself had mulled over his former officials' influence in his book, *The Turbulent Years*.

Of all Whitehall departments, the DES [Department of Education and Science] was among those with the strongest in-house ideology. There was a clear 1960s ethos and a very clear agenda which permeated virtually all the civil servants. It was rooted in 'progressive' orthodoxies, in egalitarianism and in the comprehensive school system. It was devoutly anti-excellence, anti-selection and anti-market. The DES represented perfectly the theory of 'producer capture', whereby the interests of the producer prevail over the interests of the consumer. Not only was the Department in league with the teacher unions, university departments of education, teacher-training theories and local authorities, it also acted as their protector against any threats which ministers might pose.[6]

He referred to one senior civil servant, Walter Ulrich, as 'no particular fan of government policy'. A newspaper article in 1988 claimed that Ulrich 'relished demolishing the ideas of political advisers brought in so as to ensure that civil servants did not thwart election commitments'.[7] Baker added: 'Even Secretaries of State were not immune from Walter's formidable intellectual bullying. It was one of [junior education minister] Bob Dunn's complaints that when ministers in Keith's [Sir Keith Joseph] time eventually reached a political decision about a policy matter, Walter, if he didn't like it, would unpick it by sending a dissenting minute and then re-argue the issue face to face with Keith.'[8]

In later years, Ulrich's place as the most formidable civil servant in the Department was taken by Nick Stuart, the deputy secretary in charge of the curriculum. Baker discovered that in the 1960s and 1970s, Stuart had headed the Department's own unit to promote comprehensive schools. And in 1993, the *Sunday Telegraph* revealed that he was the convenor of a hitherto unknown group of like-minded educationists that met three times a year at All Soul's College, Oxford. This group, whose members were enjoined to keep secret their deliberations, included other officials such as Sir Geoffrey Holland, members of the curriculum and testing bodies, various chief education officers and even some education journalists. Its agenda appeared to be restricted to only one point of view. One of the few to be invited who did not conform to this orthodoxy was the industrialist Sir Robert Balchin, chairman of the Grant Maintained Schools Federation. He commented: 'They didn't approve of the current government's reforms in education at all. I asked them if they ever listened to views which were unwelcome to them, and I got the distinct impression that the answer was "no". I haven't been asked back.'[9]

It was through the curriculum and testing quangos, the National Curriculum Council and the School Examinations and Assessment Council, that the civil servants were able to exert their influence. According to the former NCC Chairman and Chief Executive,

Duncan Graham, these bodies had been set up against the civil servants' advice, but they had nevertheless managed successfully to limit their powers. According to Graham, Kenneth Baker believed the NCC had wide executive powers, but his civil servants knew that its role was mainly advisory. 'All the evidence suggests that they did not want the NCC because they wanted to run the curriculum themselves with some help from what would effectively become a subordinate HMI,' Graham wrote in 1993.[10]

The civil servants' motivation appears to have been power laced with ideology. They had a strong interest in preventing the quangos from taking policy initiatives that otherwise would have been theirs. And, despite the fact that their own children were as often as not educated at top independent schools, they were sold on the programme of comprehensives, child-centred education and the movement from teaching to learning. Opinions vary about the reasons for this. 'They were simply the captives of their producer interests, just like the civil servants at the Ministry of Agriculture were in cahoots with the farmers,' said one insider. 'It was a strange mixture,' said another; 'class guilt maybe; a streak of paternalism; perhaps even the desire to stop state schools from being as good as the schools their own children went to. Whatever it was, they were utterly out of touch with what ordinary schoolchildren were going through.' A third insider was even more damning. 'They were very *de haut en bas*; they really thought poor children were stupid and so there wasn't any point teaching them anything.'[11]

The officials are said to have employed a wide variety of ruses to divert ministers from their intentions, principally through the civil servants' domination of the quangos and their subcommittees. Officials relied upon ministers being either uninformed or uninterested, they used unintelligible jargon to bamboozle them and they employed delaying devices, last minute policy suggestions and open argument to confuse ministers or wear them down.[12] They were also highly influential in appointing the membership of the quangos. It was not by chance that so many members subscribed to the new de-education orthodoxy. When ministers realised how the NCC and SEAC had been packed against them, they forced some of their own political appointees onto them. But despite claims made by Brian Cox and others, these remained a small minority. The NCC, SEAC and their successor body the School Curriculum and Assessment Authority (SCAA) have never had a majority of the Tory right, nor of educationists who understood what was happening and were committed to reversing it.

The Education Quangos

In any event, according to members and former members of these earlier bodies, the civil servants effectively ran the show. Duncan Graham observed that it was difficult to know when officials were speaking on behalf of the politicians and when on their own account. They were ubiquitous, sometimes operating individually, sometimes in packs. 'They spoke at every meeting, frequently upsetting the council with what some saw as arrogance and a dictatorial manner if they believed somebody was stepping out of line,' he recalled.[13]

He objected to them trying to take over. All papers had to go to officials two weeks before NCC meetings for their views. There was crisis after crisis about this, ending in a compromise: officials were allowed to see papers but not to alter them. But if they didn't see anything intended for the minister's eyes, they implied it would be thrown out. Minor civil servants often wrote misleading reports because they understood little; yet they would lay down the law to better qualified people. 'Junior officials would come to NCC meetings to tell a group of teachers how, for example, geography should be taught,' recorded Graham. 'They often used the word "minister" when they meant their own senior official [required them to do something] and when they said that the minister would not be happy with one of our proposals one could be fairly sure that the minister had never been consulted. The clear message given to members of the council – all of them appointed by the Secretary of State – was that they were there to rubber-stamp papers and minutes that had been written in a way which best represented the views of the DES or were, in the eyes of the department, most acceptable to ministers.' Graham wondered whether their papers actually got through to ministers or whether they merely read the officials' summaries. 'As a result I would sometimes find that when we did meet ministers we were working from quite different pieces of paper.'[14]

One early member recalled what it was like.

The civil service observer at the meetings ran them. He set the agenda, determined what would be in the council papers to go to the Secretary of State, guided the council and the officers. Papers were presented by officers, not by members. Members had very little say over the writing of these papers and the detail of what went into them. We would get documents several inches thick by Red Star the evening before the meeting. If the observer didn't like what we wanted to discuss, he'd sit there saying it wouldn't be acceptable to ministers.

For example, whenever we talked about reforming the curriculum we came up time after time against the ten-level scale which didn't make sense to some of us. At one of the last meetings of the NCC we had a discussion about the scale against the advice of the then observer who didn't want the discussion to take place. It was the only time on the NCC that there was unanimity. Everyone was against that ten-level scale. But the Minister of State, Emily Blatch, was told that the NCC supported the scale; and the council was told that the Minister supported it. But neither was right. As it happened, she telephoned and asked why the council hadn't supported the abolition of the scale. We told her we had; and she told us that the official had said that only a few extremists had opposed it.[15]

The situation was compounded further by the inadequacy of some of the members who might have been expected to take a different position. One former member commented in exasperation:

There were industrialists on the NCC who just didn't understand the system. Take technology. Parliament had voted technology onto the curriculum because it thought the country was technologically illiterate. It didn't occur to them the reason for that was that maths and physics weren't strong enough to enable good people to do technology at university. So now we're turning out thousands of children year after year who can put holes in tin cans but who can't do maths or physics to the required standard. One industrialist on the NCC said we'll never catch up with the Japanese unless we teach technology. But the Japanese don't teach it. They just teach maths and physics very well. He hadn't got the faintest idea what was being taught in the schools or the difficulty with its implementation.[16]

John Marks was one of the few members who were brought onto the NCC and SEAC by ministers anxious to redress the balance. He instantly realised the testing programme was fundamentally unworkable and that it would take up huge amounts of teachers' time by requiring tests on every statement of attainment. 'The civil servants who dominated the committee told us it was a legal requirement to test on all the statements of attainment that were written into the Act and that was that. I read the Act and didn't think that was right. It was just their interpretation. But,' he went on to claim, 'the officials – Nick Stuart, David Forrester and Eric Bolton – told me I was wrong and that they had a legal opinion to back them up. I asked to see it but they told me it was confidential to ministers. It seemed to me they were deliberately creating a banana skin.'[17] Bolton has denied this version of events, maintaining that officials were constantly telling members that the attainment targets were inoperable.[18]

Marks continued to nag away but got nowhere. In 1993, Sir Ron Dearing was brought in to run SCAA, the successor body to the NCC and SEAC, and to sort out the catastrophic mess caused largely by the unworkable nature of the curriculum and its testing arrangements. Marks, who was transferred onto SCAA, pressed the point again with Dearing and this time struck gold. 'We then got a letter from another official which said we did *not* have to test every statement of attainment after all.[19] A gasp went round the room at this. They had come up with another interpretation to allow Dearing to slim the whole thing down.'[20] In other words, there had been no legal basis for what the officials had, according to Marks, previously told them. Slimming down the attainment targets mysteriously became possible in law when the officials decided it suited them. By then, however, the system had been allowed to descend into chaos.

Eric Bolton, the former Chief Inspector who played a key role along with other senior officials in the development of the curriculum, denies that officials behaved at all improperly. 'The senior officials I dealt with played it by the book,' he said. 'When I was there, there was a genuine attempt to report to ministers what was going on. I don't remember anyone stopping anyone from discussing anything. All those members had access to the Secretaries of State and could have talked to them. The conflicts were over public expenditure constraints. The problem was that ministers could never make up their minds about anything. Duplicity runs all the time, and it's called politics.'21

Feelings, however, were running so high towards the civil servants over the role they had played in the education reforms that when Lord Griffiths, the former head of the Number 10 Policy Unit, was appointed chairman of SEAC, he said he would only do the job if the civil servants were excluded from its meetings. This development was greeted with great acclaim by SEAC's members. The civil servants were, by all accounts, furious. But they fought back. When the NCC and SEAC were merged into the single School Curriculum and Assessment Authority by the 1993 Education Act, the new legislation contained a clause enshrining in law the officials' right to attend every meeting. However, the situation then changed in a more subtle and profound way. The civil servants no longer dominate the proceedings of SCAA, members say, as much as they did at the NCC and SEAC. They don't have to. The chairman of SCAA, Sir Ron Dearing, is one of them.

Dearing was brought in to clear up a mess. In 1993, the government's curriculum reforms were running into the buffers. John Patten, the then Education Secretary, had had a disastrous tenure culminating in the collapse of his health under the strain. A man who had understood what had gone wrong with education perhaps more than any other Education Secretary, he nevertheless insulated himself from all advice and failed to grasp that the teachers' complaints about the impossibility of the curriculum and testing workload were broadly correct and entirely justified. It was Patten's misfortune to have inherited a testing system whose fundamental flaws had been left untouched by his predecessors. It was also his misfortune that the chairman of SEAC, Lord Griffiths, was part-time and often abroad pursuing his banking interests. The resulting vacuum was filled by the civil servants who were wheeling and dealing to their own, less than helpful, agenda.

The bureaucratic monster of the testing system took time to work through the age groups. Its inherent problems were relatively camouflaged in the seven-year-old tests. When the tests finally had to be applied to the complex profile of the 14-year-olds, however, disaster struck. The result was a teachers' boycott which, although it was led by the militant English teachers on their ideological crusade, was supported mainly by the middle ground of the profession which was in understandable open revolt against an unworkable system.

The political imperative, therefore, was to rescue a drowning minister and steer the curriculum ship back into calmer waters. By then, the National Curriculum Council was being run by David Pascall, a highly intelligent, muscular thinker who had previously worked in the Downing Street policy unit and had been brought in from industry. He was aghast at the shambles he had inherited. He grasped that the curriculum was absurdly overloaded and suggested it should be drastically slimmed down.

'When I arrived, I found the NCC developing a more and more complicated curriculum,' he said. 'The teachers were perfectly right to complain. The problem was that there wasn't a strong enough driver. The NCC never saw it as its role to edit; it just set up working parties which put everything in.'[22]

But he also understood extremely well that Britain was de-educating under the impact of the new ideological orthodoxy, and was determined to repair the damage. 'The curriculum according to the Act was about preparing children for adult life through cultural, moral and spiritual development but it didn't seem to me that was being followed through,' he said. Accordingly, Pascall made uncomfortable waves, causing controversy and even uproar over his attempts to put knowledge and structure back into the curriculum. In particular, he was passionately concerned to rescue the teaching of English from the unstructured prison into which it had been bolted.

'The fact that education was failing most of our children seemed irrefutable. We felt there weren't strong enough connections between what was needed and the working groups who represented the vested interests of the subjects. There was no intellectual rationale to the curriculum in terms of its aims or the requirements of society. They just set up a quango, farmed it all out to specific subject groups and then wondered why it didn't work.'

All too predictably, Pascall was denounced as a political appointment, a right-winger who was determined to move the NCC's centre of political gravity. He denied this vigorously. 'Because I had worked in Mrs Thatcher's policy unit I was seen as right wing but I'm not; some of the things they do are far too extreme. I wanted on the council one-third practising heads, one-third educationists and one-third consumers, so that two-thirds of them knew what they were talking about. I put on teachers who were very middle of the road. At no stage were the right-wing in the majority.'

Pascall believed ministers wanted him to run the new School Curriculum and Assessment Authority. But other plans were afoot. As we've seen, John Patten announced that SCAA was to be run by Sir Ron Dearing, a former head of the Post Office and a civil servant. Upon hearing the news Griffiths, pronouncing himself both delighted at the appointment and too busy to chair SEAC any longer, suggested that Dearing should take over forthwith the chairmanship of both the NCC and SEAC pending the establishment of the new body.[23] After a struggle with Pascall, this came to pass. Pascall was out, Patten was terminally wounded and Sir Ron Dearing became the Pooh-Bah of British education.

The problem was, however, that the crisis which Dearing was drafted in to defuse was seen by ministers and by him as a problem of politics rather than pedagogy. Dearing had a deservedly high reputation as an arch-fixer. The quintessential civil servant, he believed in doing deals and working out compromises behind the scenes to achieve a consensus which would allow programmes to proceed. The problem he was asked to solve was that of teachers who were in revolt against a curriculum which was so overloaded that it was unworkable. He was not asked to take on the infinitely more serious and difficult problem understood by Pascall, that the curriculum was failing philosophically as well as in its practical application. The way the establishment viewed the crisis in education was illustrated by their choice of Dearing as the man to deal with it. He was not appointed in order to win the culture war and rout the enemies of civilisation. He was appointed instead to bring peace in our time.

Curriculum Revision

The enormously high regard in which Dearing was held could be seen from the fact that, almost immediately, he was given a personal remit to review the curriculum and work out where it was failing. As a result, the curriculum was re-written. Its structure was simplified, the burden of excessive bureaucracy was lifted from the teachers and peace returned to the classroom. But it was a Pyrrhic victory. It was achieved at the expense of the content of the curriculum where the pass was sold. Dearing had seen his task principally as an editing job, whereas what was actually needed was a fundamental rewriting from first principles. What he produced was a massive compromise which, while producing some improvements, fell far short of the task of pulling British education back from the brink. Indeed, by yet further institutionalising some of the flawed orthodoxies, it even made the problem worse.

The most ferocious battles over the revised curriculum centred on maths, English and history. All resulted in a painfully crafted compromise; all, to a greater or lesser extent, sold the pass.

The Maths curriculum

The most clear-cut defeat for the traditionalists took place in mathematics. The fundamental argument was over whether children should be taught mathematical knowledge as a series of rules to be learnt and codes to be cracked, or whether they should be encouraged to devise the problems and work out the solutions for themselves through experiment and local observation. The traditionalists, including an increasing number of maths professors, believed that far too little emphasis was being given to the elementary stuff of maths such as number work or algebra. Their opponents believed that children should learn maths largely through its practical application in everyday life. They had a horror of children being taught by rote, or even being taught any firm rules at all, despite the fact that many children – perhaps the majority – only master maths by repeating exercises until those particular principles have been absorbed almost as second nature.

The particular argument over the curriculum, therefore, boiled down to a struggle over whether children should have to show they knew how to use and apply mathematics. The traditionalists wanted this requirement to be absorbed into the main body of the curriculum where it could effectively be lost. Their opponents wanted 'using and applying' to comprise a discrete attainment target, of equal value to the separate attainment targets of number work, algebra, shape space and measures, and handling data. They won.

The result has been an inevitable devaluation of the fundamentals of mathematics. Because so much prominence is given to using and applying maths – at every age level, the curriculum requires teachers to make sure pupils cover between 13 and 17 different ways of 'using and applying' – there is obviously correspondingly less time available for everything else. Thus children of all ages are expected to 'use and apply mathematics in practical tasks, in real-life problems and within mathematics itself'. From ages 7 to 11, they are expected to 'develop their own mathematical strategies and look for ways to overcome difficulties'; 'try different mathematical approaches'; and 'identify and obtain information needed to carry out their work'. From ages 11 to 14 they are expected to 'find ways of overcoming difficulties that arise; develop and use their own strategies'.[24] Such an approach involves projects and investigations which are immensely time-consuming. By the time the children have done all that 'using and applying' just how much time have they got left to devote to the mastery of fractions, decimals or algebraic equations?

In any event, the value of 'using and applying' is highly dubious. It is founded upon the visceral distrust of teaching and the corresponding expectation upon children to teach themselves by working out different approaches and solutions for themselves. As in so many other areas of the curriculum and of the new pedagogic orthodoxy, this embodies the belief that children can somehow run before they have learnt how to walk. It expects every child to be a mathematical prodigy and in consequence leaves those who are not to struggle along in untutored ignorance and error.

Children are expected to master sophisticated concepts before being introduced to more elementary work. The equal emphasis given to shape, space, measures and the handling of data imposes similar pressures. The curriculum expects children, for example, to master the application of statistics, understand probability and 'choose appropriate standard units of length, mass, capacity and time, and make sensible estimates with them in everyday situations'. In other words, the curriculum might enable children to become opinion pollsters or play the National Lottery or buy sufficient wallpaper for the spare bedroom – even though they still can't quite remember the product of eight times seven.

The History curriculum

History was less clear-cut. Once again, the argument was essentially between facts and subjectivity, between knowledge and skills. The battle was bloody and at the end of it the traditionalists were themselves divided. Some believed that the curriculum had been pulled back from the worst excesses of cultural relativism, the overvaluing of the imagination and the devaluation of British history. The curriculum was now stuffed with facts, they claimed, and British history was well to the fore. However, most of the facts that appear in the curriculum are not compulsory, merely advisory examples. Thus, between the ages of 7 and 11, children must be taught about either the Romans, Anglo-Saxons or the Vikings but not necessarily all of them; between 11 and 14, they must be taught about Britain between 1750 and 1900 but there's no mention at all of George III.[25] The history curriculum is still deeply incoherent and illogical – except for the deep vein of ideology which still courses through it.

It still embodies a clear bias away from key personalities and events towards social history and the lives of ordinary people. So, for example, Queen Victoria doesn't figure at all except in the way she illustrates 'family life at different levels of society' in Victorian Britain; the American War of Independence and the French Revolution need only to be taught to the extent that they influenced British politics, and the Stuart kings have vanished altogether. There is a huge amount about the ordinary lives of everyday people. Of course it is desirable to study social history. But with only two history periods per week on average in the schools, it appears perverse to load the curriculum with sociology and not teach the skeleton of historical knowledge which is vital if social history is to make any sense.

Yet even with these notable gaps, the curriculum still manages to be seriously overloaded. In an average two periods of history per week, no teacher of 11- to 14-year-olds could possibly do justice to all six compulsory units ranging from 1066 to 1500, 1500 to 1750, 1750 to 1900, the main events of the twentieth century, a significant turning point in European history before 1914 and a past non-European society. At the same time, the curriculum dictates a methodology of eight equal perspectives through which history must be taught. So alongside the political perspective – through which *facts* are taught – equal weight must be given to economic, technological and scientific, social, religious, cultural and aesthetic perspectives.

The result is not only that teachers must skim through it all superficially, but there is ample opportunity to turn every history lesson into a vehicle for propaganda. As one highly experienced history teacher commented:

> There's a very clear agenda running through it: an assumption that the only thing that matters is progress towards democracy. So movements of popular protest are all presented as legitimate, we're all supposed to condemn the part played by social class and we're all supposed to know whose side to be on over the Peterloo massacre.
>
> And because the whole approach is through these themes, there are huge gaps and the opportunity for propaganda because there's no clear chronology, no firm structure to hang onto. So it can easily be hijacked by any political perspective. And because change is the only thing that matters, ordinary politics aren't in there. So George III isn't there because there weren't any dramatic changes under him; he's apparently not thought worth thinking about. But what about the political institutions themselves, the role of the monarchy? If children have no knowledge of politics between 1688 and the US revolution, won't they have a rather peculiar view of chronology? There's no logic to it.[26]

According to Chris McGovern, there *is* no logic to it because of the way the curriculum was revised. After the Lewes Priory debacle, McGovern and his fellow dissident Anthony Freeman were invited onto SEAC and McGovern was subsequently involved in the revision of the history curriculum. Both, however, were eased out of curriculum development because of the uncompromising nature of their views. McGovern, indeed, wrote a minority dissenting report when the history curriculum was revised, a process which he claims was not rigorous and was ideologically motivated – a claim that has caused serious irritation, it must be said, to other traditionalists who were involved.

> When the curriculum was revised, we had three meetings to slim down history. The agenda and the writing were done by the officers. But they didn't want to cut the methodology. The perspectives were thought to be the central reason for teaching history. So they cut the content. But they didn't cut out peripheral things like non-European history because they argued that children were entitled to learn about West Africa, for example. The solution was to put a lot of the stuff in italics [the stylistic device used to indicate the material was optional]. They'd say they didn't want Florence Nightingale because she was too middle class, so they put her in italics. They'd say they'd teach the 'influence' of the French Revolution or the Napoleonic Wars but not what actually happened in those events. 'Influence', after all, takes you into economic and social history. The whole process was haphazard and done by people with no knowledge of the subject. And now most schoolchildren drop history because they are not engaged by it.[27]

Topics were dropped or included, said McGovern, without any proper rationale. For example, he tried to insert the American Revolution and was offered instead as a crumb Lord Shaftesbury as an option under Victorian Britain.

McGovern's published minority report was scathing. Most history in the curriculum listed under Britain, he wrote, was not peculiar to Britain at all but common to most of Europe. Pre-Roman history had no part to play, the Roman Empire and most of Roman Britain were optional and most of the thousand years from 55 BC to the eleventh century would have to be taught at the rate of about 50 minutes per century. There was no requirement to teach Henry V, Admiral Blake, Marlborough's wars against Louis XIV, Clive, Wolfe, the American War of Independence, Nelson, Wellington, the Crimean War, Gordon, or the Zulu or Boer Wars. The military events of World Wars I and II were part of world history, with the atom bomb prescribed but the role of Churchill optional. Nelson and Wellington were only included as 'options' alongside 'arts and architecture' or 'industrialisation in a local area'.[28]

The English curriculum

As might have been expected, the most bitter defeat for those fighting the doctrines of un-teaching occurred over the English curriculum. The final version was, like everything else, a compromise in which a little ground had been made up. But the concessions to the traditionalists were minimal. A canon of classic English texts to be taught, over which there had been such explosive arguments between those who had different views over what constituted classic texts and those who thought the whole notion of a canon was élitist, was included for 11 to 14 year olds. But it was pretty minimal even so, requiring merely two Shakespeare plays, two classic novels written before and after 1900 and four major poets before and after 1900 to be read between the ages of 11 and 16, with no canon at all prescribed before age 11.

On the teaching of Standard English through grammar, punctuation and spelling it remained vague. Occasional bows in the direction of Standard English were carefully phrased so that pupils were to be 'encouraged' or 'given opportunities' to use a range of styles of English, including Standard English about which they were to become 'aware'; they were to be taught *about* Standard English, rather than taught explicitly *how* to use it as the national language, the entitlement of everyone. The key attainment target was that 'pupils talk and listen confidently in a range of different contexts', a bland formulation designed to be all things to all people.[29] The trick worked. After all the sound and fury over the teaching of English, there was scarcely a ripple when the final document was published. Despite holding his nose at the mauling given to his original proposals, Brian Cox commented: 'This peculiar document opens doors. Most restrictions in the 1993 and 1994 drafts have been removed. What is left is often bare, boring and brief, but good teachers are allowed to develop their own initiatives.'[30] In other words, there was little in the revised English curriculum to trouble any English teacher who wanted to continue with the retreat from structure into spontaneity that was depriving thousands of children of their democratic entitlement.

The English compromise came as a particularly bitter blow to David Pascall. It was shortly after the NCC had published far more rigorous proposals for revising the English curriculum in April 1993 that he was unceremoniously dumped from the curriculum body. By the time his successor Sir Ron Dearing produced the first of several revisions to those proposals five months later, it was clear that it was not merely Pascall but his rigour that had been dumped as well. His document had set out a detailed prescription of the way Standard English should be taught through the 'correct' use of grammar, punctuation, spelling and vocabulary; it had identified phonics as an essential skill in initial reading; and it had laid down a detailed list of good books to which children should be introduced from the age of five. Five months later, he saw this approach diluted out of recognition in direct response to the pressure that had been exerted by the English teachers – pressure to which the government had caved in.

The outrage of the English teachers was scarcely surprising. Pascall's proposals would have spelled the end of the un-teaching of English, the overturning of an orthodoxy which many had never thought to question and which many others knew they dared not question for fear of losing their reputation or their jobs. It would have taken rare political courage to have stood firm against a protest on this scale. A letter published in the press revealed the magnitude of the forces Pascall had been up against. First published in *The Times* in 1992, the letter was subsequently reprinted in *The Times Higher Education Supplement* shortly after Pascall's consultation paper was published. Drafted by the Marxist Oxford English professor Terry Eagleton, it was signed in the *THES* by no fewer than 576 English dons, many if not most of them from the political mainstream.

It said:

As university teachers of English, we view with dismay the government's proposed reforms to the teaching of English in schools. Like all academics, we expect sound grammar and spelling from our students; but the government's doctrinaire preoccupation with these skills betrays a disastrously reductive, mechanistic understanding of English studies.

Similarly, its evident hostility to regional and working-class forms of speech in the classroom betrays a prejudice which has little or no intellectual basis, and which is seriously harmful to the well-being and self-esteem of many children. We are all committed to the study of Shakespeare; but to make such study compulsory for 14-year-olds, as the minister intends, is to risk permanently alienating a large number of children from the pleasurable understanding of classical literary works.

Even more disquieting is the plan for a dictatorially imposed canon of supposedly great works, in gross and wilful ignorance of more than two decades of intellectual debate among literary academics over questions of literary value and the literary canon. These philistine, ill-informed proposals would strip English of much that we and many of our colleagues regard as most precious and educational about it. They threaten to reduce a living language to a dead one, and a vital literary heritage to a mummified relic. They would do serious damage to the moral and social development of our children, and to the cultural life of society as a whole, and all who are concerned with such matters should oppose in the strongest terms.[31]

This landmark letter should stand as a monument to the state of degeneracy into which British intellectual life fell towards the end of the twentieth century. Through its weasel words, it is possible to glimpse the venomous fury of an élite whose hitherto unchallenged power to destroy a culture and inflict upon the population its own self-destructive alternative was for the first time under serious threat.

The letter suggested that these academics claimed to expect grammar from their students but would take to the barricades to prevent them being taught it. Indeed, teaching grammar or Standard English would be an act of 'hostility to working class speech'. So they would rather want people who didn't have Standard English to be denied access to the one truly national language and to remain trapped in a particular form of speech that would disadvantage them by excluding them from the mainstream. Their desire to deprive 14-year-olds of access to Shakespeare to which they themselves were 'committed' was similarly exclusivist and patronising to all those thousands of 14 year olds who – properly taught – do enjoy Shakespeare, and to all those others who would enjoy it if they were only given the chance.

But the next paragraph was the most revealing. 'Two decades of intellectual debate' was code for the totalitarian ideology which has ordained that the canon of literature is taboo because it doesn't conform to politically correct notions of race, class and gender and because, according to the tenets of literary theory, all texts can be deconstructed into meaninglessness – including Shakespeare, to which such academics are all so 'committed'. The description of the canon as 'mummified' illustrated their desire simply to pull up the drawbridge of English literary history and obliterate the past.

But opposition to the English proposals was not confined to the fulminations of dons safely corralled within their ivory towers. It constituted instead a well-developed network which had become so integrated with the political institutions that reform had become impossible. The English teachers boasted they would subvert the reform from within, and they were correct. It was not the first time, after all, that this had happened.

Compromised reform

Their symbiotic relationship with the establishment was in fact very well established. That relationship had undermined all attempts to get to grip with the problems of literacy for decades. Concern about literacy is certainly not a recent phenomenon. The proportion of the population who cannot read has long been felt to be unacceptable in an advanced society. As long ago as 1972, the Bullock committee was established to report on the problem; in 1975, it confirmed indications of a downward trend in reading standards, with a growing proportion of poor readers among the children of unskilled and semi-skilled workers.

Yet far from addressing the problem it had correctly identified, the Bullock committee provided a degree of official impetus to those New Literacy ideas which were already doing damage, and provided substance and direction for them in British schools.[32] According to the admiring words of the movement's guru John Willinsky, Bullock provided a 'degree of official impetus' for the ideas of James Britton, the New Literacy acolyte who was a member of the Bullock committee, 'to a surprising degree for a report commissioned by the then Secretary of State for Education, Margaret Thatcher'.[33]

Bullock advocated that 'personal writing' should arise from the experiences of pupils and the class, not from what they were taught, and that the teacher must recognise that 'a writer's intentions are prior to his need for techniques'. In other words, it advocated the characteristic back to front approach that children were only to be taught how to write after they had written. Willinsky wrote: 'The report offered a series of what might be taken as New Literacy responses to the literary crisis which had sparked the official inquiry; it provided both substance and direction for the British schools in a manner which is not possible in Canada or the United States, where education is a provincial and state matter.' The British establishment, with its fondness for depositing any problem into the laps of the great and the good, had unwittingly set a lethal trap for themselves and for the schoolchildren whose interests they were supposedly attempting to protect.

The Bullock report of 1975 was so compromised that one member, Stuart Froome, published a Note of Dissent pleading for more structured teaching in schools. In comments that were to have resonance twenty years later, he wrote that some schools believed children would develop language by being encouraged to speak and write and that any critical intervention would stem the flow. 'Sometimes work of very poor quality is displayed in such schools because it is believed that the child's spontaneous effort is sacrosanct and to ask him to improve it is to stifle his creativity.' He believed there was no longer rigorous marking of spelling, punctuation and grammatical errors. 'If teacher-students are as deficient in basic writing skills as our evidence suggests, it can be fairly assumed that the standards of those who do not aspire to be teachers must be correspondingly dangerously low.' He also opposed the committee's endorsement of mixed ability teaching for which there was no supporting evidence. 'The move towards mixed ability grouping in British schools is a recent one and in my view smacks more of social engineering than of educational thinking. Like the movement to abolish grades, class position and pupil competition it is really a manoeuvre to ensure that no-one is seen to excel.' Like the words of warning from David Pascall, Chris Woodhead and others twenty years later, Froome's comments fell among minds that were closed.

Anxieties, however, rumbled on. In 1981 a Schools Council project sounded the following warning: 'One of the dangers associated with an advocacy of psycholinguistic theories about the strategies children use in learning to read is that they may cause teachers to think that any direct teaching of certain skills which formerly took place is now no longer either valuable or relevant.' In 1984, the HMI published a report which warned that if literacy wasn't achieved by age seven, the effects persisted to under-achievement at 16.[34] The HMI pamphlet caused uproar in the education establishment because it dared to suggest that children should be taught something about the systematic aspects of language. The teacher Jennifer Chew recalled: 'I remember attending a meeting at which an inspector held up the pamphlet and said, "We all know who's responsible for this, don't we? We don't want anything to do with it." The incident and the words remained in my mind because everyone else nodded knowingly and I felt that I must be the only person present who did not understand what was meant.'[35]

Ruskin College conference

The truly revolutionary nature of the new orthodoxy in English teaching was finally revealed at the seminal meeting of English teachers at Ruskin College, Oxford in June 1991. This was the meeting – as we saw earlier – where the English adviser Peter Traves had spelled out the difference between 'proper' and 'improper' literacy. The conference, attended by about 450 English teachers and lecturers, was treated to a series of statements which made it clear beyond doubt not only that English teaching was now the chosen battleground for the culture war but that people of influence had the means to take that battle into the heart of the Conservative government machine – and win it.

Ken Jones, a lecturer at the Institute of Education in London and a former schoolteacher, boasted of this influence. He told the conference: 'Well developed tactics of constructive engagement were used to get hold of the rudimentary position of government ministers through the working groups and the National Curriculum Council to make the National Curriculum more friendly.' The Cox report, he said, had endorsed and validated the progressive position: in other words the importance of speaking and listening as opposed to reading and writing. And he recommended that the government's principles should be converted into 'a critique of present social arrangements' – in other words, the precise opposite of their intentions. It was at this meeting that the critical role in the strategy of the London Institute of Education itself, the most prestigious teacher training institution in the country, became apparent. There was agreement that responsibility for spreading the message would rest with the Institute.

'We've got to target the heads of English departments,' said Jones.[36]

Peter Traves continued with his principled denunciation of reading by attacking Martin Turner. The assault on real books by Turner and the media, he said, had been deeply damaging. 'What Martin Turner knows about reading could be put on the back of a postage stamp. It is a general attack on alternative sources of power, a denial that there are literacies. It is a bourgeois concept to say, little Dorothy's a great reader. [Children are taught to] read in order to control their access as consumers. As children learn to read, they feel less powerful: they are divided. The alternative has the potential to liberate. There is a political function of literacy. Children need to place their values against those of the text.'

Professor Terry Eagleton, apparently not yet having arrived at his later insight that it was government advisers who were philistines, described himself as a 'barbarian within the citadel'. University departments of English literature, he said, were 'part of the ideological apparatus of the capitalist state'. English teaching had emerged as a form of class struggle. 'Bourgeois society believes in reason, freedom, truth and justice but stands for oppression and domination. For all their high-falutin' talk about God and the family, they've got their fingers in the till,' he observed.

Great literature, the conference was assured, was 'class-restrictive, heterosexist and ethnocentric', drawn up by the ruling class to preserve its ideological dominance. Gemma Moss, a writer, complained that schools were fixated on the book culture, ignoring what children knew about TV. 'There seems to be a view that Jackie Collins is degrading and Shakespeare is morally uplifting: they're different, that's all. There's no reason why Mills & Boon shouldn't be taught alongside *Jane Eyre*', she opined. Jane Miller of the London Institute added contemptuously: 'Shakespeare, the apostrophe and the difference between uninterested and disinterested – they're just ways of separating the sheep from the goats.'

This was no idle revolutionary chatter. The conference was told how this ideology had been penetrating the establishment for years largely through the National Association for the Teaching of English (NATE) and its London counterpart (LATE). John Dixon, a former London teacher, said NATE had infiltrated the Schools Council English committee years ago. Terry Furlong, the NATE chairman, said: 'With GCSE, people realised they needed to man the barricades: to get inside groups and turn the whole thing round. We did this through the London Association, the National Association, English centres and advisers.' It had been done, he said, through a concentration on media education and multicultural and anti-racist teaching. Speaking in the wake of Cox, Furlong said: 'The National Curriculum does embody much of what we consider to be good practice. It avoids the narrow teaching of grammar. Assessment has been steered towards our ideas.'

National Association for the Teaching of English

A book published in 1992 acknowledged the impact that NATE and LATE had made on the teaching of English. Keith Kimberley wrote that they had undertaken pioneering work 'which informs much current practice in assessment . . . Over the years English teachers with others have been instrumental in bringing about: the demise of the barren comprehension task in examinations and the multiple choice test (Sir Keith also helped this one on its way out); a broadening of the range of kinds of writing thought worthy of assessment; the inclusion in examinations of the assessment of speaking and listening; the replacement of subtractive by positive forms of marking; the advent of reliable forms of impression marking; coursework alternatives to the timed test; a major rethinking of the principles for choosing literary texts; and so on.'[37]

As the struggle developed in the curriculum and testing bodies to steer the curriculum away from such ideas, the English teachers' associations NATE and LATE went into action. The education activist John Marks went along to a meeting of LATE in October 1992, shortly after David Pascall had first given notice of the way he wanted to revise the English curriculum. Before the meeting started, however, Marks was recognised and asked to leave. He did so, to loud applause. Ken Jones of the London Institute then revealed the reason for holding the meeting. The think-tanks, he said, had taken over. The government had packed the curriculum committees, rejected the research evidence and shown paranoia at the slightest criticism. There would now be continual challenges to the curriculum. And Terry Furlong, the former chairman of NATE, added that the important point about 'this degree of aggro' being planned was to choose the right moment.[38]

Furlong's comment was particularly illuminating. For it was none other than Furlong – when he had been leader of the Consortium for Assessment and Testing in Schools – who had been awarded the contract by the School Examinations and Assessment Council to draw up the original English curriculum tests which had caused the teachers so much grief. It beggars belief that such a man, who had boasted at Ruskin about manning the barricades, had been awarded such a contract by the government's own quango. This was the man who had told the *Guardian* that year, 'Teaching Shakespeare is arse-achingly boring.'[39] In 1992, commenting on the NCC's suggested list of classics including Eliot, Brontë and Swift, Furlong described it as 'another go at perverting the collective psyche of the nation'.[40] Furlong's comment at Ruskin about turning the whole thing round from the inside had clearly not been an empty boast. In the summer of 1992 Sue Horner, treasurer of NATE, became the National Curriculum Council's Professional Officer for English.

Ann Barnes, general secretary of NATE, spelled out the strategy of infiltration that had been aimed at taking control of the curriculum tests. In 1993 she wrote that other teachers felt that 'since we had to have this sort of assessment, we should make sure we had the ownership of it and make it sit as comfortably as possible with our principles. So when the contracts went out in 1989 they felt there was a lot to play for.'[41] But it was the NATE and LATE activists who that year were spearheading the teachers' boycott of the tests, which in turn led to the political demise of John Patten and the rise to power of Sir Ron Dearing. Yet NATE and LATE activists had been the very people who had helped draw them up in the first place.

In March 1993, the LATE *Bulletin* urged members to set new dates for meetings to counter the threat of the revised English curriculum and the tests. It issued battle orders.

The main issues are: the demands being made from KS1 [five- to seven-year-olds] onwards about the use of spoken standard English; the unnecessary emphasis on phonics; the imposition of a literary canon; the deletion of all references to critical study of the media (and we know whose interests that would serve!); the use of spelling, punctuation and handwriting to create ceilings on pupils' achievements. The good news of course is that if the National Curriculum is changed the way the Centre for Policy Studies wants it to be, then we will have even more arguments for not doing the KS3 SATs [Standard Assessment Tests for 11- to 14-year-olds] in 1994 and beyond.[42]

In the spring of 1994, an editorial in NATE news showed that, far from 'the New Right' having taken over the curriculum, the revolutionary English teachers knew they actually held the cards in their own hands.

On the SCAA English working group we have the vice chair of NATE, Alastair West, as well as other NATE members in the form of John Hickman and Sarah Anstey . . . Another rather double-edged element in this volatile scenario is the transformative power of English teachers. We aren't now teaching the Cox curriculum so much as what we, in our individual schools and classrooms, have made of it . . . In the same way, the new curriculum, even if implemented as it is, might in time become transformed – even, if it deserved it, sidelined, marginalised. We have the power to do this – but at a cost of time, stress, and most importantly distraction from the real business of collaboratively improving our practice [Not, note, that the most important business was teaching the children.] . . . as the ever more etiolated National Curriculum for English shrinks and freezes, other versions of the curriculum, out of a sense of critical response or out of sheer disdain for the official version, proliferate. The London Institute's new journal represents an academic consensus about just such a curriculum, rooted in specific forms of cultural and social literacy. The BFI [British Film Institute] inquiry into English will represent another.[43]

Language in the National Curriculum

And there was still one more. In the 1980s, the Education Department commissioned a literacy project called Language in the National Curriculum – or LINC. This project was designed to train teachers to handle a greater emphasis on language in the curriculum, with the intention of influencing practice throughout the country. The government promptly dropped the project in

1991, however, when ministers realised to their horror that it was little other than a Trojan Horse for the New Literacy. But it was not so easy to get rid of it.

The LINC materials made the familiar claim for a balanced mixture of methods to teach children to read. But in fact they displayed the usual depressing ignorance of how children *do* learn to read, ignoring most of the reputable research and drawing heavily instead on the gurus of the New Literacy. Accordingly they emphasised learning rather than teaching, they aimed to teach children language structure by casually slipping into the conversation words like 'adjective' in the hope that the child would work out what the rules were without being taught they were rules, and they extolled illiterate pieces of work by nine- or ten-year-olds as 'very positive', holding them up as evidence of progress when in fact they were rather evidence of educational failure. In more than 350 pages of rambling prose, LINC provided little that would actually help a teacher to teach and a great deal to support the further unravelling of the teaching of English.

Despite the fact that the government blocked publication of LINC, the materials have been widely disseminated in samizdat form and are now in common use. As Brian Cox wrote, the training pack was printed with the support of local education authorities in a desk-top published version distributed for use in in-service training sessions. HMSO, he wrote, originally planned to print 18,000 copies. By 1995, more than 30,000 copies had been ordered.[44] Yet despite Cox's praise for the invaluable good sense of the LINC materials, he revealed that for him their importance lay in a rather different sphere. 'Many teachers suspected that some Conservatives wanted to ban the LINC materials because they instinctively opposed classroom activities which helped children to analyse the underlying assumptions of political rhetoric,' he wrote. In other words, LINC wasn't about literacy at all. It was ideological propaganda.

De-Education and the National Curriculum

The National Curriculum has incorporated a disastrous series of own goals. It attempted to change a culture; but instead it was itself sucked into that culture and became used in large measure to reinforce it. The improvements that it may have fostered have to be set against the way in which – in so many important areas – it has consolidated the terrible damage done to the very idea of education by an ideology which used the education of children as a means to a political end.

This is not a situation that can be remedied by the superficial rhetoric of contemporary political debate about such issues as bad teachers or grant-maintained schools. In any event, this disaster is not confined to state-maintained schools. It is also now engulfing the independent schools, which are trying vainly to stem the tide. The National Curriculum sets the tone for the general education culture. It has a direct impact upon the public examination system. Its underlying assumptions about the place of formal structures such as grammar or mathematical rules, the pre-eminence of creativity and the imagination, the priority given to scepticism and doubt of all transmitted authority, together with the undermining of knowledge and the transfer of authority from teacher to learner, have all influenced the GCSE examination. Its influence is thus spreading within the independent schools sector whose children take the same public examinations as everyone else.

It is not merely that the curriculum and the public exams have to march to the same tune. The exam boards consist of individuals who are drawn from that same secret garden and subscribe to the same educational orthodoxies. And gradually, the A-level exam – or whatever may succeed it – will be dragged down the same path, responding in the only way the boards know how to pupils who may be equipped with the paper grades that entitle them to seek higher qualifications, but whose lack of knowledge has somehow to be accommodated within a public examination system that cannot admit that its own criteria of excellence are becoming increasingly meaningless. Indeed, Sir Ron Dearing's review of post-16 qualifications, published in 1996, proposed the dilution of A-level with vocational elements and the gradual blurring of the differences between disparate qualifications to produce an illusion of equivalence.[45] The national education currency is being devalued.

Britain is now de-educating. The whole of the British education system, from infant classes to degree courses, has been corrupted by these ideas. The collapse in knowledge among schoolchildren has meant that universities are having to water down their own degree courses to adjust to this new situation. This collapse has combined with the impact of the free market, as we shall see later, to undermine the very *raison d'être* of the universities and to unpick the academy. In this retreat from culture, the National Curriculum has institutionalised bad practices and in some ways made matters much worse. School teachers and university professors, who recognise with despair exactly what is being thrown away with such ignorant abandon, are keeping themselves going by wishing away the time to their retirement when they will no longer have to confront every day the evidence of the systematic dismantling of everything they have believed in.

The reason that even the ideological Conservative administration of Mrs Thatcher and John Major was unable to gain the ascendancy in education was not simply because of the views of some of their civil servants or the superior organising skills of the educationists; nor was it principally because the political élite generally does not educate its children in state schools (although this no doubt played an important role in their thinking).

The government lost the battle because it was fighting the effects of a culture it did not begin to understand; a culture that has spread way beyond education into every corner of social and political life; a culture that has reached into the heart of the establishment itself and subverted its institutions; a culture of libertarian individualism to which this government itself was such a notable subscriber, along with those it viewed as its political enemies. The development of that culture, its history and resonance for Britain, and the way it has undermined most crucially and disastrously the whole spectrum of relationships between individuals, between parents and children as well as between teachers and pupils, goes a long way towards explaining why there is now such a strong sense that everything of value is falling apart.

Chapter Seven

THE UNRAVELLING OF
THE CULTURE
A Broader Picture

Until now, this book has been looking in detail at what has gone wrong in British education. The cultural shift that has taken such a toll of British schools and universities, however, cannot be understood in isolation. It is part of a far wider breakdown in the concept of parenting that should be fundamental to a society: the discharge by adults of their primary responsibility to care for and control the children in their charge in order to bring about their healthy growth and development and to ensure the transmission of their culture. It is the central contention of this book that this parenting role is now foundering through profound changes in the relationship between adults and children. The damage this has done to formal education is as hard to believe as it is extensive. This is why previous chapters have described it in such detail.

But the collapse of external authority that lies at the heart of the breakdown in education has also caused the disintegration of the traditional family and the erosion of discipline and social bonds between the generations. Moreover, all these linked weaknesses feed into and exacerbate each other. So the book now broadens out from education to paint the rest of the picture of social dislocation of which formal education is such an important and prominent part, both cause and effect. Later, it will show how the breakdown in family relationships and social order are closely linked to the retreat from teaching in the classroom; and how, in turn, these are symptoms of a deeper collapse of national self-esteem and of belief in the desirability of transmitting British culture to the next generation at all.

Before that, however, the book explores the complex development of historical ideas that gave rise to the breakdown of authority and the revolution in the relationship between teachers and pupils, parents and children. The ideas about education which brought about the developments explored in previous chapters are rooted in those same historical currents of thinking which have profoundly altered the way we see ourselves and our relations with each other and are the cause of such general social and political unease.

Historical Influences

The patterns of thought that subverted education in Britain did not suddenly materialise out of nowhere during the 1960s and 1970s, as some people have imagined. These ideas had been around since the middle of the last century, from when they became increasingly popular among a bohemian fringe which acted upon the erosion of external authority by setting up progressive schools such as A. S. Neill's Summerhill or by practising the free love philosophy of Havelock Ellis.

The puzzle, however, is why the attitudes of this bohemian fringe suddenly became the norm: why and how a set of views associated with an unrepresentative élite suddenly became the received wisdom among the establishment. In both family life and in education, traditional structures began visibly to implode during the 1960s and 1970s. This process appeared to accelerate during the eighties, notwithstanding the dominance of Mrs Thatcher's government. In education, there was a complete collapse on three fronts virtually simultaneously: pedagogy, curriculum and structure. In the family, although marriage remained popular and most people stuck at it, a rising tide of divorce, cohabitation and extra-marital births relentlessly eroded the value of previously understood rules of behaviour. In all these areas, the duty of the adult world to transmit a given set of values to its children was steadily being replaced by another agenda: an egalitarian social project which not only transformed the way in which adults regarded their own and each other's behaviour, but also involved a complete change in the way childhood and its relationship with the adult world were viewed.

The progressive erosion of respect for other institutions such as the church, the civil service, the police, Parliament and the Royal Family, stemmed from the same cause: the collapse of external authority and its relocation in the individual. And at the heart of this profound change into a liberal individualist society lay a paradox: that liberal values came to turn upon themselves and began to threaten that very order and liberty on which they must depend.

How far back in history the development of these attitudes can be traced will probably always remain a matter for dispute. Liberal individualism is the defining philosophy of the modern era in western civilisation. But when does one date the start of modernity? The Reformation, when the authority of the Pope was first exploded? The Renaissance, when mankind first began to develop its all-consuming interest in itself? And how does one rank the milestones in its more recent development: the devastating impact of Nazism, for example, or of the First World War in this century; or the shattering effect of Darwin or Nietzsche in the last? Or the influence of Freud, Einstein or Marx?

The Enlightenment

It seems reasonable to regard the Enlightenment as the defining moment for the collapse of external authority, in terms of the thinkers spawned by that movement and the impact they had on the subsequent development of modern thought. Indeed, many of the attitudes expressed at that time, especially in the field of education, uncannily anticipated much of the rhetoric of today's approach. But the Enlightenment also embodied a set of paradoxes which led directly to the highly ambivalent legacy with which we are struggling today. The Enlightenment gave us freedom and liberal values; but it also gave us the French Revolutionary Terror and the Holocaust. It gave us a reverence for reason and scholarship; but it also paved the way for the flight from knowledge. It gave us individual aspirations, egalitarianism and human rights; but it also undermined the traditional family.

The Enlightenment was the movement of thought, broadly starting in the late seventeenth century and extending as far as the nineteenth century with political economists such as David Hume, Adam Smith, Jeremy Bentham and John Stuart Mill, which self-consciously set out to liberate human reason from medieval atavism, superstition and error. Rooted in the scientific discoveries of Renaissance Europe and the religious controversies of the Reformation, its intellectual genesis lay in those earlier communities of scholars such as Bacon in the sixteenth century and Descartes in the seventeenth who were the pioneers of modern scientific methodology.[1]

Until the Enlightenment, most Christian writers had seen education in terms of the transmission of truths necessary for salvation. With the Enlightenment, however, this approach was demolished. If the Enlightenment had a defining motif, it was the replacement of past authority and tradition with a new emphasis on individual experience. Francis Bacon, for example, opposed the foundation of Charterhouse school on the basis that its curriculum was to be founded on the ancient classics – sentiments of which any modern professor of cultural studies might approve.

This emphasis on reason founded on experience led to the questioning of all received authority. The essence of the Enlightenment was, in the words of Kant, 'man's release from his self-imposed tutelage'.[2] Everything was now in doubt and the unquestioned authority of revealed religion was corroded. The *Encyclopédie*, the eighteenth-century compendium of advanced knowledge compiled by the French thinkers Diderot and d'Alembert, defined the Enlightenment thinker as one who 'trampling on prejudice, tradition, universal consent, authority, in a word, all that enslaves most minds, dares to think for himself'. It was a resurgence of belief in the qualities of virtue and reason and their capacity to create a better and more just society, and it spawned numerous theories of progress. However, as the historian Roy Porter has pointed out, many leading Enlightenment intellectuals argued that experience and experiment, rather than reason, were the keys to knowledge. Man was a feeling as well as a thinking animal.[3] In this lay the roots of the subjectivity which was eventually to undermine the exercise of reason and the pursuit of knowledge itself.

Right from the beginning, liberal values contained the seeds of their own destruction. Indeed, some of the leading thinkers of the Enlightenment, along with some of their successors including Voltaire, Kant and Nietzsche, recoiled from some of the devastating implications of the thinking they were in the very process of setting in train. They wanted to redefine order, not overthrow it. They were not revolutionary nihilists, even though their philosophy paved the way for precisely such people to lay claim to their inheritance.

Voltaire, for example, believed no known religion was acceptable. He crusaded against Christianity and wanted to replace it with 'natural' religion, a non-dogmatic belief in the existence of a rational, benevolent deity. But he appeared to understand where that train of thought would lead, because he refused to allow his friends to discuss atheism in front of the servants for fear that morality would then collapse. 'I want my lawyer, tailor, valets, even my wife to believe in God. I think that if they do I shall be robbed less and cheated less,' he remarked.[4]

For such thinkers, individual freedom was most highly prized. But liberalism was not a philosophy which originated in the European Enlightenment. It was an English development which arose directly out of the 1688 Revolution and whose apostle was the philosopher John Locke. Locke produced a blueprint for English society: a liberal regime based on individual rights and natural law, rational Christianity, the sanctity of property, a liberal economic policy, faith in education and an empiricist attitude towards the advancement of knowledge and progress through experience. Locke's influence on the Enlightenment in France, however, was immense, largely through the admiration of Voltaire who was ecstatic about the English culture of individual liberty.

The impact of Locke's thinking on subsequent developments well illustrates the capacity of the Enlightenment to provide the basis both for subsequent anti-social consequences which would have horrified their progenitors and for apparently contradictory political philosophies. In this way, the Enlightenment was the source both for the social, egalitarian individualism of the political left in contemporary British politics *and* for the economic individualism of the political right. It is important to grasp the common nature of this inheritance. Otherwise, it would seem baffling to find a deeply ideological Conservative government conniving at the social attitudes of its political opponents, to which one might have imagined it would have been diametrically opposed.

Locke's belief in the governing principle of happiness – that 'things are good or evil only in relation to pleasure or pain' and that 'the necessity of pursuing true happiness [is] the foundation of all liberty' – paved the way for the view that self-interest and the general interest coincide.[5] Later in the eighteenth century Bernard Mandeville was able to claim in his *Fable of the Bees* that private vices benefited society as a whole by setting the wheels of commerce in motion. (Government ministers promoting the National Lottery might well have taken this as their text.) The reasoning was that man was programmed to seek pleasure and avoid pain and that therefore policy should ensure that enlightened self-interest ruled in order to realise the greatest happiness of the greatest number. Similarly, Adam Smith argued in *The Wealth of Nations* that the selfish behaviour of individual producers and consumers through the workings of the free market would result in the common good, helped along by the invisible hand of providence. This trajectory of thought was to undergo a renaissance under the Thatcher/Major government, with its belief in *laissez-faire* economic liberalism, and it was to provide the economic motor for the 'me-first society' of 1980s Britain.

Rousseau

The social, egalitarian individualism of the political liberal left had derived from precisely the same classical liberal roots. But another, quite different strand of thought had cut across the Enlightenment. This was to be a critical influence on the left's thinking which played such a large part in undermining education and the family. This thinking was embodied in the writings of Jean-Jacques Rousseau, the movement's most paradoxical thinker of all. Rousseau was considered to be part of the Enlightenment because he hated the *ancien régime* and had established links with Diderot and the Encyclopédists in Paris. Nevertheless, he himself was most unenlightened, arguing that reason, civilisation and progress would make mankind less happy and free. A forerunner of the thinkers of the Romantic movement, he extolled the emotions and 'uncorrupted' feelings of people in their natural (that is, uncivilised) state. In his view, it was civilisation that corrupted mankind and adult society that corrupted children.

Rousseau drew upon the existing cult of sensibility in eighteenth-century France, which appealed to the heart rather than to the head. For him, truth lay in the emotions rather than the intellect. The result was that he denied the highest forms of human endeavour, he ascribed to human beings in their uncivilised state a set of desirable attributes which were wholly fictitious and he repudiated the essence of humanity. Man, he said, was naturally good but society stripped him of his innocence. His whole philosophy represented a negation of culture and a rejection of rationalism.

It was Rousseau's book *Émile*, however, that was to have such an impact on modern theories of childhood and education. Philosophy had had no real part to play in education until the beginning of the nineteenth century when the ideas of Rousseau and his followers Pestalozzi and Froebel started to arouse interest and create a following.[6] *Émile*, published in 1762, consisted of a sustained criticism of civilisation. 'God made all things good, man meddles with them and they become evil,' he wrote. The message was stark. Children were innately creative beings; society made them evil. *Émile* combined nature worship, child-centred philosophy, an emphasis on doing and discovery at the expense of being taught, and a pervasive hatred of the existing order of things from which the child must be protected.

The message of the book was that education had to help a child's innate goodness develop. It completely rejected the Lockean approach to education with its emphasis on didacticism and the belief that children needed to be filled with knowledge by their superiors. Instead, it promulgated the belief that a good tutor relied on intuition, empathy and kindness to nurture and guide the child's natural curiosity through personal discovery and problem solving. Many of its precepts can be heard today in teacher training institutions up and down the land. Knowledge had to be acquired through the child's experience. Émile was to learn even though he was not to be taught. 'It matters little what he learns provided he understands it and knows how to use it . . . Émile knows little, but what he knows is really his own . . . The only habit the child should be allowed to contract is that of having no habits . . .' And he was to read no books, 'the curse of childhood'.[7] Above all, the child had to be respected as a child – a principle to which Rousseau happened to display his personal loyalty in real life by consigning all five of the children borne him by an illiterate servant girl to an orphanage.

The influence of *Émile* was immense. As the philosopher Anthony O'Hear has written, 'despite an ineradicable lack of clarity about what Rousseau means by nature and uncertainty about the benefit to be gained by following nature's impulses, *Émile*'s influence can be seen in every primary school in the Anglo-Saxon world.'[8]

Rousseau was preoccupied by freedom. But given his conviction that all societies were essentially repressive and corrupt, it followed that the whole social order would have to be transformed before men could be truly free and equal. Then education would be vital in consolidating the moral foundations of such an order. Education was thus the *cordon sanitaire* against a hostile social environment. It was hardly surprising that his theories of education were to become the main source of inspiration to educational theorists who thought that education meant both personal fulfilment and social change. Like Plato before him and the American philosopher John Dewey after him, Rousseau perceived education as part of an over-arching political and social project.

But his drive for equality was characterised by profound totalitarian instincts, expressed in his theory of the general will to which individuals had to give up their rights. He argued tortuously that this was supposed to protect freedom on the grounds that the 'general will' was not some external force of society but something synonymous with individual wills provided those were not led astray by transient and selfish desires. If individuals obeyed the general will, they were therefore merely obeying themselves. This was patent nonsense, made clear by his famous claim that 'whoever refuses to obey the general will shall be compelled to do so by the whole of society, which means nothing more or less than that he will be forced to be free'. In *The Social Contract*, Rousseau envisaged moral censors who would oversee the private lives of adult citizens and children as the ostensible means of enabling people to act in ways that ensured their liberty.[9]

This was clearly an early blueprint for totalitarianism, and Rousseau is credited with a line of thought that progressed through Robespierre to Hitler. Yet at the same time his thinking on education appeared to be profoundly liberal and was the inspiration for the whole child-centred philosophy of education and child development. This contradictory legacy, inspiring both democracies and despotisms and promoting liberty, fraternity, equality *and* the Terror, is often written off as a strange and inexplicable conundrum. But to understand properly why modern child-centred theories are so pernicious and have exercised such an iron grip, it is important to recognise that these tendencies in Rousseau, the 'liberal' and the totalitarian, were not opposites at all but symbiotically linked.

Rousseau was not actually a liberal at all. Classical liberalism, as expounded by Locke, relied on order and authority to safeguard individual freedom. In other words, it was a blueprint for civilisation. Rousseau, by contrast, was explicitly hostile to civilisation and valued no order or authority that was man-made. Accordingly, the freedom he envisaged for the child was not so much an expression of liberalism as of anarchy. One of the misapprehensions that bedevils contemporary debate is the false equation of 'liberal' with 'libertarian', of 'liberty' with 'licence' and of 'freedom' with 'anarchy'. The result is that anarchic impulses are defended on the mistaken premise that they are the bulwark against repression. But the very opposite is true. The destruction of order leads directly to repression. Indeed, even in *Émile* itself the freedom enjoyed by the eponymous hero is distinctly ambiguous. He may have been freed from lessons and books and allowed to roam about in a state of nature, but he is increasingly controlled by his 'liberal' tutor who manipulates his actions. So it has proved in the modern 'facilitating' classroom and among the 'liberals' in the universities and other institutions, who can countenance no alternative thinking to the prevailing orthodoxy.

This paradox greatly troubled Alexis de Tocqueville, the nineteenth-century French thinker whose masterpiece *Democracy in America* gave eloquent warning of the dangers that equality posed to freedom. He noticed that America's fundamental 'habit of mind' was 'the rejection of authority and the assertion of the right of private judgment'. More than anywhere else, it was the country where inherited authority counted for nothing; and yet 'in no country does there exist less independence of thought'.[10] This notorious conformity, he observed, was to be accounted for by that very absence of authority. Whereas 'older societies have found [it] in the traditions of antiquity, or in the dogmas of priests or philosophers, the Americans find [it] in the opinions of each other'. Accordingly, collective opinion was treated as holy writ and 'faith in public opinion becomes in such countries a species of religion and the majority is its prophet'.[11] There is such a thing as a despotism of the majority.

Rousseau, then, was a powerful influence behind the modern emphasis on feelings and experience as opposed to external rules.

However, despite their antithetical belief in civilisation rooted in order and authority, the early classical liberals laid the groundwork for a parallel collapse of authority into moral relativism, the subjectivity of values and the suspension of moral judgments. They certainly never envisaged such a development. Locke, and his two most prominent eighteenth-century followers Berkeley and Hume, never dreamed that their philosophy paved the way for the subjectivism and moral relativism, the denial of external authority, which would eventually undermine the moral order upon which their doctrines relied. But they never recognised the fatal subjectivity coursing through their own utilitarian attitudes, which replaced the question of whether something was right or wrong with the question of whether it was useful. So according to David Hume, the importance of the family, for example, wasn't based on religion or duty or an innate moral sense but on its utility. Chastity and fidelity were similarly merely useful because the 'long and helpless infancy of man' required parents for his survival. But, as the American thinker James Q. Wilson inquired in 1993, what if people come to think these things are no longer useful – as indeed they have done?[12]

The looming disaster of the loss of moral authority was recognised by philosophers such as Kant. He tried heroically to construct an absolute morality constructed on rationality and the authority of the individual, who was to be the measure of all things. Kant believed that in doing so he was shoring up the moral order; but like so many others, his contribution was a powerful force for destroying it. If the source of authority was located in the individual, there was no reason why any individual's morality should have precedence over that of any other.

Similarly Nietzsche, who was bitterly hostile to the Judaeo-Christian tradition and who was credited with announcing the death of God, famously warned of the terrible consequences of uncoupling religion from morality, when mankind would be 'plunging continually' and 'straying as through an infinite nothing'. Yet despite his anguish, he more than any other thinker did most to bring about precisely that state of affairs. He asserted that moral codes merely brought about the imposition of power by the weak upon the strong, and he poured scorn on the idea that morality could be founded upon inner sentiments. The fact that he recognised the dimensions of the destruction this implied did not detract from the scale of that destruction his own pronouncements had caused. If there was one single philosopher for the modern age, it was Nietzsche; and the doctrine he bequeathed us was nihilism.

The great paradox of the Enlightenment was that, in liberating human thought in order to enhance civilisation, it lit a slow fuse beneath it. In the words of Jonathan Sacks, the Chief Rabbi, the devastating effect was to focus simultaneously upon both the individual, detached from any historical context, and the universal.[13] This meant that power was located in both the individual and the state, with no role left for intermediate institutions such as families or voluntary associations. It set in train a line of modern thinkers from Nietzsche to Sartre and the thinkers of the British analytical tradition who all held that moral meanings were manufactured by the individual. Thus among philosophers in this century, A. J. Ayer dismissed moral talk as mere emotion and G. E. Moore believed that nothing mattered except states of mind. The finest intellects engaged in philosophical inquiry in Britain all agreed that moral choices were of as little consequence as choosing between chocolate and vanilla ice-cream. What mattered was only what-mattered-for-me.

Within three centuries, wrote Sacks, morality became a matter of individual will, preference, emotion or decision. Law became uncoupled from morality through Mill's dictum that everything should be permitted unless it did harm to others. The problem with that was that the definition of harm was subjective, and soon came to be redefined from what-was-wrong to what-was-right-for-me-and-what-I-could-get-away-with. The concepts of right and wrong became meaningless because they no longer applied to objective principles. There were no moral standards any more, only choices.

The Victorian Era

The curiosity was that this decline of morality took so long to occur. The Victorian period, after all, was subjected to repeated shocks and turmoil: the shattering impact of Darwin's *Origin of Species*, which destroyed the literal truth of the Bible; the Industrial

Revolution, the march of scientific discovery and the invention of mass democracy all shook the established order to its foundations and led to deep fears of political insurrection and moral anarchy. And yet as religious faith took a dive, the Victorians clung to rigid codes of morality. It was as if morality itself had become the new religion. Partly, this was because of that very fear and anxiety generated by these huge shocks to the system, the perception of the danger that now existed to the social order. But it was also the case that nineteenth-century doubt was quite different from late twentieth-century doubt. Darwin and Huxley made the Victorians doubt their beliefs in an agonising manner, but they then searched for different beliefs. They never denied the mind as a valid instrument for seeking the truth. The modern age, by contrast, does not believe in belief at all and understands that the only truth is that there is no truth. In the nineteenth century, even a non-believer and radical like the novelist George Eliot shared certain Victorian moral values. As the historian Asa Briggs recorded: 'She once deeply impressed a Cambridge don by telling him "with terrible earnestness" in the evening light of a college garden that whenever she heard talk of the three powerful Victorian words "God, Immortality and Duty", she felt that the first was inconceivable, the second unbelievable and the third peremptory and absolute.'[14]

Nevertheless, the attempt to hold back the tide of moral collapse was doomed. Nietzsche himself had mockingly predicted it, pouring scorn on those 'English flatheads' and 'little moralistic females *à la* Eliot' who thought it was possible to have morality without religion. 'When one gives up the Christian faith, one pulls the right to Christian morality out from under one's feet,' he wrote.[15] He was correct. Absolute but secular moral codes proved no match for an unholy trinity of pressures: the growth of free-thinking and rationalist movements connected with the rise of science, the increase in material benefits and comfort, and the fatal weakening of the Church of England caused by the schismatic Oxford movement led by John Henry Newman and Edward Pusey who departed for Rome.[16]

The result was the rise of the 'free spirits' of the 1880s and 1890s, the *fin de siècle* bohemians who behaved in a shockingly avant-garde way, particularly in sexual relationships. Their immediate guru was Havelock Ellis, the sex researcher and advocate of sexual freedom;

but their philosophical mentor was Nietzsche. As the historian Gertrude Himmelfarb has remarked: 'It is fitting that one of these new men, Havelock Ellis, should have written the first serious appreciation of Nietzsche in English, commending him for making the most determined effort to "destroy modern morals" and hailing him as the "greatest spiritual force since Goethe".'[17]

These 'new men and women' were the sexual anarchists of the late Victorian counter-culture who repudiated marriage, monogamy, conventional morality and, in many cases, conventional sexuality. They were self-consciously in revolt against the Victorian mainstream. At the same time, however, another movement of thought was providing an intellectual and moral challenge. This was the Romantic movement, a reaction against the ugly and brutal realities of Victorian materialism. Descended directly from Rousseau, it celebrated the beauty of nature and the cultivation of sympathy and the imagination as a moral response to a society that was becoming more and more 'scientific in method, rationalistic in spirit and utilitarian in purpose'.[18]

This cultivation of the feelings was helped along by the decline of both church and parental discipline. As life became more comfortable, there was a gradual disappearance of the old ascetic moral training designed to strengthen the will by weakening affection and desires. For the same reasons, parental discipline also declined, even before the arrival of Freud dealt it a mortal blow. Educating the desires took the place of educating the will.[19] The aim, as the historian of the period W. E. H. Lecky remarked, was 'to make the path of virtue the natural, the easy, the pleasing one' by appealing to the imagination and the emotions with ideals of action and character.[20]

The Twentieth Century

It was against this background that ideas of child-centred education began to flourish towards the end of the Victorian age and in the early years of the twentieth century. Descended directly from Rousseau, the idea grew that the performing arts were superior to academic knowledge and that children's creativity should be stimulated, not burdened by abstractions. At the end of the last century, these ideas began to accumulate critical mass. They revolved around a group of bohemians called the New Educational Fellowship. Abbotsholme, the first progressive secondary school was founded in 1889; it was followed in due course by Bedales. These were schools for wealthy fee payers and at that time the influence of their ideas upon state schools was nil. The great movement of thought into the educational mainstream happened after the Second World War. But according to Phil Gardner, a research fellow at Cambridge University who has studied the history of these ideas, the seeds of that post-war revolution were sown in the early years of this century, and for a variety of reasons.[21]

The first factor was the shattering disillusionment after the First World War. Everything before the war was thought to be wrong and everything after it had to be remade from scratch. In particular, there was a reaction against Prussianism which was regarded as militaristic. This developed by extension into a hostility towards elements of our education system which were seen as embodying these militaristic characteristics, as demonstrated by the precise and orderly rows in which schoolroom desks were arranged.

The second factor was the anti-urbanism of many progressives: the reaction of people like John Ruskin and William Morris against the industrialisation that was seen as blighting human potential, and the consequent rejection of the city in favour of simple rural pursuits. The third factor was the revulsion against child labour in the late nineteenth century, which led to a romanticisation of the child and a consequent receptivity to the ideas of Rousseau. The strong sense developed among writers of the early years of this century that children had to develop naturally because they were innately good while the influence of adults could only be bad. Adults were held to corrupt children when they interfered with their education. The analogy was drawn of the teacher as a gardener, with the only agent of growth the child itself.

However, elementary schoolteachers in those years weren't touched by any of this thinking. How could they be? They were too busy teaching classes of sixty or more. It was left to progressive thinkers like Bertrand Russell or A. S. Neill to start up their own little private schools. But then this ideology began to capture not the elementary school teachers but the education establishment, the teacher training colleges and the HMI, the school inspectorate. This was happening at a time during the inter-war years when teacher training was moving away from its apprentice base in the classroom to a new model in which the college was becoming the repository of training.

This new model of training was driven along by the increasing professionalisation of teaching. For most of the nineteenth century, the government had been happy with the stratified system of elementary and public schools. But in the 1890s concern started to grow that education in Britain was failing to promote national efficiency, particularly compared with America and Germany. To help combat this, it was thought that the social class of the elementary school teacher needed to be raised, and to do this education had to be made more attractive to the lower middle class. So the expansion of secondary education after 1902 was driven by the increasing number of children who stayed on to 18 in order to go to teacher training college. Teaching became professionalised.

The capture of those teacher training colleges by progressive values was therefore critical. Those teachers went to these colleges for two years during which they had a maximum of 6 to 12 weeks in the classroom. When they started work properly, however, they rubbed up against old-fashioned elementary school teachers who had been apprenticed and who had no time for these new-fangled ideas. So schools and colleges were running on quite different philosophical tracks. The college graduates had to put their training on one side and start teaching in grossly overcrowded classrooms where it was impossible to implement the ideas they had been taught.

A gap therefore developed between the classroom teachers and the educational establishment, which was in turn becoming more and more deeply convinced by these advanced ideas. In 1911, a book was published, *What Is And What Might Be* by Edmond Holmes, which was to have an enormous impact on that establishment. Holmes, who until 1910 had been the Chief Inspector of elementary schools, was a bohemian, a Buddhist and very sympathetic indeed to child-centred ideas. His book was a sensational repudiation of all those educational values embodied in the schools over which he had presided for the past 30 years. Holmes used his book to rip the schools apart. They ran mechanical, brutal regimes; they were dull and unimaginative and driven by pain and results. That was how things *were*. One school alone had impressed him, an anonymous prototype of those fringe establishments which allowed children to roam about in freedom and choose how and what they wished to learn. That was how things *might be*. He called this school Utopia.

In the book, Holmes outlined his view that the function of education was simply to foster growth. 'The process of growing must be done by the growing organism, the child, let us say, and by no-one else ... The forces that make for the child's growth come from within himself; and it is for him, and him alone, to feed them, use them, evolve them.' By contrast, many schools suppressed children's spontaneity in an attempt 'to turn the child into an animated puppet'. The child's natural activities were suppressed by being told what to say, think or do by the teacher and kept in place by punishment.

And Holmes inveighed against the practice of hearing children read aloud in terms that sound all too familiar. 'They are not reading to themselves, not studying, not mastering the contents of the book, not assimilating the mental and spiritual nutriment that it may be supposed to contain ...' They were reading aloud solely so that the teacher knew they could pass the inspector's test, 'being the outcome and expression of that complete distrust of the child which is, and always has been, characteristic of the popular religion and philosophy of the West.'

Checking on the child's progress was therefore, apparently, an expression of innate distrust of the child. But Holmes wasn't actually interested in the child's welfare at all. His philosophy purported to assert the primacy of the individual. But this was merely the means to a greater goal, which was nothing less that the complete repudiation of western civilised values. He believed that allowing a child's soul to develop would end individualism and materialism and bring about a society run on collective impulses. 'Blind, passive, literal, unintelligent obedience is the basis on which the whole system of Western education is reared,' he wrote. He wanted to end what he saw as the mechanical obedience to the Judaeo-Christian legacy, based on a distrust of man's nature, which stultified the inner life. He wanted to rid the whole culture of the 'virus of Occidentalism'. It was the West's externalism, its emphasis on results which had neglected what was inward and vital. 'The only remedy for those defects is the drastic one of changing our standard of reality and our conception of the meaning of life.' This involved a sustained attack on the 'false assumptions of Western philosophy, on the false standards and false ideals of western civilisation'.[22]

This explicit call to arms against western civilisation, using children as the weapon and the schools as the battleground, struck a powerful chord among the educational *cognoscenti*. As the foremost historian of progressive education, R. J. W. Selleck, has remarked, Holmes's book achieved an astonishing impact as a conversion of Pauline proportions.[23] This was largely because at that time elementary education was in a state of flux and uncertainty anyway, with the feeling that it was failing but with disagreements about what form it should take. So Holmes's views inspired a number of pioneer schools such as Summerhill, Beacon Hill and Dartington Hall. These establishments went in for teaching the arts and crafts, liked Freud and freedom and abhorred corporal punishment. The Malting House Garden School in Cambridge similarly stressed self-expression, the creative imagination and finding out rather than being taught. In 1933 the woman who ran it, the psychologist Susan Isaacs, was made head of the new Department of Child Development at the Institute of Education, London from where her ideas could be widely disseminated.

In 1920, Thomas Nunn, professor of education at London University, wrote *Education: Its Data and First Principles*, a textbook which put these ideas into a theoretical framework. It was a measure of the popularity of these ideas that his book was reprinted no fewer than 14 times by 1930. The theme common to all these different educational thinkers was the need to reduce the teacher's authority so as not to dominate the child, and to give children some say in the disciplinary powers of the schoolroom. Nunn's book aimed at reasserting 'the claim of individuality to be regarded as the supreme educational end'. Reading, writing and arithmetic were to be given less prominence in favour of creative subjects, and the most sterilising influence was considered to be 'rigorous conformity'. Selleck remarked: 'Individuality, in this sense, was opposed to, or at least not easily reconciled with society. The individual represented the free spirit which society might even curb or destroy.'[24]

So by the 1930s, progressivism was the orthodoxy among the universities and the inspectorate, but not in the schools which still pragmatically resisted such cranky views. They became institutionalised, however, in the 1938 Spens report on grammar schools and technical high schools, which embodied a clear repudiation of the principles of liberal education. The report said:

> It has been represented to us that secondary education is still strongly influenced by the discredited conception of an all-round training of the faculties and by the idea of a liberal education which corresponds neither to the circumstances of the pupils nor to the needs of modern civilisation . . . that there is a strong tendency to adjust the pupil to the curriculum rather than the curriculum to the needs and abilities of the pupil . . . We wish to reaffirm a view expressed in our Report on the Primary School (1931) in which we urge that the curriculum 'should be thought of in terms of activity and experience rather than of knowledge to be acquired and facts to be stored' . . . To speak of secondary school studies as 'subjects' is to run some risk of thinking of them as bodies of facts to be stored rather than as modes of activity to be experienced . . .

Fifty years later, a Conservative government was to discover that the views expressed in the Spens report were still almost impossible to dislodge.

But there was also another, immensely powerful strand of thinking that contributed to the development of progressive education. The Romantic influence was backed up by psychometrics, the science of measuring the mind, which held that children's intelligence was fixed. In the early years of this century, the new science of psychology was eagerly adopted by an education establishment wishing to improve the status of teaching by an infusion of scientific principles. Child psychology was thus rapidly incorporated into teacher training courses. Susan Isaacs herself was a child psychologist who, like many Bloomsbury intellectuals, was much influenced by Freud and the Swiss psychologist Jean Piaget whose views she did much to popularise. It was easy to see why psychology held considerable attractions for the progressive education movement, concerned as they both were with understanding and promoting the welfare of the individual. As Adrian Wooldridge commented in his book about psychometrics: 'Both psychologists and progressives started from the same assumption; that the mind to be educated, not the tradition to be transmitted, is the proper starting point of all instruction.'[25]

The work of Piaget, or perhaps to be more accurate a misreading of his work, played a key role in buttressing child-centred theories. From the early Enlightenment until the turn of the century, the plasticity and adaptability of children had always been stressed. Piaget destroyed this plasticity by asserting that mental development took place in fixed and specific stages. Educationists adapted this theory so that children in schools were only taught what it was thought they were developmentally ready to understand. Piaget thought that children actively constructed their own development. This led to the direct involvement by children in their own learning, given concrete experiences in order to find things out for themselves. The teacher's role thus became indirect. Instead of transmitting knowledge, the teacher merely facilitated the best experience and environment in which to foster the child's natural capacity to develop and learn.

When Piaget's theories were applied literally to the curriculum, they had a dramatic effect on teaching and learning, turning the emphasis onto how to learn rather than what to teach. The Nuffield Maths Project 1967, for example, said: 'The stress is on *how* to *learn*, not on what to teach. Running through all the work is the central notion that the children must be set free to make their own discoveries and think for themselves, and so achieve understanding, instead of learning off mysterious drills.'[26] But that was based on a mistake. Piaget's stages were an attempt to explore children's thinking, not a formula to promote their learning. The fact was that Piaget had kept at arm's length from education and had little to say about the educational implications of his work.

As one commentator has written:

There is another, rather large issue in Piaget's story which leads one to wonder why teachers ought to even contemplate taking him seriously and that is the issue of language. Without language, classrooms would be very strange places indeed. Teachers rely on language to communicate with children, explain things to them and probe their understanding. Yet Piaget did not really consider that much could be gained from purely verbal conversation, since he held that language merely reflects cognitive achievement rather than controls it. Again, this could be seen as minimising the teacher's role.[27]

But child-centred educationists had adopted Piaget whole-heartedly because they had come to see their own role as promoting general child development rather than initiating children into a cultural tradition. Piaget came in very handy. The fact that these educationists were doing violence to his theories as well as to the education of children was, of course, never acknowledged.

Progressivism

The influence of psychology on progressivism, however, went much deeper than Piaget. It is hard now to realise that the pioneers of intelligence testing, who have been excoriated and vilified since the 1960s for their 'élitist' views and their invention of the 11–plus, were originally lauded by the liberal left intelligentsia and were themselves firm supporters of child-centred education. It was only much later that their ideas came to be reviled as inimical to such values, a development which helps explain why, after the Second World War, these hitherto arcane theories finally became the mainstream orthodoxy.

Early in this century, liberal progressives favoured the measurement of intelligence as a means of securing social justice for everyone. Wooldridge has written about the IQ: 'The concept embodies a revolt against patronage and particularism and a plea for individual justice, demanding that everyone should be judged according to their natural worth. A society based on merit makes room for personal initiative and mobility, allocates social positions according to measured individual ability, and breaks down all bonds of patronage and tradition.'[28] So it was hardly surprising that meritocracy was embraced by the liberal intelligentsia such as Huxley, Darwin, Keynes and others, for whom the concept fitted naturally with philanthropy and social reform.

Before the Second World War, the meritocratic ideal was supported by the political left and opposed by the political right. After it, however, the political left turned away from upward mobility to equality of outcome or egalitarianism, a movement that was actually supported by the Conservative Party. In 1963 the Conservative Education Secretary Edward Boyle said in a speech: 'none of us believes in pre-war terms that children can be sharply differentiated into various types or levels of ability'.[29]

The War was a watershed. Earlier in the century, the meritocratic ideal had been supported by an evolutionary theory which entailed a hierarchy of success, by a faith in progress and in the scientific method to bring that about, and by a commitment to national efficiency. The psychologists, who wanted to spread middle class values to the slums, had enthusiastically supported child-centred theories of education and learning through play as the means of helping each child achieve its full potential.

But then, after the War, the cultural climate shifted dramatically. Gradual dissatisfaction with IQ testing and the 11–plus on the grounds that they weren't very accurate was translated into a repudiation of meritocracy itself. One of the reasons for this was the new receptivity of thinkers on the political left to Marxist analysis. When the British Communist Party under its education spokesman, Brian Simon, mounted an attack on intelligence tests, journals such as *The Times Educational Supplement* and the *New Statesman* gave this critique an enthusiastic reception. The view took hold that what had been presented as a means of opening doors for children had actually been a means of closing them. Intelligence tests served the needs of capitalism instead of the needs of children and were thus inherently anti-educational. More, intelligence tests themselves were held to be a weapon of the ruling class.[30]

In Britain, a new sociology of education developed which was positively anti-meritocratic. In keeping with the rest of sociology, it was heavily influenced by a Marxist, class-based analysis. Educational failure therefore lay in the class bias of the schools and tests were nothing more than a social construct. In getting grammar or spelling wrong, working-class children were resisting middle-class control. Failure was thus elevated to a weapon of the class struggle. Wooldridge commented in his book about psychometrics: 'For both intellectual and practical reasons many educators were convinced that the solution to apparently intractable educational problems lay not in perseverance and meritocratic classification but in the abandonment of the syllabus and the celebration of working-class culture, and sometimes even of counter-cultural rebellion.'[31]

The death knell of the meritocracy was sounded, however, not by a Marxist but by an activist from the ethical socialist tradition, Michael Young (now Lord Young of Dartington). In 1958 he wrote in his highly influential squib *The Rise of the Meritocracy* that this offending doctrine transformed so-called equal opportunity into unequal opportunity because it created an élite formed by those who had made the grade leaving the unsuccessful behind them. Thus the fateful elision was made between equality of opportunity, through which by definition some people rise above others, and equality of outcome. By the time the Labour government of Harold Wilson took office in 1964, the shift in the intellectual climate meant that not just the 11–plus but also the goal of a meritocracy itself was doomed. In *The Future of Socialism*, Anthony Crosland dismissed the established selective school system as unjust and divisive because it separated 'the unselected goats and the carefully selected sheep on the basis of tests which measured home backgrounds as much as innate ability'.[32] Crosland swallowed Young's analysis and intended that a common school would create a common culture and a classless society. In the spirit of Rousseau forcing people to be free, Crosland intended to force people to be equal, and the comprehensive school was to be the means to that end.

But this thinking was significantly muddled. The confusion between equality of opportunity and equality of outcome was never sorted out, not even, it seems, within Crosland's own mind. Professor A. H. Halsey, Crosland's influential adviser, has said: 'I was never an egalitarian. We wanted grammar schools for all.'[33] But Crosland famously declared: 'If it's the last thing I do, I'm going to destroy every fucking grammar school in England. And Wales. And Northern Ireland.'[34] More formally, however, his government circular 10/65, the turning point in the changeover to comprehensive schools, stated that its aim was 'to preserve all that is valuable in grammar school education for those children who now receive it and make it available to more children'. Two years earlier, indeed, Harold Wilson had said: 'The grammar school will be abolished over my dead body.' Despite his assurance, the comprehensive school programme set out to do precisely that.

Within a few years, moreover, Crosland's intention to preserve streaming inside the comprehensive schools lost out to mixed ability classes as egalitarianism marched relentlessly forward. A common culture was the aim; and child-centred education was the perfect educational creed to create it. So, although by now the psychologists, who had been the apostles of intelligence testing, were being run out of town, ironically the progressive ideology they had done so much to foster was being promoted as the orthodoxy for the primary and comprehensive schools by the ideologues who were in the very process of burying those psychologists' professional reputation as deeply as they could contrive.

The question still remains, though, why the War was such a watershed. After all, the educational élite had been enthusiasts for child-centred theories for decades. Why was it now that the classroom teachers finally embraced these ideas in their classrooms? There were undoubtedly many reasons which came together. The War had created a new sense of social solidarity. There was the desire to create a new society, a new social order based on fairness and social justice, accompanied by a deep sense of guilt among the middle classes that social class had deprived working-class children of their chance of success. There was the increasingly widespread impact of psychological theories of child-rearing which inspired deep guilt among adults about the harm done to children by repressing their personalities. There was the increasing influence of teacher training colleges as teaching struggled to give itself higher professional status. There was the panic among teachers at having to control classrooms of children who didn't want to be at school at all but were captive pupils courtesy of the raising of the school leaving age, a panic that made the teachers receptive to any suggested techniques that claimed to hold children's attention. And there was their losing struggle to maintain that attention in the classroom in the face of the collapse of parental authority and the rise of a peer culture that flouted all authority.

According to the educational historian Phil Gardner, there may have been another reason. Before the War, there was a significant distance between teachers and their pupils. There was no contact between home and school, no contact with the parents and no reports home. Most teachers didn't know their pupils very well. Sometimes, because the classes were so large, the elementary teacher might not even know a child's name. During the War, however, about one million children were evacuated along with their teachers. The children were separated from their parents. Those teachers found themselves therefore for the first time completely *in loco parentis* and had to sort out a range of social problems associated with these children's situation. They developed a pastoral role for the first time. As a result, they came to see themselves as central to their pupils' lives. They became much more aware of their welfare role. And because of this change in the relationship they became much more receptive to the theories that placed the development of the child at the centre of their concerns.[35]

These and maybe other factors came into play after the War to activate the progressive agenda that had been fermenting among the educational élite throughout the century. Whatever the reasons, from the 1950s onwards classroom teachers thought differently about themselves and their pupils and were then wide open to the heady cocktail of Freud, Piaget and above all the American philosopher John Dewey that was to revolutionise classroom teaching and leave children struggling in the ideological backwash.

John Dewey

It wasn't surprising, maybe, that Dewey came to have the influence he did on education in Britain. He was, after all, an American philosopher of education and Britain had consciously opted after the War to reconstruct its secondary schools along the American comprehensive model. Moreover, Britain was peculiarly susceptible to the influence of American culture through the closeness of the Special Relationship and the bonds of common language. But Dewey's influence on education, from his writings around the turn of the century, was malign, revolutionary and destructive, a fact he appeared belatedly to recognise when in 1938 he attempted to distance himself in his book *Experience and Education* from the wilder excesses of the educational doctrines he himself had set in train.

By the time he died in 1952, Dewey had become the most influential philosopher in the United States as well as a key thinker in education around the western world. It was, perhaps, not without significance that Dewey was born in 1859, that remarkable year which not only saw the birth also of the sexual revolutionary Havelock Ellis but which saw in addition the publication of three seminal texts: Darwin's *Origin of Species*, Mill's *On Liberty* and Samuel Smiles's *Self Help*. And although Dewey's educational philosophy was explicitly linked to an egalitarian, socialist agenda, its intellectual roots owed much to a way of thinking which viewed human beings as ahistoric members of an animal species, adapting themselves to their environment and their environment to themselves. Under his egalitarian banner, what he institutionalised in education was at root a savage doctrine of individualism which would effectively abandon children to a world without culture.

Dewey was the intellectual heir to Rousseau. With his combination of the extremes of pragmatism and relativism, he provided educationists with little other than a blueprint for philistinism. Children were to be taught to discount their culture and tradition. Instead, life was seen as a succession of practical problems to be solved by practical means. Knowledge and thought were therefore never matters of contemplation. Process was emphasised over product because no solution could be final, merely the prelude to a new set of problems to be solved.[36] So what counted was skill, not knowledge. 'Honesty, industry, temperance, justice, like health, wealth and learning, are not goods to be possessed as they would be if they expressed fixed ends to be attained,' he wrote. 'They are directions of change in the quality of experience. Growth itself is the only moral end.'[37]

The past was an unnecessary distraction. Anything that didn't bear on present problems should be ignored. So the idea that education should transmit the spiritual and intellectual heritage of a culture was dismissed out of hand. Values were to be reinvented with every fresh discovery by every individual child. Nothing else had any validity. 'The great advantage of immaturity, educationally speaking, is that it enables us to emancipate the young from the need of dwelling in an outgrown past,' he wrote.[38] But, as the philosopher Anthony O'Hear has written, most of what we do 'has far more to do with the discovery and recovery of meaning in our lives; and for this task acquaintance with and respect for the traditions through which meanings are channelled to us is essential'.[39]

Dewey insisted that a child's education was only meaningful when the child attempted to solve problems arising from his own experience. What was taught had to be relevant to the immediate environment. The class had to be a communal group solving problems from life outside the school, with the teacher merely leading group activities. All thinking was research and all research was original. Children had to think out all solutions for themselves, with the teacher becoming as much a learner as the child. Although Dewey stressed that the teacher had a role in arranging these activities and also in directly teaching reading and number work, he never understood that he was knocking all the props of authority itself from beneath the teacher's feet. He explicitly rejected the idea

that the teacher had knowledge to impart. On the contrary, to do
so would be to act as an 'external boss or dictator'.[40] 'In the last
analysis, all the educator can do is to modify stimuli,' he wrote.[41]

Dewey's doctrines were based on the idea of personal growth.
But as the historian Richard Hofstadter argued: 'Since the idea of
growth is intrinsically a biological metaphor and an individualistic
conception, the effect of this idea was of necessity to turn the
mind away from the social to the personal function of education;
it became not an assertion of the child's place in society but rather
of his interests as against those of society. The idea of growth invited
educators to set up an invidious contrast between self-determining,
self-directing growth from within, which was good, and moulding
from without, which was bad.'[42]

In 1938 Dewey, apparently horrified by the extremes of the
progressive education he had set in train, tried half-heartedly to
row back from what he had written. 'Many of the newer schools,'
he complained in *Experience and Education*, 'tend to make little
or nothing of organised subject-matter of study; to proceed as if
any form of direction and guidance by adults were an invasion
of individual freedom, and as if the idea that education should
be concerned with the present and future meant that acquaintance
with the past has little or no role to play in education.' But his
protestations that this was simply the result of extremists distorting
his work didn't wash. What he saw happening in the schools was a
faithful translation of his own precepts into practice. Just like many
teachers of today, he represented himself as a moderate, steering
a narrow course between extremes. But just like those modern
teachers, he never grasped that he was in the process of moving
the goalposts of extremism into the disappearing middle ground.

He was, in fact, an extreme reductionist. He couldn't tolerate any
distinctions between individuals; in his view, any division between
the educated and the uneducated would lead to unacceptable
divisions between them. All classifications and stratifications were
to be ruled out, as were any inner or spiritual qualities which
couldn't be shared. Not surprisingly, this train of thought led
directly to the destruction of learning and of culture, not to
mention the idea of moral authority itself, the understanding
of common and eternal values that needed to be transmitted
through that culture. What was astonishing and tragic was that
this appalling, amoral philistinism was so eagerly swallowed by

the education establishment, which to this day cannot understand its implications. They appear to have been seduced by the rhetoric of democracy in which Dewey clothed his reasoning, and the false equation of democracy with an egalitarianism which ordains that all individuals must be the same.

For Dewey, education was an explicitly political project. It was therefore perhaps inevitable that, when the British political and educational élites decided to turn schools into an explicit political project to restructure society to egalitarian ends, Dewey's philosophy offered a ready-made orthodoxy for the pedagogy and the curriculum that such a project required. What was thus set in train by this fateful marriage of interests was not merely the abolition of meritocracy, the prohibition on human striving and the outlawing of high expectations. The stage was set for a wholesale meltdown of education and culture through the mechanisms of moral and cultural relativism which – to everyone's subsequent professed surprise – was to leave both adults and children floundering in a moral and intellectual quagmire.

Carl Rogers

The disciple who did most to translate Dewey's vision into practice was one of his graduate students, Carl Rogers; and the main conduit in Britain for transmitting the Dewey/Rogers philosophy into the classroom was the Schools Council, a predecessor of the curriculum authority of today. Despite the fact that it was legislatively toothless, the Schools Council during the 1960s and 1970s had a huge effect on education in Britain by disseminating these American ideas. Its humanities curriculum project remains influential to this day in having established a moral relativist agenda. The ostensible aim, particularly after the school leaving age was raised in 1973, was to make subjects more exciting and controversial to hold children's attention. But Lawrence Stenhouse, the director of the Council's humanities project, drew on American works which said it was impossible in a pluralist society for one generation to transmit any settled values to succeeding generations. Stenhouse himself realised that the attempt to transmit many different values was not a value-free project. It was relativist.[43]

It was Carl Rogers who developed the notion of the teacher as facilitator and insisted that learning had to be fun. Conventional education he saw as the illegitimate exercise of power. Its strategies were the rewards of grades and vocational opportunities and the penalties of aversive, punitive and fear-creating methods such as exam failure and public scorn. He caricatured the relationship between teacher and pupil as one marked by distrust, ridicule and fear of failure. In its place Rogers proposed trusting pupils to learn for themselves, with the teacher merely facilitating the process. In 1983 he wrote:

> The facilitative teacher shares with the others – students and possibly also parents or community members – the responsibility for the learning process. The facilitator provides learning resources from within herself and her own experience, from books or materials or community experiences . . . The student develops her own programme of learning, alone or in co-operation with others. Exploring one's own interests, facing this wealth of resources, the student makes the choices as to her own learning direction and carries the responsibility for the consequences of those choices . . . Learning from each other becomes as important as learning from books or films or work experiences. It can be seen that the focus is primarily on fostering the continuing process of learning. The content of the learning, while significant, falls into a secondary place . . . Self-discipline replaces external discipline. The evaluation of the extent and significance of the student's learning is made primarily by the learner . . . The learner is the centre. This process of learning represents a revolutionary about-face from the politics of traditional education.[44]

In 1968, Rogers went so far as to predict the disappearance of teaching. Writing in an academic journal, he also denigrated subject matter, promoted learning at a child's own pace and supported the exploration of feelings in the classroom and suggested that students might never graduate. The student would

> learn to be an individual, not a faceless conformist. He will learn, through simulations and computerised games, to meet many of the life problems he will face. He will find it permissible to engage in fantasy and day-dreams, to

think creative thoughts, to capture these in words or paints or constructions. He will find that learning, even difficult learning, is fun, both as individual activity and in co-operation with others. His discipline will be self-discipline. His learning will not be confined to the ancient intellectual concepts and specialisations. It will not be a preparation for living. It will be an experience in living. Feelings of inadequacy, hatred, a desire for power, feelings of love and awe and respect, feelings of fear and dread, unhappiness with parents or with other children – all these will be an open part of his curriculum, as worthy of exploration as history or mathematics . . . [45]

The outcome of Rogers's teaching was the development of the 'values clarification' movement under one of his disciples, the influential American educationist Sidney Simon, who preached in essence 'if it feels good, do it'. Under this banner, teachers were to abdicate all moral responsibility for what they taught their pupils. As one American commentator explained: 'In discussing value-rich areas . . . (eg sex or personal habits) the teacher accepts all answers and does not try to impose his or her own views on the students . . . Responses are not judged as better or worse; each student's views are treated with equal respect . . . choosing freely is considered better than simply yielding to authority or peer pressure . . . if we uphold free choice, then we value autonomy.'[46]

What a child chose to do was not as important as the manner in which that decision was arrived at. As another American commentator noted: 'Health educators must not be concerned with the particular behaviour of their clients, but rather with the process used by their clients to arrive at that behaviour. For example, if a client (student in a school, adult in a nursing home program etc) chooses to smoke cigarettes, but has made that decision freely, the health educator has been successful.'[47] It would appear to follow, then, that since obeying any authority was intrinsically bad, it would be considered desirable for someone to decide freely to burgle a house or rob an old lady, or willingly impregnate seven teenage girls and abandon them all. Freedom had become the value that trumped all others. But it was not the kind of freedom that would have been understood by the early liberals. This was not freedom; it was licence.

William Coulson

When these anarchic and nihilistic concepts were actually applied to American schools, the result, not surprisingly, was a collapse of learning and a complete breakdown of order. The evidence so horrified Rogers's associate for two decades, the psychologist William Coulson, that he eventually came to repudiate the whole Dewey legacy in the strongest possible terms. In 1994 Coulson wrote a paper in which he denounced the visible effects. He pointed to deplorable standards of work arising from the retreat from substance. Students were taking nothing seriously, including themselves. Children were running wild because their teachers had said the main lesson was 'do your own thing' which led to situations in which they called each another names such as 'crazy stupid nigger' and covered the walls with obscene graffiti. And the justification for this mayhem, wrote Coulson in horror, was that it was all in the name of 'personal growth'.

'Dewey's experimentalism – the idea that the fund of accumulated knowledge should no longer control education – has turned out to exact a terrible price from American schools and their patrons,' he wrote. He went on to demolish the value choices that schools were now offering children over sex and drugs. 'The chaos that could be expected to follow from teaching that the proper side of the road is a matter of choice can be expected to follow in other subject matter areas where choice is emphasised. Decision-making lessons waste classroom time. In the case of drug education, it must be said that the essential task is to persuade children to recognise that they must *never* take drugs. The techniques of persuasion are well known and they, not decision making, ought to be applied,' said Coulson.

'In matters of importance, including in a democratic society, we teach obedience – if we love children. For they compare what we insist upon – school attendance, stopping at the red light, wearing protective equipment in football – with what we seem to allow. (In recent years we seem to have specialised in allowing them to choose whether or not to partake of sex and drugs.) Making a comparison between what we command or condemn on the one hand and what we allow them to choose on the other, they draw the false conclusion that what we allow must not be important – or we wouldn't allow it; we would tell them the truth; we would be insistent, not permissive.

In other words, decision making is for adults. Loving parents and good schools have always tried to improve children's vision – in this sense, that we have wanted our children to become reliable in distinguishing right from wrong . . . children deserve to be told what is the case. They deserve a factual education. Time is wasted teaching them to make decisions they shouldn't even entertain.'[48]

Coulson had understood the crucial point that has still not been understood in Britain. Democracy is for adults. Choice is an adult behaviour. Expecting children to make choices about life not only before but *instead of* teaching them what life is all about is not merely an irresponsible abandonment of children by the adults involved but imposes upon children an adult responsibility for which they are simply not ready. The consequences of a child's 'choice', whether it is a choice to skip a tedious mathematical exercise, smoke cannabis rather than tobacco or sleep with a boyfriend, may cause that child harm through ignorance – harm it was once considered an adult's unambiguous duty to prevent. Child-centred education entails the destruction of childhood.

Lawrence Stenhouse

Despite such obvious objections, the revolutionary deconstruction of education and discipline being promoted by these American educators was swallowed by the Schools Council and enthusiastically promoted as the answer to the British schoolteacher's prayers, infusing curriculum guidance in projects that were then taken up by the schools and are still in use today. In a paper delivered to the Fourth International Curriculum Conference in New York in 1969, the Schools Council's Lawrence Stenhouse renounced the teacher's role as an authority on 'controversial issues'. These he defined as war, education, the family, relations between the sexes, people and work, poverty, living in cities, law and order and race relations. He said:

> For us, understanding means more than a sum of information, affective responses and skills. It implies a structuring of these appropriate to the situation of the person who is studying it. Understanding is the achievement of an interpretative map answering both the needs of the situation and the needs of the person who is attempting to understand it.

It must be made clear that the project is not value free . . . the changes which we specify are not changes in terminal student behaviour but in the criteria to which teachers work in the classroom. These changes are defined by enunciating certain principles of procedure or criteria of criticism which are expressions of the aim. They are, if you like, specifications of a form of progress. The pattern of teaching must renounce the authority of the teacher as an 'expert' capable of solving value issues since this authority cannot be justified either epistemologically or politically. In short, the teacher must aspire to be neutral.[49]

Thus, teachers trying to give moral guidance were to be castigated for indoctrination. But of course it was these so-called 'liberals' who were doing the indoctrinating. Stenhouse devised packs of materials on each of his controversial topics that he had judged too dangerous for teachers to teach. His race pack included extracts from Hitler's writings and was blocked by the teacher unions. Taken to its logical limits, his philosophy meant that no teacher could teach that racial hatred was a bad thing, or that poverty was undesirable, or that lying or stealing were wrong. In short, it meant destroying the teacher's ability to teach right from wrong.

Chapter Eight

THE DESTRUCTION OF MORALITY
Individualism

The extreme, value-neutral doctrines handed down to the schools in Britain and America were an expression of a profound cultural shift in which to be judgmental was to be accused of attempting to limit the perfect entitlement of individuals to make up the rules for themselves. It was the apotheosis of individualism, and it had become the defining value of the culture. And since education is the process by which a society hands down the values of its culture from one generation to the next, that was what was now being handed down. Individualism was indeed the one thing that *could* be handed down in a culture which no longer believed in the validity of handing down any kind of tradition. Paradoxically, the only certainty that could be allowed was that there were no certainties, and the only judgment was that there must be no judgments. The individual self was the summation of all that was true and valuable, choice was sacrosanct and rights trumped all. The concept of a common culture, common bonds and a shared story that we needed to tell each other as human beings in order to survive as a co-operative enterprise became synonymous with oppression. The individual stood entirely alone. The 'me society' became synonymous with democracy and freedom itself.

The result is that morality has now become a subject to be discussed only by consenting adults in private. We no longer even possess the language in which we can talk about it as a society. Morality is a modern boo-word, or at best considered irrelevant. So it has become difficult even to discuss the principles by which we live. To do so immediately invites contempt, ridicule or censorship

– all of course in the name of freedom. During 1995, the BBC held a seminar to consider the difficult issue of where to draw the boundaries on taste and decency in broadcast programmes. The dominant view was that judgment was all but impossible because of the collapse of a consensus about what was or was not acceptable. Society was now too pluralistic and fragmented to allow for any one moral point of view to hold sway.

The consequence of that attitude was the BBC1 programme *Confessions*, which awarded prizes to contestants who confessed to past indiscretions. Complaints about this programme which were upheld by the Broadcasting Standards Council were directed at items featuring rewards for people who admitted setting fire to a young woman's hair; reversing a car into a policeman; stealing cutlery from a supermarket; placing an able-bodied child in a wheelchair to avoid a queue; burying a tortoise alive; and stealing clothes from a launderette. Does pluralism really mean that there can be no consensus that it's wrong to reward people for boasting about illegal or anti-social activities? Or that such activities are in themselves wrong? Of course not. Yet the BBC could not admit either category was wrong because it could not admit that *anything at all* was wrong. To say that anything was wrong would break the 'liberal' code of absolute non-judgmentalism. So its staff had to use other vocabulary to outraged viewers who complained: words such as 'unwise', or 'inappropriate', or 'a mistake'. But the word they couldn't use, even at the seminar, was 'wrong'.

A common view among young people on both sides of the Atlantic is that there are no absolutes and that morality is a matter of personal taste. In the introduction to his book *The Closing of the American Mind*, Professor Allan Bloom wrote: 'There is one thing that a professor can be absolutely certain of: almost every student entering the university believes, or says he believes, that truth is relative.' Such students, he wrote, came from every background imaginable. 'They are unified only in their relativism and in their allegiance to equality. And the two are related in a moral intention. The relativity of truth is not a theoretical insight but a moral postulate, the condition of a free society, or so they see it.'[1]

In Britain, the same phenomenon applies. Marianne Talbot, a philosophy lecturer at Brasenose College, Oxford, has written that students believe that no-one can ever claim to be right.

Many of the young have been taught to think *their* opinion is no better than anyone else's, that there is no *truth*, only truth-for-me. I come across this extreme relativist view constantly – in exams, in discussion and in tutorials – and I find it frightening: to question it amounts, in the eyes of the young, to the belief that it is permissible to impose *your* views on others.

The young have been taught, or so it seems, that they should never think of the views of others as false, but only as *different*. They have been taught that to suggest someone else is wrong is at best rude and at worst immoral: the truth that one should always be alive to the possibility that one is wrong has become the falsehood that one should never be so arrogant as to believe that one is right.[2]

As Talbot pointed out, this view was based on two fallacies. The first was that even if religious authority had been destroyed, there was no reason to think we were not bound together by our common humanity and the dictates of human survival to recognise common moral codes. The second fallacy was to think that respecting the right of others to hold different views from ourselves meant we must not say that they were wrong and that we were right. But as she also observed, the reason young people had fallen into such grievous and potentially catastrophic error was because they had been led there by their teachers.

Among teachers, there is an overwhelming reluctance to 'impose' a point of view upon their pupils. An example of this attitude among teachers appeared in an article written in 1993 by Graham Haydon, a lecturer in the philosophy of education.

It still must be said forcefully that accepting uncritically what someone tells you because they are seen to be in authority is not a good thing. . . . Doing what is right cannot be a matter of doing what one is told. Schools must produce people who are able to think for themselves what is right . . . It will not take an exceptionally clever pupil, or an exceptionally bolshie one, to ask: 'How do we know this is right or that is wrong?' Any pupil who is being taught to think ought to be asking such questions. And the same pupil ought to see that 'Because I say so' is not an acceptable answer. Nor is 'because these are the values of your society'. When exposed to a little more teaching of history, perhaps, this pupil will see that by such an argument the values of slave states and Nazi states would have to be endorsed.[3]

But this reasoning was specious and dangerous. Of course pupils should be taught to think for themselves and should understand the reasoning behind the moral rules they are taught. But the answer to the pupil's question was surely: 'Because these are the values of our common humanity and are the basis of human flourishing.' Not to answer in that way leaves it up to the pupil to decide that sometimes the end may justify the means, for example; or that stealing may be permissible if you are poor; or that lying to Parliament is justified to protect the sale of arms. Far from preventing us against succumbing to totalitarian regimes, it provides the means to endorse them. Quite contrary to Haydon's own example, it might legitimise Nazism because it would say in effect that the Nazi view of the world was merely what the Nazi thought was right and was therefore as valid as the pupil's own view. If there is no absolute right, it follows that there can be no absolute wrong, just as if there is no absolute truth there can be no lies. If rightness is simply what is right-for-me, then who is to say that Nazism was an absolute wrong? But of course it was just that, because it offended against common moral codes of humanity. Haydon's attitude, however, opened the way for his pupils to say that racial prejudice was no less right than tolerance; or that it was permissible to kill people because they were genetically imperfect. Moral relativism leads directly to despotism and tyranny. It was no accident that Nietzsche, in whose long shadow our relativist society was formed, represented a significant milestone on the road to the Final Solution.

So unacceptable has judgmentalism become that it is now considered to be worse than the behaviour it seeks to condemn. Thus we cannot now condemn adultery, say, without being in turn condemned for presuming to pass such a judgment, regardless of the harm the adultery may have caused. To pass a judgment has thus become more unacceptable than breaking up a family. But as Jonathan Sacks has observed, most people who seek to pass judgment do so not to condemn but to guide, to give people a map for their journey through life. Since we can all now choose between moralities, and behaviour is ascribed to a variety of causes other than human will, it is now assumed that our behaviour is no longer a proper subject for praise or blame or judgment of any kind and responsibility is put in doubt.

It was John Stuart Mill, in his book *On Liberty*, who argued against the 'tyranny of the majority'. Liberty depended on setting limits to government; it may only pass laws to prevent harm being done to others, not those which intrude into private life. But he also argued that censure could be oppressive. 'Protection, therefore, against the tyranny of the magistrate is not enough: there needs protection also against the tyranny of the prevailing opinion and feeling; against the tendency of society to impose, by means other than civil penalties, its own ideas and practices as rules of conduct on those who dissent from them.'[4] In other words, judgment was not permissible. But without judgment there can be no moral teaching and no moral community. Lifestyle choices become immune to criticism, and the way is open for every minority to insist that its lifestyle must enjoy equal rights, whatever the social consequences.

But in fact, non-judgmentalism is actually a most partial and particular creed. Scratch the most resolute non-judgmentalist and you will almost invariably find a very strong set of moral prohibitions – against sexism, racism or prejudiced behaviour of any kind. Even among Marianne Talbot's students, it's a fair bet that they would all agree that prejudice against black people, women, disabled people or homosexuals is absolutely wrong. The explanation for this apparent conundrum is that out of the relativist vacuum has developed what the historian Gertrude Himmelfarb has called 'telescopic morality'.

'Telescopic morality,' she explained, 'disdains the mundane values of everyday life as experienced by ordinary people – the "bourgeois values" of family, fidelity, chastity, sobriety, personal responsibility. Instead, it embodies a new moral code that is more intrusive and repressive than the old because it is based not on familiar, accepted principles but on new and recondite ones, as if designed for another culture or tribe.'[5]

Thus, promiscuity is not condemned, rather those men who fail to obtain consent for every phase of sexual intercourse; nor is drunkenness, rather those who take advantage of it; and crimes such as burglary or theft are not held to cause as much irritation as infringements of speech codes or rights of association. And through the other end of the telescope, the values that are extolled are those associated by definition with the status of the victim, people who for one reason or another have been marginalised from the mainstream, and have often had legitimate cause for complaint as a result. Such details as what may actually have happened to them, however, are now considered irrelevant. The values they represent as a class are judged to be worthwhile simply because as a class such people were once excluded from the mainstream. They were once considered to be 'deviant', in one way or another, simply because they did not correspond to the norm. But moral relativism meant that there were no standards any longer by which the norm could be measured, and so the very concept of deviancy was no longer permissible. So what was previously considered deviant became regarded as normal. Conversely, what had previously been regarded as 'normal' – Himmelfarb's bourgeois virtues – became unacceptable. Thus the traditional family, for instance, is excoriated as oppressive, while clubs promoting sado-masochistic sex are advertised in listings magazines that can be purchased at any newsagent.

Not that any of this has made us into a more tolerant society: quite the reverse. For as the political theorist John Gray has argued, toleration presupposes a prior judgment. To be tolerant, we have to know what is good or true in order to put up with behaviour that does not adhere to those values. We have to be able to tell good from bad. If we have no idea of what is good or true in the first place, then what follows is not tolerance at all but indifference.

'The idea of toleration goes against the grain of the age because the practice of toleration is grounded in strong moral convictions,' wrote Gray. 'Such judgments are alien to the dominant conventional wisdom according to which standards of belief and conduct are entirely subjective or relative in character, and one view of things is as good as any other.'[6] Tolerance was thus offensive to the 'revisionist liberals' such as the philosopher John Rawls or the jurist Ronald Dworkin. For them justice – the 'shibboleth of revisionist liberalism' – demanded that governments practised neutrality, not tolerance. So governments were reluctant to discriminate in favour of or against any kind of behaviour because equality demanded equal respect for every type of behaviour. This, wrote Gray, was only likely to breed more intolerance through the creation of group rights, which made issues non-negotiable and permitted only victory or surrender, leading to the dangers of 'arbitrary privileging' or reverse apartheid.

Collapse of the Church

The collapse of moral authority is now widespread among British institutions, inspiring in one particularly memorable phrase the image of moral relativism as 'the spongiform creed of the British establishment'.[7] Nevertheless, the collapse has been sudden. Nietzsche's prophecy has taken a long time to reach maturity. Like the linked deterioration in education, the effects of moral relativism started to become apparent in the 1950s and 1960s, accelerating sharply during the 1980s. Yet its intellectual and philosophical roots have been in place since the middle of the last century, when – in circumstances analogous in many ways to those of today – the Victorians nevertheless managed to remoralise a society that was similarly beset by the chronic insecurity and instability that accompanies profound social and intellectual change. So what caused morality subsequently to implode?

One powerful contributory factor has been the sheer pace of scientific and technological change. The world has shrunk; international travel, global business and the impact of television and telecommunications have all meant that we are now exposed all the time to other people's lives and behaviour, inviting comparison and stifling judgments. The rise of the consumer culture too, which we shall examine in more detail later in the book, has contributed powerfully to the creation of unprecedented peer pressure, particularly upon young people, which has significantly weakened the authority of parents, teachers, police officers and other previously 'judgmental' figures. But although such pressures apply to everyone it is noticeable that certain minority groups in Britain have retained a clear set of judgmental values in opposition to the prevailing cultural mood. The usual reason for that is that they adhere to codes laid down by their religious leaders. The reason why the majority culture has by contrast succumbed to all those secular pressures is because of the critical collapse, at some point in this century, of the moral authority of the Church of England.

In the Victorian age, despite the shattering challenges to faith itself, churchgoing remained strong. Between the two of them, Evangelicalism and Methodism provided the main impetus for the remoralisation of Victorian Britain. Yet since the end of the Second World War, the Church of England has lost its authority and its congregations. Many reasons have been advanced for this decline into – as the sociologist of religion Grace Davie has put it – the church from which most people choose to stay away.

It's certainly not true that the empty pews indicate the state of unbelief. On the contrary, in recent times there has been something of a religious revival on the more fundamentalist wings of all religions. The Evangelical churches with their uncompromising fundamentalism are now packed to the rafters, while the liberal, temporising Anglican churches wring their hands over their disappearing congregations. But this revival is itself a fairly recent development. There is reason to suppose that it is a reaction to a growing mood of disillusionment with the promise of progress once held out by the scientific advances that sprang from the Enlightenment legacy. People are no longer convinced by progress; they have become too concerned by the corresponding damage done to the environment, and also that the promise of happiness held out by the paradigm of progress is not fulfilled. There has got to be more to life than shopping malls and the Internet. There does now appear to be a need for certainty and purpose in an age of manifold ambiguities. But the Evangelical churches are still the fringe. They do not have the impact upon mainstream culture enjoyed by the Church of England, whose pronouncements not only reflect that culture but help to form it. Yet the Church of England isn't using that influence to hold back the tide of moral relativism. On the contrary, its pronouncements upon the family in particular display a visceral horror of judging any kind of family structure as better or worse than any other.

Dr Alan Storkey was a dissenting member of a committee on the family set up by the Church of England's Board of Social Responsibility. He refused to sign the committee's report, *Something to Celebrate*. Although this report paid lip-service to the traditional family structure of two parents looking after their own children, it refused to say unequivocally that fragmented families were undesirable. Its message was blurred. Storkey remarked:

The view that prevailed was that what mattered was the quality of relationships and not the structures. Quality was defined entirely in terms of the way individuals oriented themselves to their lives and what was right for them so the definition of the family was focused subjectively. Expectations of this report within the Church were high. Discussion of a range of social issues had been put on hold for years because the committee was going to produce the definitive response. But as its deliberations proceeded, there was growing alarm about its likely conclusions. Some members of the Board of Social Responsibility were not all happy with what they were hearing and reading and early drafts were torn up and rewritten.

The authors of the report were taken aback by the outrage it provoked within the Church. It had conspicuously failed to provide the guidance for which everyone was waiting. Eventually, although the Synod approved it the report was shelved after the Archbishop of Canterbury, Dr George Carey (who in some of his own speeches had been taking a clearer set of moral positions) effectively distanced the Church from it.

The Reverend Sue Walrond-Skynner, a family therapist on the committee, remarked on radio that the Church was not in the business of making moral judgments. But there was a real sense of betrayal. 'We've had 15 to 20 years of the deconstruction of values,' said Alan Storkey. 'The agenda now should be the reconstruction of values.' But this was hardly the first time the Church had taken a relativist position on marriage. It was, after all, the Church of England report *Putting Asunder* in 1966 which had inspired the move towards the abolition of fault in divorce by making irretrievable breakdown the sole requirement.

So why had the Church fallen into the byways of moral relativism instead of asserting a clear and unambiguous leadership role? One set of answers can be found in both its doctrines and its position as the established church. 'Its doctrines are based upon non-judgmentalism,' said Storkey. 'It's a central tenet of Christian teaching that the church must not judge others. So there's an overwhelming fear of passing social judgments at all.' This tendency has been exacerbated by the fact that as the established church, it feels bound to be inclusive and to reach out to embrace a rainbow of attitudes and behaviours. It was founded, after all, on precisely those qualities of ambiguity, negotiation and compromise which the non-established churches so despise. These churches can afford the luxury of certainty which the Church of England finds so elusive.

During the century, the Church sustained a further series of blows. It had to deal with the shattering impact of the First World War which destroyed faith in authority and whose resonances are still being worked through today; and later it had to try to reconcile the Holocaust with the survival of religious faith. In addition, it lost its influence over people's everyday lives when the welfare state took over schools, hospitals and welfare for which the Church had previously assumed responsibility. But above all, it had never really recovered from the devastating damage done to faith itself by the shaking of the theoretical foundations in the nineteenth century. Weakened by all these associated shocks, it was therefore in a vulnerable position to confront the tide of individualism which swept in after the end of the Second World War. To survive, the Church of England decided to accommodate itself to the secular mood; but in doing so, it not only weakened itself but gave a powerful boost to those very relativist currents it was so anxious to contain.

Psychopolitics

Perhaps the most important development to affect the Church in the middle of this century was the rise of the therapeutic culture. As organised religion was knocked off its pedestal and morality became the subjective expression of people's own desires, faith in the self replaced faith in God. Instead of asking the question: 'Is it right or wrong?' people now asked: 'How does it feel for me?' The high priests of the new religion were accordingly not vicars and bishops but psychoanalysts, psychologists and psychiatrists.

As the American cultural commentator Christopher Lasch put it, this was a new world view. American psychiatry proposed not just to treat patients but to change cultural patterns and spread the new gospel of relativism, tolerance, personal growth and psychic maturity. The mental health movement took over morality, philosophy and religion. Heidegger, Martin Buber and

Paul Tillich had after all redefined religion as psychotherapy. Now society itself had become the patient, and the medicine was amorality. In 1946, one of the guiding spirits of the World Health Organisation, the Canadian psychiatrist C. B. Chisholm, wanted to create more inclusive identities. 'Would it not be sensible,' he suggested, 'to stop imposing our local prejudices and faiths on children and give them all sides of every question so that in their own good time they may have the ability to size things up and make their own decisions?' Morality was redefined as a disease. 'The reinterpretation and eventual eradication of the concept of right and wrong, which has been the basis of child training, the substitution of intelligent and rational thinking for faith in the certainties of the old people' were ways in which mental hygiene could free humankind 'from its crippling burden of good and evil'. With widespread support for this cultural revolution among liberals, discrediting parents, priests and lawgivers as authoritarian, the mental hygiene movement was in the vanguard of revolt against conventional morality.[8]

The bible of the new psychopolitics was a book called *The Authoritarian Personality*, which was published in 1950 by the Marxist sociologist Theodor Adorno and his colleagues. This vilified bourgeois culture and held that conforming to middle-class norms indicated nothing less than a predisposition to fascism. The pointers to an anti-democratic personality were obedience and respect for authority. Submission to the authority of teachers was linked with 'ethnocentrism', a term they invented but which is now in common use and which included identification with family or faith. Any opposition to the subjective or the imaginative was seen as a form of incipient fascism. By contrast, the liberal was defined as 'the compassionate'. Prejudice was defined as a social disease. By defining personality attributes as potentially fascist, it promulgated the belief that merely to hold certain attitudes constituted actual or potential oppression.

As commentators have remarked, the tremendous influence of this book turned the western world upside down. Everyday family life was demonised while attacks on the social structure were redefined as therapeutic. Sexual repression, the bourgeois family, religious faith and conventional morality could be equated with Nazism. And despite its Marxist antecedents, this analysis was taken up by the new liberal consensus. In *The True and Only*

Heaven, Christopher Lasch wrote that by identifying the liberal as the antithesis of the authoritarian, Adorno and his colleagues 'defended liberalism not on the grounds that liberal policies served the ends of justice and freedom but on the grounds that other positions had their roots in personal pathology. They enlarged the definition of liberalism to include a critical attitude towards all forms of authority, faith in science, relaxed and non-punitive child-rearing practices and flexible conceptions of sex roles.'[9] The egoistic liberal emerged redefined as the ultimate individualist assuming a posture of moral indignation. The emphasis was now on having the right feelings. The early identification of liberalism with the power of reason had been junked. On the contrary, the liberal's moral courage was 'often beyond his rational evaluation' and he often had trouble keeping himself 'under control'. The picture thus painted was not of a liberal at all, but of an adolescent. Yet that was the image eagerly adopted by those who then claimed the liberal high moral ground.

Common Bonds

In a lecture he delivered in 1959 entitled 'Morals and the Criminal Law', the eminent jurist Lord Devlin sounded a note of warning. 'Societies disintegrate from within more frequently than they are broken up by external pressures,' he said. 'There is disintegration when no common morality is observed and history shows that the loosening of moral bonds is often the first stage of disintegration, so that society is justified in taking the same steps to preserve its moral code as it does to preserve its government and other essential institutions.'

The problem is that we have been educated to believe we no longer have a common morality or a common culture. But in acting as individualists, we have created a society that contains the seeds of its own destruction – or at the very least, the destruction of those very values of liberal individualism which animate it. Without common values, there can only be incoherence and a battle for supremacy between vested interests. If the only meaning lies within ourselves, on what basis can there be any kind of community? Who then decides what norms should govern? In the ensuing vacuum, selfishness is transformed into legitimate self-interest as a basis for morality.

Self-interest has been translated into rights, and their defence has become the supreme moral cause. But this selfishness has little to do with true liberalism. Back in 1911 the great liberal thinker L. T. Hobhouse wrote:

> There may be a tyranny of custom, a tyranny of opinion, even a tyranny of circumstance, as real as any tyranny of government and more pervasive. Nor does liberty rest on the self-assertion of the individual. There is scope abundant for liberalism and illiberalism in personal conduct. Nor is liberty opposed to discipline, to organisation, to strenuous conviction as to what is true and just. Nor is it to be identified with tolerance of opposed opinions. The liberal does not meet opinions which he conceives to be false with toleration, as though they did not matter. He meets them with justice, and exacts for them a fair hearing as though they mattered just as much as his own. He is always ready to put his own convictions to the proof, not because he doubts them but because he believes in them.[10]

Hobhouse's liberalism, real liberalism, was intimately bound up with judgment. He also believed that freedom and community could only be constructed through self-discipline and a notion of the common good. No-one could act as an individual in isolation from everyone else. Morality, after all, is about how we live with other people.

As Jonathan Sacks has written, moralities are like languages embodying ancient and living social processes; we are born into them and must learn them in order to communicate with others.

> Like language, morality testifies to the paradox that only by yielding to something which is *not* individual can we become individuals . . . It takes a long apprenticeship in the rules of grammar and semantics before we express alone what we wish to say. Only by a similar apprenticeship in the rules and virtues of a moral tradition can we shape the life that we alone are called on to live. Like languages, moralities are not universal. But neither are they the product of private

and personal choice. We can no more sustain relationships without shared rules of fidelity and trust than we can sustain communication without shared rules of grammar. And without a stable framework of relationships we are left confused, vulnerable and alone.[11]

It is no coincidence that we have turned away from passing on the moral codes of our tradition just as we have turned away from teaching the rules of grammar to our children. But in repudiating the very concept of judgment we have turned away from moral principles that are common to all of us. The innate attachments between parents and children, for example, testify to a deeply rooted moral sense. So do universal values such as fairness, or an abhorrence of murder or incest. The sociability we perceive in young infants testifies to the desire in all of us to form bonds with other human beings, which is what morality is all about. Bonds mean human attachments; and attachments involve commitments.

Commitments bind us to each other through fairness and self-discipline. Our moral sense depends on commitments of various kinds: to employers, to community, to friends and above all to our families. But commitment has in large measure been displaced now by choice. Our commitment to our children cannot be expressed as rights but as duties. Self-interest in families has to give way to the interests of others. But the priority given to choice means that family structures have become a matter of personal, private preference, and commitment to one's children or spouse has become a negotiable commodity. If we redefine commitment in this way, we are throwing away a blueprint for the survival of our way of life that has been transmitted down the generations for centuries. Once it has been destroyed, it cannot easily be put back again, not least because few will recall precisely what it is that has been lost. Morality and attachment are umbilically linked; but our most crucial and deepest attachments of all are under attack as never before.

Chapter Nine

THE FLIGHT FROM PARENTING

Family Pressure

It was once considered axiomatic that parents and teachers had an overriding duty to transmit a set of values to their children in order to initiate them into both their particular culture and into the human condition. Education and nurture went hand in hand. Parents and teachers each nourished and educated the children in their care both by explicit teaching and by example. That is how a culture renews itself from generation to generation and forges the chain linking the past with the future.

Alarmingly, that understanding is now breaking down. The concept of parenting has been diminished by the me-society. The very idea of duty has been replaced by rights: the right of a child to assert itself against the authority of adults; the right of adults to live as they please regardless of the consequences. Teachers are being undermined by the flight from parenting; parents are being undermined by the flight from teaching. The source of the damage in both cases is the same: rampant individualism which detaches the individual from all traditions, which declares that attachment is tyranny and which has privatised meaning so that all external authority has been repudiated: the authority of teachers and of parents, the rules of mathematics and moral behaviour, the validity of British culture and of human society itself in a western democracy. The evidence of the damage this is causing has been systematically denied, proof of the tenacity with which the egoistic individual is fighting to retain that culture of irresponsible hedonism, the twisted legacy of the Enlightenment, in the face of any attempt to restore a proper concept of duty and responsibility.

Schools are where children are taught to move beyond the confines of their own immediate experience. But the education supplied by the family is the most fundamental and crucial of all. The family is where children form their own identities and learn through their attachment to their parents to be secure and to trust the world sufficiently to play an equal part in it. It is where they learn from the way their parents behave towards each other and towards their children the elementary truths about commitment, love, caring, negotiation, compromise, sharing and putting the interests of others first. The family is where the child learns what it is to be a human being. It is where children assimilate the complex messages that form a moral sense. And it is the accumulated impact of children brought up in families where there are secure attachments that collectively makes a co-operative and altruistic society. If the family is functioning badly or if it fractures, the effect upon the child is always serious and often devastating. If there is a widespread breakdown in family attachments, then the implications for society are extremely grave.

That is what is happening in Britain. There is a widespread breakdown in the whole parenting culture – revealing itself not just in the burgeoning numbers of broken families, but in the concept of parenting itself and the understanding of the responsibilities it entails. Something has gone very wrong indeed with the way we now view the status of children and their relationship with the adult world. In family life, we are now equably tolerating the intolerable, both in changes in family structure and in the quality of the relationships between parents and children – regardless of what type of family they inhabit. It is no accident that paternalism has become the principal bogeyman; notions of care, control and protection have been inverted to represent an unwarranted intrusion upon the freedom of the individual. They interfere with the sacred right to make our own way without anyone telling us what to do.

As we have seen earlier in the book, children figure prominently in this 'liberation' movement. The child-centred philosophy is an explicit attempt to democratise and equalise children's relationship with adults. Yet, despite the rhetoric in which it has been clothed, the outcome is the very opposite of child-centred. Democratic rights belong properly to adults. Children do not have adult 'rights' to freedom because they are still life's apprentices. They have needs, not rights. Freedom for them is another word for neglect. They are primarily the recipients of duties because their principal need is to be parented, which means being looked after with love, commitment and discipline. Letting them roam freely with 'rights' and 'choice' and with no fixed boundaries effectively abandons them to ignorance, error and often harm. And it has also produced a dramatic reversal of roles. Adults, who have licensed themselves to behave as if their actions have no consequences, have become infantilised. Their children, by contrast, have been saddled with a burden of adult responsibilities well beyond their years. What child-centred theories have done is to destroy the very concept of childhood itself.

Britain simply doesn't value children in the same way that other European countries do. It doesn't appear to understand what families are primarily for. A *Eurobarometer* survey carried out for the European Union in 1990 asked respondents what they thought was the most important role of the family in society. In Portugal, 68 per cent of respondents said it was about bringing up and educating children. In Greece, the figure was 63 per cent; in Spain, it was 58 per cent. In the United Kingdom it was just 24 per cent, the lowest figure of all. British respondents thought instead that families were about personal adult fulfilment.[1]

Such an attitude is all too plain in day-to-day family patterns. British culture no longer appears to want to replicate itself. People are having fewer and fewer children. As in other western countries with an individualistic ethos (even in those where children are apparently better valued) the population in Britain is declining. A stable population needs to produce 2.1 children per woman; Britain is producing 1.8. As the sociologist Professor A. H. Halsey has written: 'The individualised as distinct from the socialised country eventually and literally destroys itself.'[2] British marriages in 1993 fell to their lowest level for 50 years, while rates of cohabitation are going up all the time. Between 1983 and 1993 the marriage rate fell by one quarter, while the number of divorces rose by 12 per cent to reach 165,000, the largest number ever recorded.[3] Although the number of divorces fell slightly between 1993 and 1994, on present trends it is still estimated that one in four children will experience their parents' divorce by the age of 16, and four out of ten new marriages will end in divorce.[4] And one birth in three takes place outside marriage.

Nevertheless, it remains the case that most marriages do not end in divorce, that most children will be brought up by both their natural parents and that most people are still committed to heterosexual unions and stable marriages.[5] Some people think, therefore, that alarm about family breakdown is nothing more than 'moral panic' and that nothing much has changed in people's behaviour. Such complacency, however, is seriously misplaced. The numbers involved in the fragmentation of the family are very substantial and rising. The damage, however, is also incremental, with the children of broken families themselves by and large less able to form stable relationships when they reach maturity. And in any event, the problem is not merely one of family structure. Fragmenting families are but one part – although an important part – of a complex web of authority relationships between adults and children which are going badly wrong. They are the outcome of, as well as a powerful contributory factor to, the collapse of tradition and authority which is creating an atomised society of isolated individuals.

Evidence for this is not merely to be found in Britain. Throughout the western world, there is a rising tide of juvenile distress and disorder, ranging from depression and eating disorders through suicides, educational under-achievement, drug and alcohol abuse and finally to crime. In an authoritative and exhaustive survey of trends over time in western Europe and America, the child psychiatrist Professor Sir Michael Rutter and the criminologist Professor David Smith discovered that since the end of the Second World War, the incidence of *all* the disorders studied in young people had shown a sudden increase. These trends, they reported, could not be accounted for by rising poverty or unemployment, not least because they had increased during periods of low unemployment and rising living standards as well as periods of high unemployment. They suggested that increasing levels of family discord and breakup may have played a part in creating these disturbing trends, along with the growth of an alienating youth culture, earlier engagement in sexual relationships, greater peer group pressure and greater personal autonomy combined with heavier financial dependence upon parents.[6]

The research confirmed what is apparent from empirical

observation, that children are increasingly showing the distressing personal effects of being subjected to a range of intolerable pressures. Of course, every child reacts to pressures differently; and very often, these pressures operate not just alongside but upon each other, so it becomes hard to separate out the cause of the problems. But it seems clear that at the root of it all lies the fact that children are no longer being adequately cared for by the adult world. Preoccupied by their own self-interest, adults from all social classes now routinely display a high level of indifference to their children. This is made all the more shocking by the fact that neither they nor apparently other 'responsible' adults even recognise such indifference for what it is, let alone pass judgment upon it.

Adult Indifference

In January 1994 a seven-year-old child, Stacey Queripel, was found strangled in woods near her home in Bracknell, Berkshire. She had run away three times because she thought nobody loved her. Her mother Gilliane had been married twice and had children by three different men. She blamed her boyfriend for making Stacey unhappy by saying he was going to be her proper father. She also blamed the men in her life for building up relationships with Stacey and then leaving. Stacey and her sister Leanne had been sent to bed early while a group of men watched television and smoked cannabis. The door was left on the latch while men came and went as they pleased. The inquest heard Gilliane frequently put her hands round Stacey's neck and smacked her hard when losing her temper. The coroner said: 'What parent at some time has not been driven to distraction by children playing up? Life as a single parent must have been very difficult . . .'[7] This implied that the coroner was moved more by the mother's situation than by the plight of a young child who was so unloved, so neglected and so full of grief that she ran away three times.

Magdalene Bush, aged 16, left home after a row to live with her boyfriend. Her mother was an academic administrator and her stepfather was a university lecturer. She was reported as saying about her mother:

She said a lot of stuff about my father. I knew Chris was my
stepfather and I always believed that John was my natural
father, but she told me he wasn't. I was very shocked and very
upset . . . When I went, I waited around the village and after a
while I went back down the steps to my house. I was thinking
about going back in but they were inside having supper. At
that moment I just felt that they didn't even care. If they cared
at all, they would have been looking for me. That was it – I
went . . . I phoned my mum and said I was going to Paul's
tonight and wasn't coming back. She said: 'Suit yourself.' I
was very upset about my father – I eventually found out that
apparently it's not certain that he is my father. I'm not really
sure what the situation is. I was very upset and confused. But
after a week, I just thought 'sod that' and didn't really think
about it so much.[8]

If her mother, Janet Earl, recognised the grief, betrayal and hurt
in her daughter's remarks, she certainly didn't let on. Nor did she
accept a shred of responsibility for what had happened. She said:

I wanted to communicate to her that at 16, generally you
are not ready to settle down and marry, and that a sexual
relationship should wait until you have such feelings of
abiding commitment to a partner. I made it abundantly
clear that I did not approve. I said there was no question
of her boyfriend sleeping with her at our home. She merely
accused me of hypocrisy . . . We have brought them up
to be self-sufficient and considerate of others' feelings, to
have self-respect and some kind of moral perspective. Every
one of these concepts and our last vestiges of authority as
parents have been undermined. I cannot understand how
a government that professes to espouse family values can
erode them so successfully with this absurd loophole in the
benefits law.[9]

So the parents were off the hook because the government was to
blame. What the government had got to do with her daughter's
feelings about her father, or the fact that the mother could hardly
preach about family values, moral perspective or self-respect when

it wasn't certain who her daughter's father actually was, remained far from clear.

Jill Robinson wrote in a newspaper about Tom, her 14-year-old son. She said she couldn't cope with him any more and she was handing him over to his father from whom she had split up ten years previously. She now believed that Tom had been damaged by the separation. What was wrong with Tom? According to her, he was hypercritical of the house and car, he was shallow and cared only about possessions, he was critical of her shape, he blamed her for going into his room or conversely not doing his laundry, he demanded £50 for trainers and got it, he didn't help with the washing-up when she was ill and he refused to go to bed at an acceptable time. 'Are all adolescents this selfish?' she asked. 'Would he be behaving in the same way if his father were present? Have I tried to overcompensate for the split, as some friends have suggested? . . . He obviously does not love or respect me, and now that he is bigger and taller than me, I have no control over him.' Her parting shot was: 'I hope his father succeeds where I have failed. Otherwise, as I have told Tom, the next woman to abandon him will probably be his wife.'[10] Thus this mother appeared to blame her son for his own distress, with the *coup de grace* that other women would abandon him just as she was doing on account of his behaviour – regardless of the fact that this was caused, as she simultaneously admitted, by the damage that had been done to him.

In family after family now, there appears to be an almost wilful refusal to appreciate the grief and trauma caused to a child by the sundering of its parents and the fact that one of the parents is voluntarily absent from both home and child. However explicit the child's feelings of devastation may be, the adult world appears to want to minimise this impact, to explain it away by blaming a host of peripheral factors or, most frequently, by blaming the child. Children are shunted about quite routinely between step-parents, dumped by one set upon the other, as if they were no more than inconvenient bits of spare baggage whose peripatetic existence is now so common it has become entirely normal. Their teachers commonly perceive the impact of such disrupted back-grounds more acutely than the parents. The teachers observe the extremely high correlation between educational under-achievement and troubled family life and they draw the obvious conclusions. But it's not so obvious to the parents, nor to the academics whose 'research' usually 'proves' that any ill-effects are all the fault of the government.

Different children, of course, react in different ways to the breakdown in family relationships. Not all misbehave or fail to thrive. And there are plenty of intact families where parenting is wholly inadequate. But the cumulative result of chaotic, indifferent or cruel family backgrounds is the increasingly anarchic, violent behaviour of the young and, even more alarmingly now, the very young. It was reported in 1995 that children as young as four were among a record 10,000 violent pupils expelled from school during the past year. Liz Paver, a head teacher from Doncaster, said: 'They swear and kick at each other – the only thing they know is violence. These are children – not from inner cities – so poorly nurtured and cared for that they have effectively had their childhood taken away.'[11]

Many children are being brought up in violent households. A report published by the National Society for the Prevention of Cruelty to Children in 1995 indicated that many adults had grown up in homes where abusive or violent behaviour was part of family life. Almost half of the people interviewed had witnessed physical violence between those who cared for them; one-third had been beaten with a belt, strap or shoe; more than a third had never or rarely been hugged or kissed by their father, and more than one in six by their mother. And of course, many of these adults would themselves now be parents, with all that implied for the onward transmission of patterns of violence and the parenting deficit.[12]

Violence, neglect, indifference and other forms of inadequate parenting occur in intact as well as fractured family households. But all the evidence is that this kind of behaviour is more likely in irregular families. A study published in 1994 suggested a high correlation between children born out of wedlock, teenage parenthood, chaotic home backgrounds and child abuse. The lowest risk factors of all were in homes where the natural parents were married. It said that children were 20 times more at risk of child abuse if both natural parents were cohabiting rather than married. From the evidence of 35 child abuse inquiries, the child was 18 times more likely to be killed by one of its two cohabiting than married parents, and 74 times more likely if it was living with its unmarried mother and her cohabiting boyfriend.[13]

*　　*　　*

The political argument about family breakdown has tended to focus around the issue of lone parents. In fact, concern should be much more broadly focused around disruption, since the core issue is the loss of the meaning of commitment and the flight from responsibility and duty. Inadequate parenting occurs in families of all types, but the problems this causes for children are generally compounded in families after divorce, where there are transient or missing parents, where families are re-ordered through the arrival of step-parents or where a number of different 'partners' come and go in the children's lives. These problems are spread across the social classes, but there are now whole communities where the combined effects of second- and third-generation family fragmentation along with endemic unemployment – in other words, the loss of the two primary means of socialising individuals – have created alarming enclaves of incipient social anarchy.

According to a local primary head teacher in one such area in north-west England, out of about 170 families with children at her school only six had a father in residence. The children were highly sexualised, she said, not through sexual abuse – though that was there too – but through the casualisation of the mother's sexual relationships. Fathers were just written out of the script. One child didn't want her father to come to the school Christmas play because she would have been thought weird to have a dad. Boys were growing up without any sense of responsibility, said another teacher; they believed it was normal for men to father lots of different children by different women who could be beaten, abused and abandoned at will. These inadequate parents neglected their children, though they thought they loved them, because they were little more than children themselves. They indulged them by giving in whenever they demanded anything; yet at the same time they'd kick them out onto the streets to fend for themselves all day. At four or five years old, the children would be roaming the streets. There was no bedtime and they'd be up late watching sex videos which the parents left lying around. By the time they were ten or eleven, their parents had given up on them.[14]

The poignant fact about many young people who have been brought up in such a state of emotional chaos or neglect is that they want to avoid doing to their own children what their upbringing will almost inevitably condemn them to do. They know they have been short-changed and that it has done them harm. And if they have got into trouble with the law, they themselves often blame the absence of discipline caused by their chaotic family backgrounds and in particular the absence of their fathers. On a parenting course at Deerbolt Young Offender Institution, County Durham, young fathers were desperate to avoid doing to their children what had been done to them. Yet their own backgrounds made that very difficult to achieve. About 30 per cent of boys at Deerbolt aged between 15 and 21 admitted to being a father. But because of their own emotionally wrecked backgrounds, they didn't know how to cope. About 75 per cent had either been brought up with no resident father or had step-fathers. They described family backgrounds of neglect and cruelty in which their thoughts were overwhelmingly with the fathers who had left them. None was married; they regarded marriage as an irrelevance. They talked of commitment but didn't understand its implications.

Their teacher, Jane Mardon, said they thought of parenthood along the lines of *Hello!* magazine or TV advertisements. 'They do have difficulty thinking ahead and imagining the consequences of their actions,' she said. 'Before becoming fathers, they didn't think ahead at all. Some of these boys haven't been loved from day one. They have nobody. It's very difficult for them to learn to love. They come out with stereotypes. Some think it's about giving people what they want.'[15]

Mavis Grant is head teacher of Mary Trevelyan primary school in Newcastle, serving the kind of estate from which boys like these are disproportionately drawn. About two-thirds of her pupils come from disrupted family backgrounds. She said that many parents found it hard to imagine the impact of their difficulties with each other upon their children.

We're often in a situation here where one parent takes out an injunction against the other seeing the child and we have to deal with the effects on the child who often becomes withdrawn or lethargic, developing problems with other children. Some may never have known their father, or for a short time. That has an impact on their expectations of family life and their own expectations of their own role as they grow up. Their own fatherlessness means the boys assume the role is about physical protection more than caring or creating a secure home environment. It's more to do with defending your rights against someone who's given offence.

They look for attention in one of two ways: by becoming physically aggressive or very distant and switched off. Sometimes they become quite isolated, making no attempt to join in group games. Some of the boys become obsessed with trying to please you in every possible way. Their self-esteem is often very damaged. They have no-one to measure themselves against, and sometimes the extra pressure put on the mother means all her energies are spent day to day on keeping the family together and less time for nurturing the child's confidence. Sometimes boys as young as seven or eight will say, 'I'm the man of the family.'

There's less of a loss for those who've never known two parents than for those whose parents have split. They don't have that sense of bereavement or witnessing the anger and so have fewer problems adjusting. But children who've never known their father at all have problems with insecurity, their sense of identity or belonging. All these children find it hard to make attachments; or else they become excessively attached. Often, children who've left to go to secondary school come back here day after day when they should be at their new school because here is the only security they've ever known.[16]

These accounts, from the mouths of the children themselves and from their teachers, speak of a terrifying desolation and a stunting of human potential. A deeply alarming situation is steadily building up, both for individuals and for whole communities, because of the breakdown in organised family structure. It is exacerbated by the related erosion of parental authority which affects all families and which we shall examine in more detail later. But the increasing fragility of the conventional nuclear family – two parents bringing up their own children together – is itself a major cause of children's lasting distress and damage. Yet despite such testimony, and despite the weight of research evidence from both Britain and America, opinion formers have gone to extreme lengths to pretend that no such damage is being done and that the only problems which arise in these circumstances are caused by poverty. One has only to observe a child psychiatric clinic, or visit an independent school, or look at the British Royal Family to see plenty of well-heeled children whose lives have been wrecked by family breakdown.

The evidence of the harm being done is in fact overwhelming. In every area of their development, children from fragmented families do worse, relatively speaking, than children from families which are intact. Of course, some thrive perfectly well; but for the majority, the cards are stacked against them and they suffer as a result.

A study by Dr Kathleen Kiernan, *The Impact of Family Disruption in Childhood on Transitions Made in Young Adult Life, 1992*, using data on more than 17,000 children born in 1958, found that children of divorced parents were more likely to leave school at 16 and home before 18. In stepfamilies, girls were at twice the risk of becoming teenage mothers and having a baby outside marriage, at twice the risk of leaving school at 16 and at three times the risk of leaving home through ill-feeling before 18. Dr Martin Richards of the Centre for Family Research at Cambridge University wrote that the consequences of divorce upon children were even worse than if a parent died. Poverty couldn't be the whole explanation since when the remaining parent remarried the financial situation improved but the outcomes for children got even worse. Children's self-esteem suffered after divorce, he wrote, through the loss of the parent, usually the father. 'It is not just that he has gone after the divorce but that he has chosen to go – or at least that is how children will often see it. The fact seems often to lead to difficulties for young people (both men and women) in forming relationships and a consequence of that may be premature marriage, cohabitation or parenthood.'[17]

In the United States, a great deal of research has come to broadly the same conclusions. The American researcher Judith Wallerstein looked at middle-class families after divorce and concluded that divorce was a chain of events, relocations and radically shifting relationships. Five years after divorce, more than a third of the children involved were suffering moderate or severe depression; 10 years afterwards, a significant number were troubled, drifting and underachieving; 15 years afterwards, many were struggling to establish romantic relationships of their own.[18] Step-parents seemed to make the situation even worse. As the researchers Andrew Cherlin and Frank Furstenburg put it, through divorce and remarriage children were related to more and more people to whom they owed less and less.

Yet the *bien-pensants* of the British intellectual élite, in the media, in the universities and in the pressure groups, first ignored and then distorted these findings. Family life was not disintegrating, they said airily, merely changing. And changing for the better, since happy parents meant happy children; divorce merely brought conflict in marriage to an end, to the benefit of the children. A moment's thought would inform one that this cannot be so, since divorce very often greatly exacerbates conflict because of the fight that then takes place over both children and property. Of course, in some cases divorce will end the misery, to the relief not only of the parents but also of the child. In this of all subjects, one cannot generalise about particulars; there are, and always will be, instances where a break in the family is the best outcome for all concerned. But to say that generally divorce marks an end to the misery of the children is wishful thinking. Indeed, it often happens that at divorce the conflict is for the first time centred specifically on the children themselves, who are then emotionally torn apart.

Wishful thinking is typical of the anti-evidential mind-set that characterises this whole issue. Evidence is roundly denied, just as it is in education. There has been a flight of reason into ideological propaganda and distorted thinking. The effect of this has been that thousands of parents, who were divorced on the assumption that this was indeed the best outcome for their children, have done those children harm that many of them have subsequently come to recognise and regret. Yet people who point this out are virulently denounced in a crude attempt to marginalise them or intimidate them into silence. And woe betide any British researchers who produce more of such evidence. They are likely to find their funding is promptly cut off by the people who purport to be defending liberal tolerance against the reactionary revival.

But then, as the American liberal commentator Barbara Dafoe Whitehead observed, this debate is not about evidence as such but about values in an intense culture war.[19] Since the earliest times and across cultures, the most successful means of ensuring physical survival and promoting the social development of the child has been the family unit of the natural father and mother. Those primitive societies where this has not been the pattern have remained primitive societies. The nuclear family has been associated with northern Europe as far back as records go. But in recent times, this fact has been denied. Among liberal circles, the view is commonly expressed that the nuclear family was 'invented' during the industrial revolution; or as one Labour peer airily remarked: 'Fathers only started to be thought necessary in the late Victorian period.'[20] One does not have to read the authoritative research on family formation to realise the patent absurdity of such observations.[21] But such thinking is the product not so much of ignorance as ideology. Parental separation, once regarded as inimical to children's best interests, has been re-packaged as essential to children's happiness. That happened because of a shift from concern for the well-being of the child to the well-being of the adult. In order to present parental separation as normal, the two-parent family had to be re-sold as deviant.

As Whitehead commented, researchers were markedly reluctant to say that these changes to the family represented social regress rather than progress. It was just too risky to make normative statements any more. 'This reflects not only the persistent drive towards "value neutrality" in the professions but also a deep confusion about the purposes of public discourse,' she wrote. 'The dominant view appears to be that social criticism, like criticism of individuals, is psychologically damaging. The worst thing you can do is make people feel guilt or bad about themselves.'[22] In other words, the worst thing one could do was to require a moral response.

A representative sample of this ostrich mode of moral discourse was furnished by an article published in 1995 by the influential left of centre think-tank, the Institute for Public Policy Research, in which feminist ideology masqueraded as objectivity. The central grievance was that concern about family breakdown, dismissed as 'moral panic', sought to pin the blame on women. Countless children, it claimed, were brought up in stable, loving one-parent families; greater numbers were blighted by unstable, unloving, irresponsible two-parent families. No evidence was adduced for this last assertion because of course there is none; the evidence indicates, on the contrary, that most harm accrues to children in fractured families. This article accused its opponents of nostalgia for an idealised past, woven round a 'fantasy of cementing men into families'. Why it should be considered so outrageous to believe that fathers should be expected to remain with their families was not clear. The article repeated the canard that 'the traditional nuclear family model took shape in the industrial era and suited those conditions', and it asserted boldly: 'What counts is not the biological connection but the quality of the intimate relationships in a child's life.'[23] But biology does matter greatly to a child. Presumably these authors would be less quick to argue that biology didn't matter if anyone were to suggest that the bond between mother and child was of no consequence. Children are devastated if either parent leaves them. It affects their very identity. It gives a child the message that 'one of the two people who made me has chosen not to look after me and share the home that will shape me'. That is a devastating message and it is why the 'quality of the intimate relationships' that follow is so overwhelmingly difficult for such a child to cope with.

The routine distortion of evidence about family breakdown has been a pronounced characteristic of the debate about the family. At the heart of this otherwise baffling irrationality lies a passionate unwillingness to acknowledge the facts for fear of the consequences that may then ensue. This was illustrated by a conversation with a leading social scientist who had denounced the ethical socialists A. H. Halsey and Norman Dennis as wrong when they claimed that children who came from broken homes did worse, relatively speaking, than children who did not. Telephoned to discover upon what research he based his criticism, this academic proved reluctant to answer the question. Instead, he released a stream of invective and asked furiously: 'What do these people want? Do they want unhappy parents to stay together?' Pressed to identify the research which would prove Halsey and Dennis wrong, he eventually said, in summary, this: of course they were correct as far as the research went, but where did that get anyone? Nowhere! Was it possible to turn back the clock? Of course not! And why were they so concerned above all else for the rights of the child? What about the rights of the parents, which were just as important?[24]

That, surely, is the crux of the matter. If the evidence were accepted that family breakup harms children, it would mean that adults could no longer drop or swap sexual partners in the belief that they were doing no harm to their children. The culture of individualism would take a knock; the doctrine of individual rights upon which it is based would be revealed as promoting adult freedom only at the expense of the best interests of their children. It would no longer be possible to maintain the cardinal doctrine that everyone's lifestyle is equal in worth to everyone else's. As with the ideologies of the classroom, facts are simply too inconvenient to be allowed to stand in the way of behavioural licence free of moral censure.

The lengths to which the liberal intelligentsia will go to neutralise or marginalise such inconvenient facts were demonstrated in the reactions to a study carried out by two researchers in the Exeter University Department of Child Health, Dr John Tripp and Monica Cockett. This reported that children in re-ordered households were more likely than children in intact families to have health problems, to need extra help at school, to encounter friendship difficulties and to suffer from low self-esteem. But what made this report so politically explosive was its findings on conflict. Conflict has been used as the great justification for marriage breakup. There is no dispute that children suffer greatly when their parents are at war with each other. The argument has been, though, that since divorce brings that conflict to an end it is therefore in children's best interests if their parents do divorce – and quickly. Despite the fact that in many cases divorce patently does *not* bring conflict to an end, this theory has been trotted out over and over again to justify the assertion that family structure was 'not deteriorating, only changing'. But what Cockett and Tripp so inconveniently discovered was that although children from conflict-ridden intact families fared worse than children from peaceful intact families, they did worse still *after* their parents divorced than before. Moreover, separation and divorce often initiated conflict or made it worse.[25]

This research enraged the intelligentsia, which launched a whispering campaign against the researchers and threatened their continued funding. From the point of view of such critics, it is crucial that evidence like this is marginalised because it exposes the fact that what they claim to be reality is in fact little other than ideology. Most people do not share their implicit starting point, that the conventional family is tyranny. But it becomes very much easier nonetheless to justify a relaxation of self-discipline if the intellectual climate is telling you constantly that there will be no ill effects as a result. The attitudes of these élites are corrosive because they attempt to bamboozle people into believing that the cultural goalposts have shifted by redefining the deviant as normal, thus encouraging more people to behave in an irregular fashion.

Most people, though, remain conventional in their attitudes. In

1986, the British Social Attitudes survey found that 71 per cent of those questioned thought more should be done to safeguard the institution of marriage, and 74 per cent agreed that most people took marriage too lightly. And children, for obvious reasons, are often the most conventionally minded of all. The journalist Linda Grant reported in the *Guardian*: 'I asked the kids of Highgate Wood School what was the thing that scared them most and they said, something happening to my mum and dad. My parents splitting up. What was the most important thing in their lives? "My parents, family. Being wanted. Family." '[26]

In the same article, Monica Cockett said of the Exeter study she had co-authored: 'Why our study was an uncomfortable message was that we don't listen to what children say they want. To a child, the family is still total security. To a child, the natural parents are still the real and closest members of the family, and that's why we don't have closed adoption any longer. Biological ties are so strong that it's hard for the children of Fred and Rosemary West [the Cromwell Street 'House of Horror' murderers] to accept the full evil of someone who is supposed to care for you, and it's very hard for a child who had had a wonderful father who has walked away to reconcile those two things. It's an awful lot to take on board, and we're asking children to function like adults without the adult sense of internal security. They grow up with their emotional world unattended to and hollow.'[27]

Commitment

The point of detonation for the individualists in the family debate is the concept of commitment. The whole point of commitment is that it is an undertaking that restricts one's freedom of action. But of course, to the unbridled individualist there can be no restriction at all on freedom. Any restriction is a form of oppression. Commitment has therefore been redefined so that it becomes negotiable. So the commitment of adults to each other, or to their children, now lasts just as long as they are prepared to adhere to it. This, of course, is not commitment at all but freedom and irresponsibility camouflaged beneath a distinctly flaky undertaking. Yet it is called commitment by people who maintain that it is as secure a basis for the upbringing of their children as the commitment of marriage. Not surprisingly therefore, in view of this semantic sleight of hand, the practical expression this non-commitment now takes in the form of cohabitation breaks down more frequently than does marriage (which itself has been eroded by the new definition of commitment). Family relationships have moved from being a covenant to a contract, breakable unilaterally when one party decides it's time to move on.

As Norman Dennis has written:

> Parents as 'partners' instead of 'husbands' and 'wives' means that the whole system of kinship has been deinstitutionalised, not just the marriage bond. The commitment of unconditional and irrational affection to the children of kinsfolk cannot but be lessened when increasingly the question of whether the parents of those children stay together becomes a conditional and rational decision, depending on how far staying together suits the partners in terms of their personal right to happiness and self-fulfilment. Not only does lone parenthood, therefore, necessarily mean a diminution in the number of potential givers of unconditional, long-term affection. The increase in the chances of partnerships and marriages ending as lone-parenthood means commitment is more likely to be withheld across the board and from the start in all partnership, marriage and lone-parent situations.[28]

The paradox of the conventional family is that it is the best mechanism yet devised for promoting the integrity of the individual, the inner-directed, self-reliant personality; yet as an institution it has been chronically undermined by that very individualism it has helped to foster. The ideal of marriage remains strong; what is now vanishing is the self-discipline needed to sustain it, and a general comprehension of the reasons why it was necessary in the first place. The roots of this profound change lie in the emergence of individualism during the eighteenth century, in precisely the same impulses that gave rise to the primacy of the self that has wrought such havoc in education. Originally, the British family was intimately linked to considerations of kin, inheritance and community. Above all, it was a mechanism for transmitting property.[29] But the key concept was transmission, maintaining an unbroken chain of continuity from one generation to the next. It was pre-eminently a mechanism for the exercise of duty beyond the self. When notions of sentiment and romantic love flowed into the family, however, with their unpredictable and destabilising characteristics, the self took over and the once pre-eminent duties of transmission and continuity, with which they were in direct conflict, gradually became marginalised.

As the historian Roy Porter has observed, the family traditionally stayed together through strong legal, cultural and economic ties which after the eighteenth century gradually gave way to romantic love. This transition eventually led to the current unprecedented crisis, he said, since the traditional family was the best institution for socialising people into skills, obedience, and law-abiding behaviour. It gave people emotional stability and fulfilment, a sense of belonging, challenge, endeavour and optimism and produced a high level of social cohesion.

At the same time, however, it produced a lot of personal frustration. From the end of the nineteenth century, sexual radicals from Shelley to Havelock Ellis and the Bloomsbury Group viewed the family as a kind of cage.[30] This was, of course, precisely what happened in education, which also spent a long time gestating its own destruction and similarly spawned off-beat theories taken up by the same bohemian fringe at the turn of the century. The question poses itself with the family, then, as with education: why, after these attitudes had been confined to a fringe group of intellectuals for so many decades, did the family suddenly come apart at the seams in the 1970s and 1980s?

The obvious detonator was the permissive society of the 1960s, when the arrival of a rebellious youth culture which had money in its pockets for the first time altered the cultural landscape. But that does not explain the origin of a permissive culture which revolved around the separation of sex from marriage and which was characterised by the revolt by the young against the adult world; nor does it explain why that adult world felt so patently inadequate to confront the challenge to its authority. The answer seems to lie in that same rise of psychopolitics which had so influenced educational thinking, with a strong dash of anthropology thrown in for good measure.

Indeed, the two were linked in another very important way. The development of formal, compulsory education took away from the family much of its educative function and placed it in the classroom instead. So the family very quickly lost sight of its crucial role in socialising the young into a culture, refocusing instead on relationships within marriage and between parents and children. The result was that families became increasingly vulnerable to the manifold pressures of a culture based on individual fulfilment. There was no longer the counterbalance of an overriding duty to keep the family together in order to transmit a set of cultural values. That had been hived off to the schools. What no-one foresaw was the catastrophic effect of the very same pressures operating within education itself, so that in the end *no-one* was transmitting the culture to the next generation.

As the century wore on, the falling birthrate, rising divorce rate and the changing status of women all put pressure on the family. By the 1920s there was already considerable concern about its state of health. But during the following decades, the family was dealt a number of significant blows which greatly eroded its fundamental role as the primary agent of social continuity. Marx and Engels were critical of the bourgeois family because they believed it was essential that mankind was freed from the legacy of the biological imperative. Nothing could be fixed in our natural history because we had to be makers anew of our own destiny. Everything we did was therefore the product of our environment and so there could be no such thing as 'natural' behaviour or 'normal' family patterns. Engels claimed that marriage continually evolved through social and economic change, and that monogamy had created a patriarchal society in which the woman 'was degraded and reduced to servitude; she became the slave of his lust and a mere instrument for the production of children.'[31]

Freud had cut the already disappearing ground from under parents' feet. The rise of the therapeutic culture, replacing 'is it right?' by 'does it feel good for me?', was a hammer blow to parental authority. But social anthropologists were promoting an agenda of cultural relativism, anxious to refute the theories of Freud in order to eradicate anything that smacked of biological determinism and which might lead to racism. The anthropologist Margaret Mead accordingly produced her lyrical description of the sexual life of primitive tribes and used her findings to mount an all-out attack on the American nuclear family on the grounds that it crippled the emotions and produced maladjusted individuals. Despite the flaws in her arguments, this species of anthropological relativism did much to uncouple the family from its own western traditions.

This anthropological assault developed alongside the Frankfurt School of sociology with its Marxist critique of the authoritarian family, sexual repression and puritanical morality. These thinkers linked the conventional family to bourgeois repression: for Erich Fromm, the patriarchal family produced 'the authoritarian character' and for Wilhelm Reich 'the authoritarian state in miniature'.[32] The real reason why the conventional family became a revolutionary target, however, was that it was recognised as the building block of society. So any philosophy that wanted to transform the nature of society had first to transform the nature of the family.

According to Christopher Lasch, the 1950s were far from being a 'golden age' of the nuclear family. Instead, those years were a period of revolutionary ferment. The therapeutic culture had taken hold. American marriage counsellors no longer considered it a success to hold a marriage together; the couple had to work out their future in terms of their own values and (just like in education) with an emphasis on personal growth and happiness. There was a growing feeling that patriarchal family life should be replaced by more democratic, fragmented and individualised family patterns with an overriding need for communication, mutual understanding and respect.

It was only then that the family finally imploded, under the combined impact of the generational revolt and the new feminist critique of the 1960s and 1970s, heavily influenced by psychological radicalism. Marriage was deemed to be about companionship, not child rearing. Couples had to shake off their rigid roles and respect each other's need for individuality. An ideology of non-binding commitments evolved. Relationships became negotiable and transitory. The family could no longer be expected to transmit its own values. Parenthood was too important to be left to parents. Sex was separated from love and marriage. Divorce became seen as a creative act.

The Nuclear Family

In the 1967 Reith Lectures the anthropologist Edmund Leach pronounced an anathema upon the family which was to resonate for decades. 'The family looks inward upon itself; there is an intensification of emotional strain between husband and wife, and parents and children. The strain is greater than most of us can bear. Far from being the basis of a good society, the family, with its narrow privacy and tawdry secrets, is the source of all our discontents.'

The fact was that most people wanted, and still want, what Leach found so unendurable. The idea that had taken hold among the anti-family critics was that the 'private' nuclear family was a kind of introverted corruption of a previous ideal, an extended family form in which somehow everyone looked after everyone else with no emotional intensity and was a lot healthier as a result. But this wasn't true. The work of the family historian Peter Laslett showed that in previous centuries the number of extended families in residence was actually small. Far from a previous golden age, many families of the past had been locked into grinding poverty, ignorance and economic tyranny. Old people had been neglected, children had been filthy and uncared for; kinship relationships had harboured seething resentments, conflicts and troubles.

But the perception of the nuclear family had been twisted by the impact of Freud, not to mention the many misreadings of Freud, and by the subsequent growth of psychopolitics. By the late 1960s, the family was being presented as the enemy of personal happiness. Through the work of the radical psychiatrist R. D. Laing and others, an anti-marriage and anti-family cult caught the imagination of the media. The family was tyranny, forcing children into prescribed patterns of attitudes and behaviour which resulted in conflict, misery, disorder and psychotic illness. This was made all the more terrifying because of the privacy in which such oppression was taking place, privacy disturbed only by the superior insights of the therapeutic profession – a perception which, of course, greatly enhanced its status as the new priesthood of the age.

The analysis was not only distorted but ideological. The modern family was presented as the tool of bourgeois capitalism, grinding the individual into the shape required by an exploitative economic machine. Parental love, in Laing's opinion, was emotional death. 'From the moment of birth,' he claimed, 'when the stone age baby confronts the 20th century mother, the baby is subjected to these forces of violence, called love, as its mother and father have been, and their parents and their parents before them. These forces are mainly concerned with destroying most of its potentialities. This enterprise is on the whole successful. By the time the new human being is fifteen or so, we are left with a being like ourselves. A half-crazed creature, more or less adjusted to a mad world. This is normality in our present age.'[33] Later, Laing was to deny that he was anti-family and said families were 'one of the best relics of a crumbling system we have to hang on to'.[34]

The political New Left grabbed hold of the anti-family analysis and made hay with it. In *A Brief Guide to Bourgeois Ideology*, published in 1969, the activist Robin Blackburn said that the family was the main arena of deliberate physical violence and that one in five families had a history of schizophrenia. Even well adjusted, happy families were condemned because they excluded illegitimate children or unwanted grandparents. Juliet Mitchell wrote in *Women's Estate* in 1971: 'The family . . . embodies the most conservative concepts available; it rigidifies the past ideals and presents them as the present pleasures. By its very nature it is there to prevent the future.'

On the contrary; the family was there to ensure the future. But a newly emerging revolutionary creed swallowed the same analysis. In 1963, Betty Friedan proclaimed in *The Feminine Mystique* that American women had become slaves in the home. In the light of the contraceptive pill and the economic transformation in women's position and attitudes, marriage became seen as outmoded and responsible for their oppression. Not just marriage but sexuality and gender itself were deemed to be social constructs.[35] Everything was transitory and subjective; all tradition and custom was innately oppressive. In fact, Friedan was not anti-family as such; she was only protesting at the way it had confined women. In 1981, in *The Second Stage* she complained that she had been misunderstood and that feminism had done harm by the over-reaction against the family. But her 'recantation' was far too late.

Others had meanwhile leapt into the fray. Marriage was equated with patriarchy. In 1971 Germaine Greer called women to arms in *The Female Eunuch* by instructing them to refuse to marry and to reject marital fidelity; she also advocated communal childrearing. 'Our society has created the myth of the broken home which is the source of so many ills, and yet the unbroken home which ought to have broken is an even greater source of tension as I can attest from bitter experience,' she wrote.[36] By her second book, *Sex and Destiny*, her position had somewhat changed. 'If the Family pits itself against the power of the preachers of instant gratification,' she wrote, 'it presents itself as a stumbling block to the establishment of the consumer society, which has resources which can sweep the whole tenacious structure into the void. Among those resources are intelligentsia of every shade of red, pink and blue, who have mounted attacks on the Family from every side.' Too late she had discovered that a healthy conventional family is actually a protector against the forces of selfish individualism, which she herself had done so much to promote.

Although these ideas were driven along by revolutionary political and ideological theorists who were out to change society, that does not explain why such notions caught fire among the public. The explanation surely is that they suited the general mood and gave expression to feelings that a lot of people shared. There was a feeling of optimism, of the limitless nature of human potential, of an opportunity to create a new society based on freedom and social justice to replace the old culture of subservience through gender, class and race. People's attitudes had changed: mass education, the advent of television, young people with money in their pockets, the rise of the consumer culture and the advent of the contraceptive pill all contributed to a new cultural climate of voracious individualism. People measured what was being said and written by their own experiences and found enough there with which they could identify. Some families *were* oppressive. Many women *did* feel trapped. Racial and sexual minorities *were* discriminated against. What was less apparent, however, was that dealing with these often legitimate grievances by trashing all institutions, rules and authority would promote a pathological victim culture within a wider environment of social anarchy which would be inimical to reason and to progress itself.

Into this cauldron dropped the liberalisation of the divorce law in 1970. In the words of the principal historian of divorce Professor Lawrence Stone, no fault divorce on demand was launched just at the point when a social and cultural hurricane was beginning to blow through western society. The new law both reflected and helped to promote the change in the culture that was under way. Ostensibly passed to buttress marriage by allowing the 'empty shells' to be discarded, it had the opposite effect. The divorce rate rocketed, way beyond the number of failing marriages which had been effectively bottled up. The subsequent rise in divorce, wrote Stone, was due to the increasingly pervasive ideologies of individualism, the pursuit of personal happiness and the expectation of speedy gratification. Modern technology, the rise of consumerism and unparalleled affluence had made the idea of deferred gratification redundant, eroding the sense of obligation and responsibility within the family and society at large. Unrealistic expectations had been created of sexual and emotional fulfilment; there was more stress on rights and less on duties. 'As a result, many marriages which would have been regarded as tolerable in the past are today seen as unendurable,' he wrote.

Sexuality, marriage and procreation all became detached from each other. If the focus of adult relationships had moved over the centuries from property to love, it moved now again from love to sex. The spiritual trinity of love, sex and children was shattered and the sexual act became instead the apotheosis of self-gratification. The result of the new permeability between the state of marriage and unmarriage, the rise of cohabitation and the collapse of the stigma associated with children born out of wedlock, was that there was no longer any standard for the orderly progression from one stage of family formation to another, nor even one single socially acceptable mode of bonding. Yet the history of family relationships had been marked by a series of both informal and increasingly formal measures imposed by both church and state to protect individuals and society at large from the anti-social consequences of irregular unions. 'In this respect,' Stone observed, 'we have, quite suddenly, and perhaps only temporarily, created a situation not dissimilar to that which prevailed in early modern England.'[37] In other words an advanced industrial society, under the destabilising influence of extreme individualism, was now dismantling centuries of social progress and reinstating a form of social medievalism under the banner of enlightened attitudes.

Values were now turned on their heads. In this brave new world of moral relativism, where moral judgments were anathema and discrimination had become a term of abuse, there was a new tyranny of egalitarianism. All lifestyles were morally equal. What was deviant had become normal, so what had been normal now had to be stigmatised as oppressive (itself an ironic paradox in this apparently judgment-free universe). Prejudice was no longer attached to individuals but to whole classes; as a result, it was transformed into oppression. What mattered now was not whether behaviour was 'right' but whether it made people feel good. It became more important to spare people's feelings than to spare the consequences of their anti-social lifestyles upon others. Indeed, by definition such people couldn't behave anti-socially because they were themselves victims of oppression (with heterosexual men mainly cast as the villains of the piece). Thus any attempt to maintain that the two-parent family was still a preferable option for society was denounced as stigmatising broken families. Lone parents became a new oppressed class.

In 1980 a book appeared entitled *Splitting Up: Single Parent Liberation*, edited by Catherine Itzin. In the preface, Paul Lewis of the National Council for One Parent Families wrote of the 'cruel practical difficulties which society puts in the way of one parent families reaching their potential as whole, happy and well-balanced families' and of the effect of poverty, especially if the lone mother had a new baby. There was no acknowledgment of the part individual choice may have played in arriving at such a parlous position. On the contrary, the individual was not responsible; the state was to blame for everything. One-parent families, wrote Lewis, should be seen 'not as a growing problem but as the vanguard of a revolutionary change in the structure of society'.

It was made to appear that if it wasn't for the wickedness of the state in deliberately crushing this new victim class, British society would have basked in a kind of osmotic nirvana of personal relationships. In *The Making of the Modern Family* in 1977, Edward Shorter wrote: 'The nuclear family is crumbling – to be replaced, I think, by the free-floating couple, a marital dyad subject to dramatic fissions and fusions and without the orbiting satellites of pubertal children, close friends or neighbours . . . just the relatives, hovering in the background, friendly smiles on their faces.'[38] What was supposed to happen to the 'pubertal children' in this paradise wasn't clear. They were now so obviously an inconvenience they simply had to be air-brushed out of the picture altogether.

The family had been redefined as the sum of individual decisions uprooted from tradition, custom or obligation. It was a short step from the new short-term contract culture of marriage to regarding the family as just another consumer choice among many. If there are no binding obligations, if commitment is infinitely negotiable, then individuals are free to shop around for sexual partners with whom they may make arrangements to suit their particular circumstances. Within the family itself, children themselves came to be seen as merely products. As Halsey wrote, children had become like commodities which adults could choose to have in preference to other consumables, like cars or videos or holidays.[39]

In this universe of subjective values, children became objects. Despite all the child-centred rhetoric, they were not seen as individuals in their own right at all but as extensions of the adult ego. It was the adult woman who had rights. She had the right to prevent a child from being conceived. She had the right to get rid of an unwanted child, which was seen as having no independent claim on life at all since it was merely an extension of her own body, over which she had the right to do as she wished. And she had the right – facilitated by technological advances – to be provided with a child, to remedy the distress of herself and her partner (if there was one) caused by their childlessness. The desire to have a child became redefined as a need, a need so grave that it was translated through the good offices of medicine into a right to be granted by whatever means science could devise. Having a child thus changed from being the fulfilment of a covenant with the future to being the completion of an adult's egoistic ambitions. The perception of the child itself also changed. It was no longer an individual laying his or her own claim to a share in the human project, but rather an adjunct to the adult's own universe.

The modern child had become instrumental to the desires of the adult world: in the schools, as a means of transforming society, and in the home as a means to further the demands of adult fulfilment. Concern for the child's own welfare now had to take a back seat. The way the relativist ratchet had turned was well illustrated by the announcement by the British Medical Association in 1996 that it was no longer opposed to surrogate motherhood. It had previously objected to the practice on the grounds – among others – that predictable conflicts between the two sets of parents and subsequent confusion over identity would cause the child distress and harm. It was now dismissing those ethical objections on the grounds that such confusion in the mind of a child was now commonplace anyway. 'Complex familial relationships are now the norm for many people and cannot be seen as necessarily damaging,' it said.[40] In a few insouciant words, this pillar of the establishment conveyed the desperate degeneracy of our most fundamental assumptions about a child's need for security and attachment. Just because familial limbo was now unfortunately all too common, it appeared that anyone could now treat children like dumpable objects without a second thought.

As Christopher Lasch had observed: 'Indifference to the needs of the young has become one of the distinguishing characteristics of a society that lives for the moment, defines the consumption of commodities as the highest form of personal satisfaction and exploits existing resources with criminal disregard of the future.'[41]

It's not just that adults now see children as extensions of their own egos. It's also that they no longer see a clear or valid distinction between the status of the adult and the status of the child. The boundaries have become dangerously blurred. Equalisation has meant that just as the classroom teacher now feels obliged to take a back seat while the child gets on and educates itself, so parents have been crippled by a disintegration of their own authority. They no longer feel willing or able to tell their children what to do. Instead of adults teaching their children, they have come to believe that their children have to teach them.

The Collapse of Parental Authority

Parental authority collapsed for a number of related reasons. There was the romanticisation of childhood that started with Rousseau and progressed through Dewey and was helped on its way by the revulsion against the harsh discipline and exploitation of child labour in the nineteenth century. There was the trumpeted death of God and the collapse of the authority of the Bible, stripping parents of their ability to summon up a supreme authority to back up their dealings with their children. And then there was Freud. Parents now realised that the way they dealt with their children – whether they picked them up too much or too little as babies, whether they were open or secretive in what they said, whether they were strict or lenient with them – would have unalterable consequences and could mark their child for life. As the American author Marie Winn wrote, parental confidence began to evaporate.

'Suddenly, parents were burdened with the fearful knowledge that only they stood between their child's growing up "normal" and his growing up to be a neurotic, screwed-up adult.'[42]

The result was a change in the focus of child-rearing away from the child's socialisation and upon its mental health. The 'inner' consequences came to matter far more than the 'outer', regardless of the fact that the latter were crucially related to the former. A collaborative relationship developed between parent and child. As Winn observed:

> No longer does the parent operate from his vantage point of superior knowledge, of *adult* convictions. Rather, the child is enlisted as an accomplice in his own upbringing. At every step along the way the parent tries to discover why the child is behaving as he is behaving, why he, the parent, is reacting as he is reacting. And when the child misbehaves, the parents do not use 'child psychology' to bring him around to more socially acceptable behaviour. Instead they try to make him *understand* why he feels the way he does by providing the sort of enlightenment a therapist might give a patient . . . Everywhere today parents are explicating the texts of themselves, trying to get their children to understand, to agree, to forgive, instead of simply telling the kids what to do.[43]

Many parents fell into the trap of thinking that by being 'liberal' with their children they were acting in their best interests. In fact, these parents confused liberalism with studied indifference. They thought that by demonstrating tolerance of their children's attention-seeking or rebellious behaviour, they would effectively buy their children's approval and gratitude. The opposite was often the case. 'Liberal' parents felt that by not disapproving they were being tolerant. As has already been noted, tolerance, however, implies a prior position of judgment and disapproval; without such disapproval, tolerance is merely a polite word for indifference. In the absence of clearly defined boundaries, children had nothing to rebel against, nothing against which they could define their own identities by proving themselves to be the opposite.

The outcome is children who are rootless, unhappy and often out of control. Teachers report difficulties with even very young pupils as a result. A paper written in the late 1980s by two lecturers in

education, Jean Lawrence and David Steed, reported that a retreat
from clear-cut notions of morality, accompanied by a shift towards
more consensual and negotiated agreements between adults and
children, had damaged the authority of both parents and teachers.
Younger parents were ambivalent about authority and what to tell
their children. Parents either gave up the attempt because they
felt helpless and powerless, or took refuge in authoritarianism.
Teachers said children couldn't make friends or work co-operatively
in groups and used aggression to demand attention they didn't
know how to attract or to cover their own sense of inadequacy or
bewilderment.

Often, children from well-off homes acted up the most, said the
authors, because their parents didn't give them enough time and
attention. They had never had to conform to any rules over, say,
bedtimes or TV watching. 'Many parents so lack confidence in
giving guidance that by the age of seven they regard children as
being old enough to decide matters for themselves and give them
completely free rein. Many children, lacking clear frameworks
and guidance from supervision or lack of insistence on suitable
behaviour achieve a degree of autonomy and employ strategies
which amount to blackmail: tantrums, stubbornness and argument
are all used with tired and anxious parents to extend this area of
independence. They employ the same strategies in school with their
teachers.'

Some parents, they wrote, were guilty about saying no to anything. Others were bewildered and unsure what to say. Children
learned as a result that adults were either volatile, inconsistent,
violent or malleable. Many parents swore or were aggressive and
advised children to behave accordingly towards their teachers.
Many children were growing up lacking a consistent mother or
father figure and experienced continuous changes of adults at home.

'Teachers note that disturbances at home affect children in a
number of ways,' they wrote. 'They seem unable to concentrate
or listen; they are restive; often they will ask "what do we do?",
interrupting teachers as they attempt to explain. Many seem
incapable of carrying out even simple tasks such as changing
for P. E., dressing themselves and using a knife and fork.' They
didn't listen to instructions, answer or come when called. 'Often
they have no idea that behaviour corrected by teachers is wrong;
many can't put their ideas into words and their frustration turns

into aggression . . . Many seem to be incapable of co-operating in play and have not developed skills of extracting themselves by walking away from trouble; ganging up provides a kind of support, and many trouble-makers find mutual support by mixing with each other.'[44]

As parental authority has declined, so the influence of the media and the peer group have increased. Half of all 7- to 10-year-olds and 70 per cent of 11- to 14-year-olds have a TV set in their room. Knowledge of the outside world is mediated by adults less and less. As Lasch observed, this precocious knowledge that children are absorbing, derived from TV, films or magazines, is then used as a stick to beat their parents with if they fall short of ideal behaviour.

> Formerly it was parents who were self-righteous. Now they are unsure of themselves, defensive, hesitant to impose their own standards on the young. Forced at every turn to defend their authority by abstract standards of justice and legitimacy, they fall back on argumentation, negotiation and covert manipulation – 'stage managing'. The same thing happens at school. The distance between teacher and student narrows, the teacher becomes an 'opinion leader', and the curriculum stresses 'realism' – how to get along with others, not how to think for oneself.[45]

This exposure to the adult world has made children particularly precocious when it comes to sexuality. Youth culture, including that directed at pre-teenagers, is saturated not just with explicit sexuality but with anti-family messages of nihilism and degradation. This is particularly true of American rock music, described by the American cultural commentator Allan Bloom as a non-stop masturbational fantasy which presents sexual desire as the be-all and end-all for children who have less and less means of understanding love, marriage or family or the connections between the three. Many modern lyrics dwell on transitory sex that is cold and sadistic.

Popular culture makes a great deal of money out of the mass sexualisation of children. Their music, their clothes, the TV and films they watch envelop them in a miasma of sexuality. TV programme controllers who say that the plot lines of soap operas are moralistic are missing the point. Children who are glued to these dramas and learn through them to make sense of the world are turned into voyeurs at the peepshow of adult behaviour, drawn into a brutally reductive adult culture in which sex is divorced from spirituality or commitment and marriage is devalued or ridiculed.

Magazines read avidly by young teenage girls such as *TV Hits*, *Sugar*, *Bliss* or *Just Seventeen* are heaving with graphic 'advice' on sexual matters, such as whether to sleep with mum's boyfriend or have oral sex. Such items are justified by the publishers on the grounds that they offer 'information' to children and are fulfilling a public service. As often as not, they say, they advise against 'having sex'. This may be true. But they assume that the 13-year-old girls who read these magazines are capable of making informed adult choices about such dilemmas as whether to have sex with their boyfriends or to have a lesbian affair. They present as normal to children a brutally crude, mechanistic world in which sexual gratification is merely another commodity, available not just to adults but to them, along with the clothes, CDs and cosmetics.

Yet parents often connive at such precocious nihilism and present their prematurely sexualised children as role models – against a background of ostensible public revulsion at child sexual abuse. Thus the press featured the story of Helen Benoist, a 'wild child' who had left Whitley Bay High School at 15 to dance with a topless troupe and at 16 was stripping off in nightclubs and dating the 55-year-old nightclub owner Peter Stringfellow, who claimed to have bedded more than 2,000 girls. But it was her parents' attitude that was startling. Her mother was 'quite happy' with Helen's lifestyle and her father, an environmental health officer, said his daughter's happiness was most important. Any duty to prevent her from harm was not mentioned.[46] Helen was following in the footsteps of Emma Ridley, another teenage nightclub stripper whose mother connived at her lifestyle of under-age sex and marriage at 15 to a Los Angeles nightclub boss twice her age. And Mandy Smith was encouraged by her mother to move in with the ageing Rolling Stone Bill Wyman when she was just 14.

The number of parents who act in quite this way may be very small. But these examples illustrate a general phenomenon, that children are being treated as adults. The very notion of parenting appears to be in the process of being written out of the script. The idea that childhood and adolescence involve a gradual initiation into the human condition through graded steps appropriate to a child's age has been destroyed. Television, movies, videos, popular music, clothes: the whole popular culture industry exposes them to the full panoply of adult influences without any graduation.

Not only does the adult world now expect children to be mature, but the roles are being reversed. In 1986 a book was published to give advice to children whose parents were getting divorced. *Dinosaurs Divorce: A Guide for Changing Families* by Laurene Krasny Brown and Marc Brown was illustrated with cartoons of dinosaur parents who fought, drank too much and broke up. It offered children advice such as : 'Try to be honest if they ask you questions; it will help them make better decisions', or 'Living with one parent almost always means there will be less money. Be prepared to give up some things.' What the book was actually doing was asking children to be sympathetic, understanding, respectful and polite to their confused, unhappy parents; the sacrifice was going to have to come from the children. They were expected to be *more* mature. In the world of divorcing dinosaurs, the children rather than the grown-ups were to be the pillars of patience, restraint and good sense.[47]

This breakdown of the distinction between adults and children can cause a confusion of roles that is devastating for the child who is expected to become, in effect, the parent's parent. Instead of being looked after, the child finds it is assuming the caring or protecting role, a development which is often unconsciously encouraged by the parent.

All the boundaries have now become permeable. Marriage merges into unmarriage. Men merge into women as traditional gender roles blur. Home and outside world merge into each other through television. Children merge into adults. Pupils merge into teachers. Private lives are on public display every day. Private sexual behaviour becomes public titillation. Everyone defines their own boundaries – including children, who need boundaries to be set for them. But children are now seen as able to deal with whatever life throws at them.

According to Professor David Elkind of the Department of Child Study at Tufts University, Massachusetts, this new perception did not appear because of some new insight into children. 'It emerged because post-modern parents *need* competent children,' he wrote. 'We need children who can deal with out-of-home care from an early age, who can cope with divorce and who will be left unfazed by seeing people murdered in the streets or behaving wildly on drugs. The media reflect this new image of child competence.'[48]

So we now have our knowing, street-wise children, reflected by media that pander to the image. The notion of child competence is now also being peddled by philosophers who are keen to push the doctrine of personal autonomy to its extreme limits. An example of such a *reductio ad absurdum* was provided in an essay by the philosophy professor John Harris, published in 1996. Advocating children's 'liberation', he wrote that children were as 'entitled to the same concern, respect and protection as are adults' and that any claim that a particular child was not competent to exercise such rights should be tested just as it would in the case of an adult.[49]

But this role confusion is catastrophic for children. Artificially minimising the distinction from adulthood and bestowing upon children a body of 'rights' may appear to some to be 'empowering' children. But it actually cripples their development. Children need to be treated differently from adults because they *are* different. Giving them 'choices' before they are mature enough to know what choices mean is merely another species of neglect. A good example of this is the Children Act 1989, a pioneering piece of legislation that gave children a battery of 'rights'. These rights produced some perverse results which have caused children serious harm. Because children now have the 'right' not to be touched by a professional adult other than in severely restricted circumstances, staff are powerless to prevent delinquent children in children's homes from wandering about, getting into trouble with the law or working as street prostitutes. The Act also gives children the 'right' to have their wishes taken into account during court proceedings to determine with which of their separated parents they will live. On some occasions, the court has acceded to the child's wishes with catastrophic results – in extreme cases, leading to violence by the custodial parent against the child. In every such case, a burden of choice is imposed upon the child for which most are not ready. One child psychiatrist wrote that he wondered whether the interests of children had not been put back years by the Act. It had been written, he said, by lawyers and civil servants with little idea of the reality of disturbed children and complex family relationships.[50]

One of the reasons for the Children Act was the succession of child care scandals in which grossly unsuitable carers had assaulted, neglected or otherwise abused the children in their care. The solution to this problem should have been to reform the caring system, but this was deemed impossible. The chosen solution was instead to 'empower' children by equipping them with rights. So at the very heart of the Children Act, the view was enshrined that children had to be protected from the adult world. Adults were dangerous. Children were the repositories of wisdom. It was a fallacy that went right back to Rousseau. The adults who were most dangerous to children, however, were those who had failed to discharge their responsibilities to children by ensuring that professional carers were adequately selected, trained and supervised and who instead chose to dump the burdens on the children themselves by saddling them with inappropriate 'rights'.

Children cannot deal with the burdens of adult choices and behaviour. Far from liberating them, it makes them confused, frightened and unhappy. It is hardly surprising that they become depressed or take refuge in drugs or alcohol. Liberty and equality are for adults, not for children. Nor, despite their worldly-wise appearance, do children become precociously mature. Those children who have been deprived of their childhood try instead to cling on to that childhood longer; sometimes, their entire adult lives are lived under a shadow.

Importance of Family

The conventional family, along with the relationship between adults and children, is in the process of being deconstructed. Some of those who actively pushed this process along did so with the revolutionary aim of transforming western society. In the main, however, the change was widely embraced because of the pressures of a highly individualistic culture committed to egalitarian democracy. If this process is not checked, however, it will destroy that very democracy and individual liberty which gave rise to it in the first place.

De Tocqueville observed that an individualistic society depended upon the communitarian family for its existence. The nuclear family is not only the key transmission mechanism for a culture, but it is the building block of liberal democracy. It teaches independence, self-restraint and responsibility, all of which are essential to a free democratic society. If the family fails to pass these virtues on, democracy itself is threatened.

The conventional family traditionally fostered individualism because of its structure. Marriage was controlled by the church rather than the clan, and so couples were allowed to decide for themselves whether to marry. It was a consensual union in which the husband couldn't easily dispose of his wife for his own convenience, and so – despite the dreaded paternalism – it produced something approaching equal respect, gradually promoting privacy and a further impetus to individual fulfilment.[51]

As the French anthropologist Emmanuel Todd has suggested, family relations provide the model for political systems and define the relationship between the individual and authority. The nuclear family structure, he wrote, was incapable of renouncing individualistic values. Sexual liberty was often associated with political liberty, but in fact sexual liberation often led to a reaffirmation of state authority.[52]

The conventional nuclear family structure produced autonomous individuals who were both free and responsible, holding a balance between individualism and social responsibility, acquisitiveness and altruism. One could almost say, therefore, that the breakdown of the family helped create the social conditions for the advent of Thatcherism by creating individuals freed from those constraints upon selfish behaviour. The danger lies in the political system that will arise if the deconstruction of the family continues unchecked. Whatever it is, it won't be a liberal democracy safeguarding the rights of the individual.

The paradox of individualism is that it contains the seeds of its own destruction, and the implosion of the family structure which has nurtured it could be the catalyst for such a disaster. As Allan Bloom wrote in *The Closing of the American Mind*, the tension between freedom and attachments is the permanent condition of mankind. In modern political regimes, freedom now has primacy over family, community and even nature. Concern with self-development has revealed itself to be inimical to community. Attachment today is conditional. 'Of course many families are unhappy,' he wrote. 'But that is irrelevant. The important lesson that the family taught was the existence of the only unbreakable bond, for better or worse, between human beings.'

Bloom noticed special handicaps among his students whose parents had broken their attachments to each other. In a chilling insight, he wrote of his students whose parents were divorced:

> I do not have the slightest doubt that they do as well as others in all kinds of specialised subjects, but I find they are not as open to the serious study of philosophy and literature as some other students are. I would guess this is because they are less eager to look into the meaning of their lives, or to risk sharing their received opinions. In order to live with the chaos of their experience, they tend to have rigid frameworks about what is right and wrong and how they ought to live. They are full of desperate platitudes about self-determination, respect for other people's rights and decisions, the need to work out one's individual values and commitments, etc. All this is a thin veneer over boundless seas of rage, doubt and fear.[53]

Bloom's students feared both isolation and attachment and lacked confidence in the future; their enthusiasm was replaced by self-protectiveness. These students appeared to be a paradigm of modern society itself, embodying the spiritual and emotional isolation caused by the loss of attachment to other human beings and the natural order. The loss of such attachments, to both family and to established traditions and cultural frameworks, spells disorder, both in the lives of unattached individuals and within the environments which they then so brutally lay waste.

Chapter Ten

THE DISORDERED CHILD
Crime

Crime is above all a moral issue. It is also the preoccupation of the age. The level of public disquiet at its inexorable rise distorts the political debate as politicians dart from one short-term panacea to another, conveniently leaving office before they have to face the consequences of yet another failure to bring the true rate of crime down. It is hardly surprising, however, that their 'solutions' appear to have so little effect since they invariably fail to tackle the causes of the phenomenon. The argument divides those on the political right, who think crime is entirely the fault of human behaviour, from those on the left who think it is entirely the fault of economic circumstances. These are both distorted perspectives.

The causes of disorder are highly complex and difficult to disentangle from each other. But it appears clear that the roots of crime lie in a breakdown of the moral sense which occurs in certain circumstances, leading to a collapse of both formal and informal social controls. Individuals internalise a moral sense as they develop through childhood and adolescence. It is acquired through a secure attachment to their families and to the surrounding culture, through which they learn the elementary codes of human behaviour and the relation between acts and their consequences. But in recent years there has been a comprehensive breakdown of such attachments. Family life has become conditional and contingent; employment is either insecure or non-existent; religious belief has been eroded; schools, both in what they teach and the way they teach it, increasingly abandon children to their own devices.

Instead of authority, firm rules and fixed boundaries which define the world as something intelligible to which the child can become attached, there is now merely an endlessly shifting landscape of subjectivity and ambiguity. The child has become an autonomous and solitary individual, left alone to construct his or her own meaning from the world. Who then can be surprised to see more and more children breaking rules that the adult world tells them over and over again have no absolute validity? Who can be amazed when such children appear to have no connection at all to the social structure? Yet the facts should shock and disturb. In 1995, a report by Dr Clive Wilkinson for the National Youth Agency found that in some areas up to 10 per cent of 16- to 18-year-olds had no contact with official society at all: they were an alienated underclass who had barely attended school after the age of 12 or 13.[1]

Family and crime are symbiotically linked, and not merely because the implosion of the one leads to the explosion of the other. They are linked now by a shared erosion of certain key values. Every society needs order if it is to survive. To have order, a society must have values. Those values, expressed most vividly in the changes within education and family life, have become fragile and tenuous. There has been a breakdown in moral transmission from one generation to the next. In the adult world, both parents and their surrogates – teachers, social workers and increasingly the judiciary and other members of the establishment – are retreating from the parental role of promoting the care, control and development of children. In particular, they display a failure to recognise the need for clear moral judgments, discipline and punishment as part of a child's social learning process. Their retreat from this agenda marks a retreat from the principal duty of adults to socialise a child.

Although the official statistics for recorded crime are deeply ambiguous and open to manipulation, there is no doubt that crime is a worsening social plague. Since the middle of the 1950s, crime rates have rocketed upwards without any sustained periods of decline. In 1950, about 500,000 crimes were recorded, but this has now gone up to nearly six million recorded crimes per year. And recorded crime is merely the tip of the iceberg; the real rate is very much higher.[2]

Figures ending June 1995 showed a 5% fall in the annual recorded crime figures for the second year running, and the first fall in violent crime – by 2 per cent – for almost 50 years. But these figures should be taken with a pinch of salt. Not only did they follow a 6 per cent increase in violent crime in 1994, but they also included a rise in the number of murders from 668 to 729 and an increase in the number of robberies. Moreover, changes to the classification of various types of violent crime also put such figures in doubt. The more authoritative British Crime Survey reported a rise in all crime and in crimes of violence, which are estimated to have risen by about 15 per cent over the past 15 years. During the 1980s, the rise was particularly steep; recorded property crime rose during that period by about 100 per cent and violent crime, according to the British Crime Survey, by about 21 per cent.[3]

Nevertheless, Britain still has a very low murder rate by comparison with other countries. Random killings remain very rare. But what has caused such anxiety is a perceived change in the nature of the violence being committed. A few examples illustrate the trend.

Michael Carling, aged 60, the mayor of Tickhill, South Yorkshire, was beaten senseless when he tried to stop vandals destroying a flowerbed. The vicar reported him as saying in hospital after an operation to repair his jaw that he could not understand why a simple and minor request like 'Please don't stamp on the flowers' should bring such a savage reaction and get his head kicked in.[4]

Paul Brighton, aged 44, a father of three, had his skull crushed by a gang of youths. He had chased them after they threw beer cans through his window while he was watching TV with his family.[5]

On V.E. day a 90-year-old woman, Annie Saxton, was stabbed, beaten, sexually attacked and strangled in her house.[6]

Alan Sturgess, aged 47, died of a heart attack after his car was pelted with bricks by a mob of jeering teenagers outside his house.[7]

A 14 year-old girl was thrown out of a train in Croydon by a man after she refused to drink from a can of lager he offered her.[8]

To the horror of the nation, a south London headmaster, Philip Lawrence, was murdered in December 1995 after he went to the rescue of a pupil who was being attacked by a gang of youths at the school gates. Following this crime, a reporter from the London *Evening Standard* visited the area to talk to local youths. He found threats of violence were commonplace and the carrying of weapons was normal. A gang of 13-year-olds tried to swipe his mobile phone, pushing and swearing. One boy said: 'There's nothing to do in this area apart from the youth clubs. And they're not that good. We need some arcade games in the centres.' A 19-year-old girl, asked what she thought of youth facilities, kicked the reporter in the shin and said she didn't like people asking questions. She then punched the photographer in the face. Another youth pulled out a knife.[9]

In October 1995, the Archdeacon of Sheffield talked about society breaking down on the Manor estate in the city where after dusk youngsters smashed and burned fences, stole cars and threatened residents.

According to Dr Masud Hoghughi, the child psychologist who for many years ran the Aycliffe youth treatment centre in County Durham, crimes of violence were the most seriously under reported of all young people's crimes. They were increasing in number and gravity and the age at which they were committing them was getting younger. The under-17s were thought to be responsible for up to one third of sexual offences and up to 50 per cent of arson. Increasingly, he wrote, more young people were operating at the edge of society's tolerance. The increasing availability of weapons, the spread of drug crime, greater violence in computer games and amplification by the media were all creating an atmosphere in which violence was common.[10]

Of course, people have been alarmed by a perceived breakdown in social order for decades. That does not mean, however, that things haven't been getting worse. What has changed is that a current of violent and murderous rage appears now to be flowing just below the surface of society. An air of unease hangs over our public spaces because human behaviour has become much more unpredictable. Crimes of violence *are* increasingly random. They are also becoming grossly disproportionate to any provocation, and more sadistic. This kind of crime speaks not so much of opportunism as nihilism. It means one no longer feels safe sitting at home in one's own front room watching TV.

What also appears increasingly clear is that it is too simplistic to talk of a disappearance of the notion of right and wrong. Most of these young offenders know perfectly well that burglary or violent assault is wrong. But they know it as a kind of abstract code that has no relevance to their own lives or to the effect on other people. Indeed, the most chilling aspect of many of these crimes is that they appear to be committed with a complete absence of empathy, insight or understanding of what it means to *be* another person. These young criminals cannot make the link between acts and their consequences. There is nothing in their own lives to enable them to realise that anti-social acts cause other people pain and grief. There *is* no other. The only thing that has meaning now is the self. Is that so surprising? That is, after all, precisely what they have been taught by the adult world.

But it is not just the children who are out of control. In many cases, the parents connive at their misbehaviour, taking their part against the teachers who seek vainly to discipline them. One inner city headmaster, in a school top-heavy with poverty and broken families, described graphically the unholy alliance of children, parents and many of his teachers who were linked in the same culture of self-absorbtion and instant gratification.

A significant minority of children come here at age 11 pushing, punching and swearing and many haven't learned to read or write. The primary schools are far too informal and lacking in rigour. The parents aren't coping with their children. The children aren't coping. The children are fearless; they have no respect for their parents or for the police. They are no longer wary of the adult world. They don't connect actions to consequences. They believe they will always be baled out if they get into trouble.

Parents often walked into the school, he said, and started threatening the staff if their children had been disciplined by them. They often behaved as if they were their children's equals. One mother was discovered brawling in the school lavatories with a child who had been annoying her daughter.

Parents say the most astonishing things to me. One child had sworn at one of our teachers, but the mother's attitude was that the teacher had done something to deserve it. She said she remembered what it was like being picked on by the teachers. These parents were educated in the 1970s when the schools were in total disarray.

The teachers here tend to have different views about whether they embody authority or not. I imposed the old traditional ethos by reintroducing school uniform and putting up an achievement board on the wall. The teachers go along with it but mostly not with any enthusiasm. I'm for ever picking up pupils for their behaviour or the fact that they're not wearing the proper uniform; most of the teachers probably wouldn't bother.

In West Africa I sat in a mud hut and watched three-year-olds being taught their tables by a ten-year-old. Most of these teachers here find it a struggle just to keep these children sitting down in the class. The children don't see the value of education in the way they used to do. They look at the people who've got the big BMWs around here and they see they are drug dealers. Their families don't value education. These children are rich. Even our poorest children have got what they want already. They don't need to strive in the way my generation did. They've got no staying power. They can't even walk to school; if there's a train strike, the parents say they can't come in because they can't make a 40 minute walk.

The primary schools teach them an ethos of boredom. They're taught to stick at something only until they get bored with it, then move onto something else. These children live their lives on a high level of excitement and stimulation. They are being trained in habits of mind that are profoundly antisocial. When they first come here, sitting in a chair is foreign to them. Their perceived needs have to be met immediately. If another child annoys them, they hit him straight away.

Children misbehaving in class are generally unhappy because of their disrupted family backgrounds. The level of violence isn't necessarily greater but it's less predictable, with a greater element of randomness. We get very young mothers on their own, for whom their sons are like little brothers. The parents often see their children as social accessories until they reach puberty, when it all breaks down. These parents can't accept that anything is their fault. They always blame everything onto someone else. I'm in despair over it. I feel a stranger in my own country.[11]

Our culture now actively promotes the expectation of instant gratification. Our education system is now constructed around the absolute need to stop children becoming bored or having to make any effort which may cause them a degree of discomfort. For these children, failure is not an option. The result is that, with no challenges to meet, many regard the whole endeavour as a waste of time. This culture holds that children have no staying power and so it serves up everything in bite-sized chunks before they've had a chance to digest what they've already been given. The result is that their ability to concentrate is destroyed. In their family relationships, children are taught by example that if the other person doesn't come up to scratch they can be traded in for a new model. There is no question that anyone should have to repress their desires for a greater good. Repression is, after all, the greatest crime that can be committed against individual fulfilment. Every relationship, including that between parents and children, is negotiable. Television, videos, computers and the Internet give them immediate results with hardly any effort. Popular culture tells them constantly they have a right to have what they want and to have it *now*. Thus, the concept of discipline has almost entirely lost its meaning. Its very premise is considered an affront.

The high rates of crime and incivility, coupled with the perception that fewer and fewer people have the will or the ability to exercise judgment or authority, have created a climate of popular anxiety and fear. A MORI poll revealed in 1995 that one in five people now carried protection when leaving home, more than one in four had given up walking or public transport because they felt frightened of being attacked, and nine out of ten people felt that the likelihood of their becoming a victim of crime had increased over the past 10 years.[12] Researchers at Leicester University reported that shoppers were deserting the high streets for shopping malls because they feared crime and felt threatened by beggars, drunks and vagrants.[13] Such fears tend to be dismissed by 'expert' opinion as unfounded.

Statistics are regularly trotted out to show that one's chances of becoming the victim of a violent crime are minuscule, even more so if one falls into the most nervous categories of women or elderly people. This argument, of course, misses the point. It is because of their fear that these 'nervous' categories largely remove themselves from those public spaces where such crimes are committed, thus creating a self-fulfilling prophecy. The argument also takes no account of the perceptions of insecurity and menace fuelled by those everyday incidents and incivilities which do not figure in the statistics but are sufficiently commonplace to influence the public mood.

But the argument that such fears are unfounded is merely one in a series that seeks to deny the realities of crime. In particular, strenuous attempts are made to blame crime onto economic factors such as poverty, unemployment and social deprivation, and thus by extension onto the government. A typical example was the report written in 1995 for the Employment Policy Institute by John Wells of Cambridge University, which claimed that unemployment and poverty caused crime. Although Wells acknowledged that not all unemployed people resorted to crime, he suggested that the growing sense of illegitimacy of a social order that countenanced mass unemployment along with increasing income inequality could eradicate the normal constraints on crime.[14]

However, all the available research indicates that the relationship between unemployment and crime is very much more ambiguous. Crime took off and kept rising steeply from the 1950s onwards, throughout the so-called golden period of rising disposable incomes and low unemployment. In the late nineteenth century, crime rates fell steeply even though there was little movement on relative poverty. There was a striking decline in relative poverty between 1961 and 1979, when the average standard of living rose by about one third while that of the poorest tenth rose by one half. But during that period there was an equally striking rise among males in drug abuse and crime. The classic study for the Home Office by Simon Field in 1990 showed that, over the long term, crime rates correlated with rates of consumption. There was no correlation between rates of crime and the unemployment that after a time lag follows any fall in consumption.[15] According to the criminologist Professor David Smith, crime couldn't be associated with inequality either, because less developed countries had more inequality and lower crime rates.

In Sweden, by contrast, the crime rate had increased sharply during a post-war period of high economic growth. This was a period in which inequalities in income were being reduced and in which Sweden was introducing one of the most comprehensive social welfare programmes in Europe. 'At the national level,' according to Professor Smith, 'high crime rates are associated with wealth, economic growth and more even distribution of incomes.'[16]

In the short term, however, it's true that unemployment can influence crime, largely by combining with other key factors to create a network of circumstances in which crime is encouraged to flourish. The fairest thing one can say at present about the relationship between unemployment and crime is that it is complex and not yet understood. But the evidence to suggest that crime is primarily a disorder of affluence is batted aside by the 'liberal' intelligentsia whose political beliefs remain constructed around the doctrine that poverty and inequality are the sole causes of social ills. Poverty is indeed an ill in itself and makes many problems infinitely worse. But the simplistic and ideological reductionism of the left, mirrored by the equivalent prejudice on the right which sees man as fundamentally flawed and evil, blinds them both to the fact that crime is the product above all of a society in which attachments have become fractured, in which social relationships have been squeezed out by impersonal and anonymising forces, in which structures that once bound people together in a sense of common and shared endeavour have broken down and in which as a result both external and internal social controls have been eroded.

As Rutter and Smith wrote in their survey of youth disorders in 1995, moral standards disintegrate under certain social conditions. 'There is good experimental evidence to demonstrate that these mechanisms of moral disengagement do operate,' they said. 'If societal trends tend to strengthen some or all of these mechanisms, then the increase in crime could be explained not by a decline in personal or moral standards but by an increasing tendency to disengage these standards in a selective and self-serving manner.'[17]

In other words, moral behaviour is influenced by the prevailing culture. And that culture has come to embody distorted relations within the schools and the family. The responsibility for rearing and disciplining children has been taken away from parents and given to 'experts'. During the 1950s and 1960s, crime was relabelled 'deviancy', offending was transformed into a matter for therapeutic inquiry and punishment became yet another concept to be consigned to the relativist garbage can. As Christopher Lasch observed, during the 1940s and 1950s the therapeutic model was extended to the law in a battle against retributive justice, a trend that started in America under the impact of its massively influential psychopolitical culture. In 1949, the American judge Hugo Black declared: 'Retribution is no longer the dominant objective of criminal law. Reformation and rehabilitation of offenders have become the important goals of criminal jurisprudence.' In the same year another lawyer wrote: 'The concept of treatment has replaced the concept of punishment.'[18]

As the criminologist Jock Young has written, the period from the late 1950s to the early 1970s saw a ferment of ideas in criminology in the attempt to reconstruct the narrative of crime: labelling theory, new deviancy theory, conflict theory, subcultural theory, control theory, neo-classical theory. Despite these bewildering changes and the even more bewildering jargon, the impact of all these ideas was to overturn the way ordinary people as well as the academics viewed crime and the attempt to contain it. The overriding impression created by these debates was that nothing worked. The theory that social conditions caused crime had collapsed in the face of statistics which showed that crime had gone up while jobs, disposable income and welfare spending had also gone up. The theory that crime could be controlled by hitting the criminal harder had taken a dive when higher spending in the United States on police and prisons failed to make an impact either.[19] The effect of this double-barrelled defeat was to encourage people to deny that there was any real problem at all and to claim that it was all a case of 'moral panic'.

But one idea had taken firm and unshakeable root. This was the therapeutically influenced perception that criminals behaved the way they did because of deficiencies in their upbringing which meant they didn't stand a chance of leading law-abiding lives. Therefore they couldn't in all justice be punished for their transgressions. They needed instead to be 'helped'. Punishment was therefore barbaric, a double whammy perpetrated against the victims of circumstances. There was, of course, no place in this thinking for the victims who had suffered from these criminals' actions. And there was no recognition of the absolutely crucial fact that one of the main ways in which such offenders were indeed the victims of their circumstances was that they had never been given the discipline, including punishments for their misdeeds, which is an essential constituent of parental love. They were, in fact, crying out for discipline and for proper parenting. Instead, they were treated with an indifference masquerading as benevolence in a world where moral judgments had become taboo.

Punishment became linked to revenge. The old canard about 'an eye for an eye' was trotted out, in ignorance of its actual biblical meaning – which was not revenge but proportionality in justice. The notion of 'just deserts' is an essential component of the strong feelings about the essence of justice which we all innately possess. It is a notion which links punishment to fairness, and it means that if punishment is denied, a powerful impression of injustice and unfairness is created. Yet that is precisely what happened, aided by the growing perception that of the three-pronged rationale for punishment – deterrence, reform and retribution – deterrence and reform didn't work. Neither fines nor imprisonment, which had become the principal means of punishment, seemed to deter the majority of criminals from reoffending, and so far from reforming them, prison appeared merely to reinforce their antisocial behaviour. To a utilitarian age, convinced that the only good in an action lies in its measurable consequences, the notion of just deserts cuts little ice if the consequences are ambiguous; and if it undercuts the principles of personal autonomy and freedom that are so central to our individualistic society, it becomes intolerable.

Criminal Justice

In the 1950s and 1960s the philosophy of criminal justice was predominantly consequentialist. Social workers, psychologists and criminologists gradually transformed its character to an emphasis on rehabilitation and reform in which the notion of punishment became outmoded and unhelpful. Among academics, however, the tide has turned somewhat over the past 20 years and the idea of retributive justice has revived. Victims of crime are now taken into account. The idea has gained ground that punishment is required by justice to remove the unfair advantage that the offender has taken, as has an acknowledgment that punishment expresses condemnation which is not just important in itself but may make possible a moral dialogue with the offender. In other words, punishment has become recognised in some academic circles as a species of paternalism designed to nurture the moral autonomy of the individual.[20]

According to this analysis punishment is an educational process, teaching offenders a moral lesson they will understand and accept for themselves. It is the act not of a harsh but of a caring society. It presupposes that the offender is attached to a set of values defining his identity as a moral being. It shows him how seriously society takes the wrong that has been done; it also signals that the debt has been paid and life can go on.[21]

Punishment is thus crucial to the establishment of a moral sense, both within the individual and within society. Without a moral sense there can be no prospect of any end to disorder. But, regardless of any developments of thought in academic circles, the agencies that come into contact with offenders are viscerally hostile to the very notion of punishment. There is a deep reluctance even to use the word and bafflement at the idea that it is essential. 'Why do you have to insist on using it?' asked one chief probation officer in some irritation. The word just wouldn't be used, he said, by probation officers. 'They do accept the notion of punishment, but they don't call it that. Instead they talk about offenders accepting responsibility for what they did, or confronting their offending behaviour.'[22]

But that wasn't the same as punishment at all. Why was there such extreme reluctance even to use the word? Because, he said, it was associated with physical punishment and with cruel behaviour. But punishment needn't be physical or cruel. A child can be punished by being deprived of pocket money, or by being forbidden to watch TV. The problem with punishments inflicted by the state, however, is that they are so remote and impersonal they often fail to make the kind of impact on the offender's life that can be achieved by depriving a child of its favourite possessions or activities. That is a very real difficulty; but retreating from the concept of punishment itself is a category error.

Masud Hoghughi, the child psychologist, states:

Punishment is the deliberate infliction of hurt in retaliation for wrong done. Unless you can create circumstances whereby antisocial acts result in an immediate and unmistakeable negative reaction to the perpetrators, you don't make the behaviour you disapprove of unacceptable. The only way you can define criminal acts as unacceptable is by making the consequences unpleasant. We don't do that with juveniles because as part of our liberal development we have removed the negative consequences of actions and made them indeterminate and subject to so many interpretations they don't make sense to young people. So they are not punished for their wrongdoing.[23]

Instead, it is punishment itself that is classified as wrongdoing. The report of the Gulbenkian Commission on Children and Violence in 1995 muddied the distinction between violent or physical punishment of children on the one hand, and all punishment on the other. Its opposition to the former merged seamlessly into opposition to the latter. Retribution, it said, should have no part to play *whatever* in the criminal justice process for juveniles. It said: 'The commission does not accept that the seriousness or violence of a crime is in itself a reason for restriction of liberty of a juvenile; to allow such considerations to determine sentencing and placement is incompatible with the aim of rehabilitation. Considerations of retribution, and retributive public opinion, should play no part in sentencing or treatment.'[24]

This assumption that the criminal justice process for juveniles must be concerned *only* with rehabilitation and the protection of the public, and that there was therefore no place whatever for any notion of just deserts or punishment of any kind, was startling. Not only was there no place for the moral sense, but the Commission made it clear that morality had no place in the juvenile criminal justice universe at all. It said: 'In considering violence and how to reduce it, a concentration on moral judgements is unlikely to be helpful, acting as a distraction from issues amenable to change.' Remarkably, it appeared that for the Commission the only considerations that were relevant to juvenile crime were social deprivation, discrimination and access to firearms.[25]

There is evidence that this moral confusion appears to be now spreading from the therapeutically-influenced caring professions into the more traditional reaches of the establishment. In 1994, a boy of 14 stripped a boy of 11, physically assaulted him in a seven-hour attack, used a set of keys to scratch his throat and back, twisted his neck until it cracked and smashed his head against a fence until he passed out. The victim eventually recovered. At the age of 16, the attacker was given a 12-month supervision order by Mr Justice Sachs at Preston Crown Court which he said was to enable him to develop 'social skills'. Dr Halla Beloff, a social psychologist at Edinburgh University, said later: 'Young people who engage in property offences might be likely to get far harsher decisions from magistrates and judges, and this suggests that if you do terrible things to a little boy that's not so bad.'[26]

In 1995 Antoinette Campbell, aged 23, left her six-month-old daughter alone in her flat for 30 hours during which the child died of starvation and dehydration. She said she had wanted 'a bit of freedom and a bit of space for herself' and admitted behaving selfishly. Sir Lawrence Verney, the Recorder of London, gave her three years probation because she needed 'help not punishment'.[27] There are, of course, cases where mothers kill their babies while under the influence of depression or psychosis. But this sentence seemed to imply that mothers are simply incapable of the type of monstrous selfishness which requires a moral response – even where the offender herself has acknowledged as much.

In 1995, a 13-year-old boy who raped a girl of nine received a supervision order at Cardiff crown court from Mr Justice Jowitt who told him to 'think about purity and being a decent fellow'. The little girl told police he had been performing 'a sexual act' in front of her about 20 times and had threatened to kill her.[28]

Meanwhile, social workers attempting to deal with juvenile delinquency appear to be paralysed by the moral confusion embedded in their own professional culture and within the Children Act. In 1995, four teenagers who smashed up a house in Southampton after absconding from a council home were taken on holiday to Turkey for a week, after which three ran away. A social services spokesman commented: 'The purpose of the holiday has, disappointingly, not been achieved.'[29]

In the same year, social workers apologised for the behaviour of three boys in their care, aged between 12 and 14, who were responsible for a trail of destruction including £10,000 worth of damage done to cars at a garage, graffiti daubed on church walls, stolen Communion wine and rifled church boxes. Sonia Haywood, Wiltshire's assistant director of childcare, said the staff had been powerless to stop the children from leaving the home.[30]

What has disappeared from these children's lives is the understanding by the adult world, embodied in natural parents, their surrogates and the law itself, of what a proper parenting role should entail. This is despite evidence which suggests that the poor quality of family relationships is of overwhelming importance as a determinant of crime.

As the Home Office reported in 1995, the quality of the family relationships among delinquent children is pivotal. 'Those with a poor relationship with one or both parents are more likely to be inadequately supervised by their parents, to truant from school, to associate with other delinquents and ultimately to offend themselves,' it said. 'Young people living with both natural parents were found to be less likely to offend than those living with one parent or in a step family, but these differences are largely explained by differences in the quality of the relationships between young people and their parents and the capacity of parents to exercise effective supervision.'[31]

All the evidence points to strong links between poor parenting and delinquency. As the authority on delinquency Professor David Farrington has written, antisocial behaviour develops when the normal learning processes based on rewards and punishments are disrupted by erratic discipline, poor supervision, parental disharmony and antisocial parental models. Broken homes and early separations, he wrote, also predicted delinquency in many studies. Teenage mothers were more likely to have delinquent children, probably because several other risk factors came into play like poor child-rearing methods.[32]

A study carried out in Newcastle by Professor Israel Kolvin and others, reported in 1990, found that social deprivation on its own was not sufficient for delinquency to develop. Good parenting, it concluded, protected children against the acquisition of a criminal record. Yet despite this and much other evidence, the pathology of family life is persistently downplayed in analyses of crime. Even where the importance of family and parenting are acknowledged, such discussion tends to concentrate on external factors affecting the family such as poverty or unemployment; due obeisance is paid to teaching 'good parenting' skills which, in its failure to address what is actually going wrong with parenting, manages merely to be a glib and meaningless platitude. So what is the reason for the reticence?

The main reason is that the family is the most explosive issue of all, lying at the very heart of the conflict between the culture of individualism and the need for a cohesive society. The implications of its current deficits and disintegration present 'liberals' with such unpalatable challenges to their deep-seated beliefs and to their own lifestyles that they shy away from recognising its central importance in tackling crime.

In particular, there is a deep reluctance to accept that family breakdown contributes to children's delinquency. To believe that it plays a part is *not* to say that family breakdown inevitably produces children who become delinquent. That is clearly not the case, any more than it is true that poverty inevitably breeds disorder. It is not social circumstances *or* individual behaviour by themselves which cause crime; it is the interaction between the two. Family breakdown on its own doesn't lead to crime, any more than does inadequate parenting in intact families without any other contributory factor. It is rather that these deficiencies in family life create instead emotionally vulnerable or damaged children who react to social problems such as poverty in a different way from children who are emotionally healthy and intact. People with different personal biographies react differently to similar circumstances. That is why video nasties have catastrophic effects on some children but not on others; that is why poverty does not always produce criminogenic families. The roots of crime lie in the way all these factors work on each other. But it remains the case that the single most constant factors are family disorder and inadequate parenting, that whole constellation of problems in the families of delinquent children.

The central purpose of the family is the nurture and socialisation of its children. The families of delinquent children fail on both counts. The core tasks of parenting are the care and control of children which create the conditions for their healthy development. Both care and control fail in delinquent families. Such families find parenting extremely difficult. They have little knowledge about good parenting behaviour, not least because they themselves were often so poorly parented. They tend themselves not to be stable, sensitive, mature or communicative characters. They don't understand the desperate need of children for unconditional love and fair and consistent discipline. They don't command respect. Their attempts to set boundaries are weak and inconsistent; sometimes they collude with their children against authority, sometimes they punish them with great harshness. If they've been in trouble with the law themselves, they find it even more difficult to lay it down to their children. The harshness of their own upbringing makes them feel that they and their children are at the mercy of uncontrollable events. They think they love their children, and yet – by leaving them, for example – they don't always behave in their children's interests.

These are some of the conditions that create the damaged and vulnerable children who consequently lack a moral sense, who lack the ability to connect actions to consequences and who lack the sensitivity and ability to empathise with others. These deficiencies precondition them to crime if other circumstances – unemployment, educational failure, poverty – come into play. Yet there are still dangers in such an explanation. An analysis that says delinquent children are merely the helpless products of their dysfunctional families can lead to an ideology of emotional determinism no less pernicious in its effects than the economic determinism of the egalitarian left. It leads to the sex offender telling his psychologist that the reason he committed the rape was that he was himself the victim of child sex abuse. But of course there are many victims of child sex abuse who do not commit rape. To think that behaviour is determined by our family pathologies not only flies in the face of reality but denies individual moral responsibility.

For professionals to come along and say in effect that the delinquent child is a helpless victim of his circumstances colludes with and reinforces that vacuum in responsibility and in so doing helps seal that child's fate. Yet that's what our professional response has done to delinquent children. Children learn a moral sense first by social learning, by observation and imitation of those closest to them, but also by consequences. They learn that criminal actions are unacceptable only if they see clearly that the consequences of such actions are unpleasant. There has to be an immediate and unmistakeable negative reaction to the crime. But we don't do that with adolescents. We have now developed such a sense of guilt about the terrible lives they have led that we give them messages through the criminal justice system which are at best highly ambivalent. We have confused explanation with justification. We find the thought of punishment abhorrent. Yet without punishment – and punishment does not have to mean imprisonment – we cannot expect these adolescents to understand that what they have done is unacceptable. Worse still, we have retreated from setting any boundaries for them at all.

And then, having been so badly failed by both his natural parents and his state 'parents', having been handled with kid gloves for his delinquency rather than treated to the tough love that might give him the beginnings of a moral sense, the young adult criminal finds that the full fury of the system descends upon his head and he is thrown into a prison from which he will doubtless emerge confirmed in his criminality and possibly with a drug addiction to boot.

According to Hoghughi, the main reason why crime policies fail is the inadequate attention they pay to the fundamental laws of human behaviour. Behaviour, he believes, is shaped by its consequences. But the offender's balance sheet is complex. The anticipated rewards from crime include material gain, peer approval, revenge, pleasure, excitement, the release of tension and self-enhancement. But that balance sheet won't be unduly affected by conscience, shame or public censure because these things don't usually figure much in the adolescent's emotional repertory.[33]

And that is because such an adolescent has been made to feel there are no worthwhile connections between himself and the society he inhabits. His family, his school, his place of work, the role models in his culture and the institutions of the state all pump out the same relentless message. He is merely a commodity in a marketplace, uprooted and disconnected from any fixed and stable structures. No-one will teach him the judgment he needs to connect with the mainstream. He must make his own meaning where he can. He has been taught not only that nothing matters because it has no objective value, but that things that don't matter are therefore equal and that such equality is itself the highest good. Insecurity and unattachment are his watchwords. He is on his own in a world where only the fittest will survive.

Chapter Eleven

THE NO BLAME, NO SHAME, NO PAIN SOCIETY
'No Such Thing as Society'

Individuals have been left stranded by the politics of both left and right. The paradox is that both were ostensibly committed to granting individuals' desires. It has been a great mistake to imagine that the Labour and Conservative parties have embodied opposing political philosophies. It is more accurate to say that since 1979, despite the distracting rhetoric of adversarial combat, they have represented but two sides of the same individualistic coin. The left stood for egalitarian individualism in the social sphere, for the doctrine of equality of values and lifestyles; the right stood for libertarian individualism in the economic sphere, for the doctrine that those who could achieve wealth and success should be left alone to do so while those who lost out would have to go to the wall. Neither stood for a culture based on altruism, fuelled by a principled concern for other people. The moral relativism of the left was thus the mirror image of the debased liberalism of the right, which wasn't surprising because both sprang from the same intellectual ferment back in the eighteenth century when the rule of authority was overthrown and the autonomous individual emerged supreme.

Some time during the 1950s, the political left lost its grasp of the language of moral discourse. As a result of the rise of the individualistic, consumer culture and the collapse of the Church, altruism began to wither away and individuals handed over the duty of responsibility to the state. In 1960, the great ethical socialist thinker R. H. Tawney wrote that the world was in retreat not merely from particular principles but from the very idea that political principles existed. Morality that transcended economic expediency and the belief that the ends didn't justify the means seemed part of a 'remote and worn-out creed'.[1] Tawney's moral stance fell utterly out of fashion. The view that character and choice affected conduct was derided. Individual responsibility was considered of little significance compared to the forces of economic circumstances. So ethical socialism was left vulnerable to Marxism which destroyed it from within and laissez-faire economics which assaulted it from without.[2]

Mrs Thatcher came to power in 1979 proclaiming a return to Victorian values. There was little sign, however, that she had any understanding of what Victorian values actually were. The Victorians had achieved the quite extraordinary feat of remoralising a society in danger of fracturing under the double impact of industrialisation and the shattering collapse of religious authority. It had pulled off this achievement by rooting the individual very firmly in society and in a moral ethic which started from the notion of the common good. As Gertrude Himmelfarb has observed, the 'self' to the Victorians meant something rather different from its 1980s incarnation. It was rooted in the social norms and approbation of others, and entailed duties and responsibilities as well as rights. By contrast, today's 'self' is narcissistic and does not have to prove itself by reference to any values or people outside itself. For the Victorians, self-help was centred in family and community; among the working classes, this came out as neighbourliness, among the middle classes as philanthropy.[3]

Mrs Thatcher, however, took on the individualist agenda in isolation. In her hands, it became simply a reaction against the corporate state and against any kind of collective activity at all. For her, the individual stood in opposition to society. 'There is no such thing as society. There are individual men and women, and there are families,' she once famously remarked.[4] It was an observation which revealed her confusion between the collectivist organisation of the state and the normal relationships of a coherent culture of shared objectives and endeavour. Her political project was conceived almost entirely around a narrow, utilitarian economic model of human behaviour. The result was an atomised creed which lent itself perfectly to political opportunism, populism and consumerism. It was set up explicitly around the pivotal figure of the individual consumer. It thus promoted a culture of individualism that destroyed attachments and lent itself to relativism, by telling everyone they were all equally entitled to make their several demands. There would be no arbitration between them from any fixed position embodying the common good, but it would be left to market forces to determine which of these demands would be the fittest to survive. Choice was elevated to be the sacred principle of this religion of the self.

It was also a deeply philistine doctrine. It was certainly not respectful towards tradition or culture. It was, after all, a revolutionary creed that pinned the blame for Britain's decline on its institutions. Hostile to privilege and deference and by extension to the associated ethos of *noblesse oblige*, it was accordingly malevolently disposed to any distinctions between individuals based on the claims of education or professional training, and saw these instead as passports to feather-bedding that needed to be scrapped. It was suspicious of all élites as a conspiracy against the laity. In this, it had much in common with the left which similarly despised British institutions, except where it could control them, as for example in local government. So both left and right formed unholy alliances to level down all such distinctions and bring the élites to heel.

In medicine, for example, the internal market which transformed health care from a service to a series of rolling business contracts was introduced in the late 1980s with no trials and virtually no professional consultation. And despite the ferocious rhetoric employed to denounce its introduction, the fact was that the managerial left loathed the medical profession and was quietly delighted with a system that transferred the power of hospital consultants to themselves. Similarly, the 'cardboard cities' that aroused so much indignation among the Conservative government's opponents were also testimony to this unholy alliance. It was an article of faith among so-called liberals that the disintegrating family was an unchallengeable 'right', as was the freedom for paranoid schizophrenics and other mentally ill people to live free of institutional restraint. The Thatcher government, from the other side of the individualist mirror, cut welfare benefits and hospital beds in the interests of reducing the reach of the state. The result was teenagers who were fleeing from their fractured families and mentally ill people to whom no hospital could or would offer asylum living on the streets in cardboard boxes. Thus crude populism marched hand in hand with egalitarian ideology. And as the Thatcherite economic and political hegemony became more entrenched in the national arena, so the relativists of the left correspondingly dug themselves in more deeply behind the barricades of family, schools and identity politics in the private domain.

National politics both reflects and influences a culture. By their words and deeds, politicians contribute towards setting a moral tone and can help raise or lower standards of behaviour throughout society. Political leadership under the Thatcher/Major government, however, embodied the values of selfishness, irresponsibility and deceit. This was the government of the me-society, which in its cynical manoeuvres embodied above all the principle of self-interest – its own. It thus sent out to people from the example of its own behaviour an unmistakeable set of messages: that selfishness was desirable, that the end justified the means, that truth was infinitely malleable, that winning was its own justification, that other people were disposable and that the only duty of individuals was to themselves.

At every opportunity, the government sloughed off responsibility. It set up a vast and burgeoning quangocracy, devolving the administration of government to appointed bodies or to outposts of the civil service. The effect was that lines of command became blurred and it became difficult to hold anyone to account for anything. Escapes of prisoners from the high security Parkhurst and Whitemoor jails in 1995 were blamed on the quasi-autonomous Prison Service and on the prisons themselves; ministers never accepted responsibility. People died on hospital waiting lists because of shortages of intensive care beds; again, ministers blamed the hospitals but never accepted responsibility.[5] The report by Sir Richard Scott into the arms for Iraq scandal revealed that Parliament had been misled; no minister accepted responsibility.[6] Public service was dismembered and still no minister took responsibility for anything that had gone wrong.

The government had elevated 'dishonesty' into an art form. As Ian MacDonald, head of the sales secretariat at the Ministry of Defence, so memorably observed in evidence to the Scott inquiry: 'Truth is a difficult concept.'[7] The phrase 'economical with the truth', first coined by the then Cabinet Secretary Sir Robert Armstrong during the Australian court case in 1986 in which the British government tried to ban Peter Wright's book *Spycatcher*, became a by-word for the verbal sophistry that was standard at Westminster and in Whitehall.[8] It was no longer possible to shock the public by any suggestion that a minister had lied because the public had now come to accept that this was the way politics was conducted. No politician could be trusted; there was an automatic presumption of fast dealing.

Government was in fact constructed on a pyramid of deceit. Facts were manipulated as a matter of routine to maintain the fiction that the business ethos which now prevailed in the public service had brought about improvements in public life. Hospital waiting-list figures were manipulated downwards while patients were well aware that they were being discharged from hospital before they had properly recovered in order to bump up the 'throughput' figures. If the same patients returned to hospital (which wasn't uncommon if they had been sent home prematurely) those re-admissions bumped up the throughput figures even more. And there were other ways in which patients were double counted. Then there was the fiction of the endlessly rising standards in education; and the many changes in the way unemployment statistics were compiled to reduce the total number of people listed as out of work.

Revelations that members of Parliament were taking money from outside lobbying firms sent out the message that MPs might be bought for a price. These revelations led to the appointment of the Nolan committee into standards in public life.[9] There were also many examples of financial irregularities or incompetence by health authorities, government ministers and quangos, in which huge losses of public money had led not to disgrace but to promotion or knighthoods for the perpetrators, that the Public Accounts Committee produced a special report into unprecedented levels of public sleaze.[10]

All of this contributed to an unparalleled cynicism about public life and a collapse in the authority, not merely of the government of the day but of the entire political class. Political leadership came to excite disbelief, cynicism and irritation on an alarming scale. But at the same time, this display of values was corrosive. The message it promulgated was that everyone was out for himself. Public life and positions of leadership were perceived to be merely avenues for self-promotion. Authority itself was contemptible because it ripped off the public. The public was assimilating the impression that authority could – indeed, should – be disregarded. At the same time, however, it was also digesting authority's own behaviour which, however disapproving the public appeared to be, sent out the signal that the attitudes it embodied were normal and even tolerable. It told people they could behave badly and get away with it. 'What kind of role models are these for our young people?' asked a headmaster in despair.[11]

If the style of government promoted a collapse of authority, the substance helped reinforce the culture of rampant individualism. One of the fundamental tenets of the Thatcher/Major government was that individuals should be freed from constraints; they could only flourish if the state was off their backs. Accordingly, deregulation was a key plank of their policies. There was even a Deregulation Minister. There was, however, no acknowledgment that a civilised society can only proceed if there *are* constraints on behaviour, that regulation is therefore a necessary part of life if a society is to try to avoid harm, and that freedom must be balanced against responsibility. Deregulation brought in its wake runaway financial borrowing, dangerous railways and beef that became suspected of destroying human brains.

In 1996, the government proposed to deregulate gambling by allowing more casinos to be built, ending the prohibition on advertising, reducing waiting times for membership and allowing payment by debit card for the first time. The junior Home Office minister, Timothy Kirkhope, said: 'What struck me was that the people in casinos were all very smart and normal. It was nothing at all like a James Bond film set. The people seemed happy. Why should we stop people choosing to spend their evenings playing roulette or blackjack?'[12] But gambling had been regulated for good reasons. It had been considered highly undesirable, encouraging profligacy and irresponsibility and leading people into ruinous debt; so although it wasn't outlawed, it was considered that nothing should be done to encourage it. But now the guiding principle had changed. Instead of doing nothing to encourage a harmful activity, the government felt it must do nothing to prevent people from their right to have fun. Fun and harm appeared to be mutually exclusive, particularly if you happened to be wearing a suit at the time.

The same attitude underpinned the National Lottery. This was a way of sloughing off onto the public by way of a sweepstake some of the responsibility and accountability for raising money through taxes for public services. It furthermore did so by institutionalising not only greed for riches beyond the dreams of avarice but also the something-for-nothing culture, the dissociation of acts from their consequences, the disregard of the needy and vulnerable and the promotion of gambling addiction. It was also a good illustration of the way in which official acts influence public behaviour, however much the public may say it disapproves of them. Some of the charities whose fund-raising was said to have been badly affected by the huge sums the public was pouring into the Lottery tried to boost their incomes by adapting a similar scratch-card game – which was particularly implicated in promoting addiction to gambling.

All this demonstrated how John Stuart Mill's definition of harm to be prevented (the only criterion which justified interfering with people's behaviour) was infinitely redefinable under the impulse of individual selfishness. Indeed, politicians genuflecting to the market forces of libertarian individualism, who placed a very high value indeed on having fun, appeared to be able to redefine harm with effortless ease. In January 1994, the then Deregulation Minister Neil Hamilton made a speech to the Libertarian Alliance, a group which supported the legalisation of sado-masochism and drugs and the privatisation of the army, in which he called for the deregulation of all personal and social behaviour.[13]

Corruption of Liberalism

Libertarian Tories, like the libertarian left, simply couldn't understand that licence was not synonymous with the authentic liberal values of a free society. The unfettered market cannot produce a civilised culture because it sets citizens against each other for personal gain instead of working together for the common good. It is not underpinned by virtues such as trust, integrity or altruism but is a savage, unprincipled structure in which the weak are junked as trash. But government ministers persisted in endorsing its anti-social characteristics. In 1994 John Redwood, for example, lauded selfishness and declared that free markets enabled people to express their interests without interfering busybodies telling them how to run their lives. 'The public interest is taken care of by the private interest of wanting to make money,' he said.[14]

The most obvious illustration of the pernicious effects of this doctrine could be seen in the pockets of social devastation around the country where mass unemployment had laid whole communities waste. There was no public interest in the poverty this caused, nor in the depression and other illness that followed, nor in the erosion of the work ethic, nor in the destruction of one of the primary mechanisms for socialising young men, nor in the creation of that new phenomenon, the unmarriageable male whose prospects were so poor no sensible girl would have him. The damage done by mass and endemic unemployment was incalculable. Yet to the then Chancellor of the Exchequer Norman Lamont, unemployment was 'a price well worth paying'.[15] The message from that was that human beings were expendable.

This Tory ideology was a debased form of liberalism. It stemmed from a reading of early liberals such as Adam Smith which had never been applied so narrowly and in such a distorted manner, and certainly not in the heyday of those Victorian values Mrs Thatcher admired so much. Adam Smith, as we saw earlier, believed that self-interest was the motor of the general interest and that the enterprise of individuals, when left free of regulation, was capable of carrying the standard of material well-being to undreamed-of heights. But the Tory ideologues wrenched this doctrine out of its surrounding moral context, ignoring both Smith's observation that individuals had to apply self-restraint to control their selfish impulses and his prophetic fears about what might emerge from such an emphasis on business. 'These are the disadvantages of a commercial spirit,' he wrote. 'The minds of men are contracted and rendered incapable of elevation, education is despised or at least neglected, and heroic spirit is almost utterly extinguished. To remedy these defects would be an object worthy of serious attention.'[16]

Indeed, the defining motif of the Thatcher/Major years, the idea that economic self-interest was the principal motor of human behaviour and that crude materialism was all, was a travesty of Smith's thinking. As long ago as 1921, R. H. Tawney had sprung to Smith's defence against the same distortion. 'No interpretation could be more misleading,' he wrote, than to present Smith and other early liberals as the apostles of selfish materialism. On the contrary, they had been classical exponents of the great traditions of English liberty which dated back to the Middle Ages.[17] The fact was that these classical liberals had been rooted in traditions of moral authority upon which they explicitly depended, and which made their concept of liberty – which had been, after all, a defence against political and economic tyranny – into a noble ideal.

The corruption of liberal thinking on which the Tories drew so heavily can be traced through the *laissez-faire* economic theories of the mid-nineteenth century Manchester School of liberalism and the Victorian analyst Herbert Spencer to the post-war Austrian economist and fervid anti-socialist Friedrich Hayek, whose thinking had a profound effect on Sir Keith Joseph and through him upon Mrs Thatcher. It was in Hayek's writings that the modern individualistic ethos, based on a ruthless desire to stamp upon anything that might stand in the way of selfish fulfilment, found its perfect justification in political and economic theory, based as it was on extreme free-market ideas and a notion of the minimal state. It was also deeply and fundamentally philistine. As the political thinker John Gray has written, Hayekian liberalism suggested a view of society 'in which it is nothing but a nexus of market exchanges, such that allegiance can be secured to a liberal political order that is universal and embodies no particular cultural tradition. In this paleo-liberal or libertarian view, the erosion of distinctive cultures by market processes is, if anything, to be welcomed as a sign of progress toward a universal rational civilisation.'[18]

The outcome has been a self-undermining liberal project. As Gray has observed, the Tories' neo-liberalism destroyed the bedrock of a society which people imagined they had been elected to preserve. 'When our institutional inheritance – that precious and irreplaceable patrimony of mediating structures and autonomous professions – is thrown away in the pursuit of a managerialist Cultural Revolution seeking to refashion the entire national life on the impoverished model of contract and market exchange,' he said, 'it is clear that the task of conserving and renewing a culture is no longer understood by contemporary conservatives.' Liberal democracy itself was now imperilled as the market failed to deliver the goods demanded as 'rights' and utopia never arrived. Worse, unconstrained market forces undermined social and political stability, particularly through imposing unprecedented levels of economic insecurity and the dislocation of families and communities through the mobility of labour which in turn helped the disintegration of family life and the promotion of crime.[19]

In allowing this to happen, the Tories had reinforced those nihilistic cultural tendencies on the left to which they purported to be so implacably opposed. It was no accident. Both sides espoused revolutionary creeds. Both set out to overturn past authority in an attempt to remake society. Both systematically splintered all attachments in the misguided belief that this would set people free. Both were in the business of levelling down and eradicating distinctions in homage to the gods of personal autonomy and choice. The outcome has been a gradual dissolution of social and civic bonds.

As the political writer David Selbourne has observed of the civic order:

> To seek to 'pare down' its obligations to the citizen body to their 'inescapable core' is nothing but the mirror image of what the anti-social citizen would do with his own duties to his fellows. Thus to sell off, or 'hive off', to the 'private sector' parts of the prison and police functions, the Whitehall civil service, of the Inland Revenue, of the health service, of the Customs and Excise service, of the Schools Inspectorate, of the social security system, to turn public institutions with a long history of civic service into 'commercial delivery systems', is not merely good – or more often bad – house-keeping. It strikes blows at the nation's civic existence, at those institutions which, being in the public not the private domain, give shape, coherence, continuity, reassurance and identity to the citizen's life in the civic order to which he belongs.[20]

The neo-liberal Tories did much to further weaken those institutions which were already under siege from the secular trends of modern thought. Although professing themselves to be fiercely monarchist, they did nothing to insulate the monarchy from the balance-sheet mentality which demanded that even the Royal Family provided value for money. Following the example of the museums in her realm, the Queen was forced to turn Buckingham Palace into a business enterprise by opening it up to the public and charging admission. Aided by their allies in the tabloid press, the Tories helped reduce the Royal Family to a demotic peep-show which denied the validity of the monarchy as a public service and helped fuel the republican sentiments of the left.

In similar vein, their attacks upon the Church of England further weakened that already enfeebled institution. The Church, it is true, had entered the political fray by regularly criticising what it saw as a needlessly harsh political and economic culture. Instead of acknowledging that the Church spoke with any authority at all, the government behaved as if it had no right to speak its mind and that everything it said, whether in the field of social policy or theological doctrine, was unmitigated rubbish. The problem was, of course, that much of what the Church was saying was self-destructive, since in its anxiety to embrace cultural diversity it denied its own authority. But the government's contemptuous reaction could hardly fail to erode that authority even further.

Contract Culture

The most corrosive effect the Tories had upon the civic order was the supreme importance they ascribed to business principles, which ceased to be a means to an end and became instead an end in themselves. This meant that values were judged solely by measurable outcomes. If outcomes could not be measured, then there could be no value. So productivity, efficiency and cost-effectiveness became the guiding principles of the age. Of course these attributes are important. But other values may matter just as much, if not more; yet these were written out of the script altogether. So those principles that humanise business, such as trust and loyalty, were trampled underfoot in the rush to rationalise, downsize and privatise. A condition of permanent insecurity was now built into the system as employees were sacked or uprooted. The rise and rise of managerialism meant a corresponding loss of attachments and the erosion of professionalism. Trust, after all, lies at the heart of the social contract with the professions. Employers used to trust professionals to do their job properly and in return they received benefits such as incremental promotion and job security. That contract was shattered in the interests of flexibility, 'choice' and above all the cutting of costs.[21]

The result was the erosion of the concept of public service, that transcendental value which acts as society's invisible glue and which embodies a civic version of the common good. But because it *is* invisible, it didn't figure in the brutal utilitarian criteria that measured value and success. So the public services found that their role was redefined. The police, for example, were judged by the tangible result of catching criminals. But catching criminals is only a small part of their job. Their main role is to maintain public tranquillity, defuse problems and prevent crime. But these inconvenient facts were brushed aside. The result was that their ensuing preoccupation with clear-up rates distorted their efforts to prevent crime from happening in the first place, to the disadvantage of the public. But it wasn't in the police's advantage to prevent crime, since fewer problems in their patch meant less public money. Thus was illustrated the perverse effect of grafting a business ethos onto public service.

Public service took a similar and devastating dive in the beleaguered National Health Service. The value of health care was distorted by the high priority given to bringing down costs. So hospital beds were closed because they cost too much to keep open, and surgeons had to close their operating lists when the money ran out. The internal market meant that instead of the binding mood of common endeavour which was the animating spirit of the health service, hospital fought hospital and department fought department for money. Not only, therefore, did services deteriorate but altruism was progressively squeezed out of the system to be replaced by irresponsibility and even criminal neglect. The commercialisation of the blood transfusion service cut the service and threatened to destroy the 'gift relationship' of free blood, symbol of British altruism and of a shared commitment to the common good. As the *Sunday Times* reported, hundreds of tons of medical waste, including amputated limbs, infected blood and used syringes, were being dumped by rogue contractors in lay-bys, car parks and warehouses across the country causing an extreme public health hazard. The reason these rogue contractors were hired by the hospital trusts was that they were cheap and the trusts were looking all the time for ways to save money.[22]

Business principles caused high and low culture progressively to disintegrate. The BBC, despite the explicit public service commitment that was its *raison d'être*, found it was having to measure success in terms of its ability to win the ratings war with commercial television. The government encouraged the growth of cable and satellite broadcasting, regardless of their impact on taste and decency standards and the subsequent effect upon the BBC. The result was that the standard of BBC programmes tumbled as the ratings war was fought out on the terrain identified as the lowest common denominator of public acceptability. This led the BBC to broadcast programmes like *Confessions* – discussed earlier in the book – which were ethically unjustifiable, along with other tasteless and exploitative game shows and a relentless diet of voyeuristic sex. The press followed precisely the same trajectory. The quality newspapers one by one progressively abandoned their ethos of public service as they chased each other deeper into a market defined by hedonistic and frivolous values. But it was very difficult for anti-Thatcherites to attack these trends. The idea of public service in popular culture implied paternalistic assumptions about standards and what people should or shouldn't see or read. The notion that institutions like the BBC held the nation together was anathema to those on the left who thought it was élitist to believe that the BBC should stand for anything at all beyond the rainbow of individual choices. So it happened that the power of individualism triumphant on both left and right sustained the vandalism inflicted by commercial pressures upon British popular culture.

The disappearance of the concept of the common civic good hit the universities very hard as well. There, the ideals of liberal education had not been subverted merely by anti-education ideologues. The universities, and with them the culture of which they were the custodians, were caught in a pincer movement between left and right. The Thatcher government never understood or accepted the argument that education was important for its own sake. It made this quite clear in its 1985 Green Paper, *Higher Education Into The 1990s*, which called for higher education to serve the national economy more effectively. The Tories turned the universities into a kind of utilitarian entrepôt, in which the values of education were progressively undermined by the values of the balance-sheet.

The result was that free intellectual inquiry was all but snuffed out by business imperatives. In their new entrepreneurial mode, dependent as they had become on short-term contracts, academics increasingly inquired only into those subjects which were likely to produce conclusions that were sufficiently politically acceptable to attract funding. The crucial pooling of intellectual ideas started to dry up because of the fear that colleagues might derive from such exchanges the commercial advantage of a competing contract. Government funding increasingly depended on performance measured by indicators which had little to do with the traditions of academic inquiry or standards of excellence and everything to do with throughput and output.

As one university professor explained, the criteria meant that every three to four years everyone in a university department had to produce four refereed publications. The university received money in proportion to the number of publications and their quality:

But what criteria are used to judge the quality? The Higher Education Funding Council hires academics to judge whether they are of national or international quality. But what does that mean? It's entirely subjective, and open to manipulation and abuse. People with established reputations are more likely to get judged well than people who are unknown. There's no way these panel members can read all these scripts. And it's impossible to achieve consistency of comparison within any one discipline let alone between them all. Furthermore, it's impossible to find academics who know about these things. A professor of sociology on the panel, for example, may know everything about criminology but nothing about ethnicity which he's supposed to be judging.

It's impossible to produce four pieces of a very high standard in this time. Moreover, this means that longer pieces of scholarship which may take many years to complete and don't lend themselves to short publications don't get agreed because there's no money in them. It creates a corrupted atmosphere in which academics start to distrust each other. In this institution, our senior management wondered aloud what to do with staff who weren't publishing enough, whether they should they have their contracts renegotiated or be dismissed.

The way was also being opened, he said, to the government interfering through its assessors in what was being taught. During inspections of teaching, he said, assessors were demanding to see lecture notes. 'One academic who refused was threatened by his vice-chancellor with the sack. These assessors won't have a clue about what we're teaching. It's turning students and colleagues into informers. We've already had one suggestion that staff members should get colleagues' students to complain to them about those colleagues. This was turned down, but it was indicative of attempts to curry favour with the funding council paymasters.

'In England, people don't crush freedom by specific fiat. They do it by economic underfunding and bureaucratic action.'[23]

This 'never mind the quality, feel the quantity' approach by the government has powerfully reinforced the cultural relativism which simultaneously invaded the universities from the left. And the contract culture has operated in other ways to institutionalise the lunacies of the left in education. Universities receive money in proportion to the number and standard of the degrees they award. So there's intense pressure to award degrees regardless of the quality. One admissions tutor said that in 1994 he decided he should fail 11 out of 44 first year students who hadn't made the grade. But since this would have caused a crisis at the university because of the loss of fee income, only one was failed. The result was that the remaining ten students had to struggle to keep up.

Moreover, he said, inspections looked at whether the students were satisfied instead of assessing the quality of what was being taught. 'They ask how many firsts you offer and since the national profile is that, say, 8 per cent get firsts they want to know why you're not awarding that number. You open yourself to the criticism that you're not giving the students a fair deal. Now, our best students are probably as good as they've ever been. But we're giving 2:1s to some people on the basis of other elements of their degree while their German component may remain sub-standard.'[24]

The contract culture has undermined the public examinations as well. The examination boards tender for exam contracts with the schools. But the schools are under huge pressure from the exam league tables to show more and more passes at higher and higher grades. So the exam boards compete with each other to pass more and more students at higher and higher grades in order to get those contracts.

The London Mathematical Society, whose anxieties about maths standards were discussed earlier in the book, reported: 'League tables encourage schools to move their examination entries to what are seen as "less demanding" examination boards – and the working group was supplied with compelling evidence that this is happening. Moreover, demands on students and overall standards are slipping as the boards themselves compete to retain market shares.' Similar unfortunate effects of market forces, they said, were at work elsewhere in higher education. There were pressures to retain students for the funds they generated even if they failed their maths courses. There was also pressure to award more firsts and upper seconds, since this could be taken as a measure of effectiveness. All these pressures were driving down standards. 'The losers include the more selective departments whose traditions of excellence are undermined, and the high attainers (on whose future so much of the country's prosperity depends) who fail to be challenged and whose achievements are not adequately recognised.'[25]

So the balance-sheet obsession of the Tory right has joined forces with the relativism espoused by the left to destroy the building blocks of the culture. The university professors' dismay about the ideological erosion by the left of standards of mathematics or modern languages has been compounded by a corresponding onslaught from the ideologues of the right with their philistine contract culture which has institutionalised the very erosion of academic standards the Tories pretend to be halting. In fact, they are doing the exact opposite. They have bolted relativism into the system.

The Common Good

The factor common to both left and right in these matters is their refusal to recognise a concept of the common good. Both sides of the political divide subscribe to the firm dividing line between public and private morality laid down by John Stuart Mill, who ordained that individualism should only be circumscribed to prevent harm from being done. But apart from the subjectivity of harm as a concept, without any notion of the common good there can be no corresponding concept of common harm. And without any idea of common harm, the government feels it must not pass judgment on behaviour. The left marches behind precisely the same banner. The result is not only that it believes adultery can't be condemned, but also that the Conservative government sought to remove it as a wrong in its attempt to change the divorce law in 1996.

But public and private morality are in fact connected. Individuals' behaviour can have an impact on society which damages the social fabric. That is precisely why the state, back in the sixteenth century, moved to regularise marriage and divorce. The public commitment by two individuals to each other in a marriage does not merely bind them to each other; it binds them as a partnership into a relationship with the state and in which the state has an interest. That interest is largely connected with the importance to the civic order of children being properly nurtured and socialised into the culture. As a result, it used to be considered that it was in the public interest to lay down the conditions under which the contract of marriage could be broken – if not terminated by mutual consent – which were adultery, desertion or cruel or otherwise unreasonable behaviour by one spouse towards the other.

One of the great mysteries has been how the Tory party, which boasted for years that it was the 'party of the family', could have presided over its accelerating disintegration. Mrs Thatcher's government contributed to that disintegration through a tax and welfare system which relatively disadvantaged the two-parent family. Even more remarkably, it was a stern Presbyterian moralist, the Lord Chancellor Lord Mackay, who in 1995 introduced legislation to take the concept of fault out of divorce. This did more than remove the allocation of blame or responsibility from the divorcing spouses. Since the responsibilities of marriage itself were only defined in law negatively through the fault grounds in divorce, it followed that the removal of all such grounds would remove any legal concept of responsibility from marriage itself. This would effectively decouple marriage from public morality and make it, uniquely for a contract, an arrangement that one side could freely terminate against the other's wishes. How was it, then, that a government ostensibly committed to the family and a Lord Chancellor personally committed to traditional moral values and hostile to moral relativism could have been the architects of such a measure?

The answer was not just the desire to save public money through reducing legal aid for divorce. It also lay in the Millian split between public and private behaviour. This meant that the libertarian Tories, committed as they were to an extreme form of individualism, denied that the government had a duty to prevent a common harm. As the former policy adviser to the Thatcher government Ferdinand Mount made clear in his book *The Subversive Family*, the state could not interfere with the family because the family was itself the main bulwark against state oppressiveness. What these Tories failed to see was that unless the state propped the family up, it would have little to protect it against the huge cultural forces that would otherwise tear it apart.

The Tories accordingly took the view that the state could and should do nothing about marriage breakdown because it was merely the result of massive cultural forces. This took no account of the fact that the law not only reflects but reinforces the values of a society. The Tories also believed the law should not try to impose morality upon people. But in every other area, the law reflects, embodies and puts into practice a set of moral values. Indeed, in areas such as sexual or racial discrimination the law attempts to change people's behaviour. So it was only in the area of family law that this non-interventionist view appeared to hold. Thatcherism thought the best thing to do with the family was to leave it alone.

What was striking about the debates that took place over the Mackay divorce measure was the widespread incomprehension of how important it was to retain fault. Even where people understood that this would remove the concept of responsibility for bad behaviour from marriage altogether, they still said that fault was on the one hand an arcane abstraction irrelevant to real life and on the other that it was a major cause of pain and distress. The glaring contradiction between these two positions did not seem to occur to those who held them simultaneously.

But in every other area of the law, the prime consideration is not to avoid pain but to make people who have done wrong face the consequence of their actions. Removing fault from divorce was an attempt to sanitise proceedings by getting the parties to collude in an evasion of truth. It was claimed that matrimonial disputes were too sensitive and complex for anyone to apportion blame, and that the law had no business poking its nose into personal behaviour. But that was an argument for abolishing the whole of matrimonial law or anti-discrimination legislation. Such a premise certainly doesn't hold in the law of contract which enforces agreements and penalises broken promises.

Logic and evidence, however, were in short supply during the divorce debates. The consensus appeared to be that the main purpose of any legislation must be to avoid causing pain. Justice, it appeared, was hardly to be considered. The premise, of course was false. It was not legal 'fault' that caused pain but the fact of the breakdown of the marriage; it was not setting out the reasons for the breakdown that caused conflict, but the division of the spoils of property and children. And it most certainly didn't follow that the avoidance of pain for the adults, if this could be achieved, would mean the avoidance of pain for the children. The assumption that the one followed from the other echoed once again the more general cultural fallacy that children and adults had somehow become interchangeable.

Supporters of Mackay's divorce measure were motivated very largely by the desire to avoid pain. What was painful, it appeared, must automatically be bad. The idea that removing the painful consequences from an individual's anti-social behaviour might contribute to the stock of public harm, or to put it another way that the association of pain with anti-social activity might be in the public interest, met with widespread incredulity. Such is the state now of public morality in a highly individualistic culture, whose goals are personal gratification and the pursuit of an elusive human happiness.

Through taking up such positions and through their own behaviour, the libertarian Tories conspired to promote the moral relativism they claimed to despise. The Conservative government under Margaret Thatcher and John Major institutionalised through its political programme the no blame, no shame, no pain society. And in the process, it helped the disintegration of British culture itself.

Chapter Twelve

THE DECONSTRUCTION OF THE PAST
Cultural Frailty

A t some point during the latter part of this century, the British national story ran into trouble. The intelligentsia turned upon the culture of which they were the custodians and, like the mythological pelican which tore at its own flesh, began to consume their own *raison d'être*. As education and family started to crumble, a crisis revealed itself at the very heart of our culture and our civilisation. National and personal identity appeared locked together in a fatal embrace as they slowly started to disintegrate together.

This profound atrophy of the human spirit is the belated legacy of the Enlightenment. As such it affects, to a greater or lesser extent, the whole of the modern western world. But it has not affected every society in the same way. Some cultures appear to have resisted its full nihilistic impact. It is those countries which have particularly individualistic traditions – notably America and Britain – where the rot has set in most deeply.

Indeed, one might go further and suggest that America was the principal conduit for this cultural dismemberment. It was the American 'new man' of the eighteenth century, the rational, self-interested individual building a country free of the encumbrances of the past, who was such a focus for Enlightenment enthusiasm and ideas. It was precisely that American free and entrepreneurial spirit that caused de Tocqueville in the nineteenth century to write in *Democracy in America* of the ambiguous and damaging effects of individualism: he saw this led people to isolate themselves from society which is left as a result to look after itself. It was America which institutionalised the therapeutic culture which replaced the question 'Is it right?' by 'Does it feel right for me?' And it was America which, in a line stretching from John Dewey to Carl Rogers and Sidney Simon, decoupled education from authority and institutionalised moral and cultural relativism in the attempt to engineer an egalitarian society.

The influence of American attitudes throughout western culture has been enormous. But it has not been universal. One might speculate that one reason why France, for example, has not experienced a deterioration in its own culture was because it fended off American influence in an explicit defence of its own language, history, traditions and artistic expression. It had a strong enough sense of its culture to enable it to do that, rooted as it was in a very clearly defined French identity which found its first expression at the time of the French Revolution. That simply wasn't the case in Britain, where American influence can hardly be over-estimated. The natural ties of a common language and a shared sense of history were amplified in political and cultural terms by the Special Relationship with the United States formed after the Second World War, in which the balance of influence lay very strongly with America. It was a relationship founded, after all, upon a perception of Britain's intrinsic weakness, embodied in the Attlee government's perception of Britain's diminished strength in the post-war world.[1]

Nevertheless, despite the influence of America upon Britain as a whole, the cultural weakness that it exploited was particular to England. There are significant differences in attitude towards education and family between the indigenous English and the country's minority communities, as well as between the English and the Scots, Welsh and Northern Irish. Virtually all these non-English cultures have a much more acute sense of the importance of education than do the native English. One can suggest a number of reasons for this: particular intellectual traditions, such as the Scottish Enlightenment; the sense of being a beleaguered and struggling nation (beleaguered, as often as not, by the English); the vivid and painful experience of being an outsider community struggling to establish itself in the mainstream, evident among the Asian, West Indian or Chinese communities and before them the Jews. Whatever the particular explanations, all these originally non-English communities have one thing in common: a strong sense of their own group identity. They are therefore sensitive to the supreme duty of transmitting to their children through both word and deed the moral, spiritual, social and intellectual lessons derived from their own traditions. The purpose of the whole exercise is to keep the culture alive. Minority faiths attach tremendous importance to the family precisely because of that imperative to keep their particular show on the road. Education is therefore the very pivot of that transmission of values between generations that tells people what they are and roots them in a particular tradition.

But not, it seems, in England. In England, it is very difficult to talk about national identity *at all*. What is a source of pride to the Scots, Welsh and Irish is a target of derision or intense disapproval in England. And because of the cultural dominance by England within Britain, hostility to the idea of English culture has become entangled with hostility to the idea of British national identity. When Dr Nicholas Tate, head of the School Curriculum and Assessment Authority, advocated that British schools should transmit a sense of national identity to their pupils he was accused of something akin to cultural fascism. In the liberal orthodoxy, national identity was synonymous with nationalism and was therefore to be suppressed. It followed, therefore, that if English or British national identity couldn't even be acknowledged as something to be valued, it was not going to be transmitted through the normal channels of education. Indeed, education became the mechanism of presenting national identity in a negative light: other forms of identity were encouraged instead. So what happened in British education was nothing less than a collapse of the perceived need to transmit an agreed set of traditions and values. The desirability of keeping the English or British cultural show on the road at all was effectively denied. More, even the very existence of an English or British identity was denied. It was all got up by other people, apparently, and never had any substance. So there was nothing good to perpetuate. Or so the English intelligentsia told itself.

This meant an abandonment of the ideals of a liberal education. The idea of initiating children into a culture clearly doesn't apply if that culture hates itself and wants to melt away into the pluralistic ether. In essence, a liberal education valued knowledge as a good in itself rather than primarily as a passport to skills, a highly paid job and a successful economy. By acquainting children with 'the best that has been known or said in the world, and thus with the history of the human spirit', in the words of the Victorian thinker Matthew Arnold, it taught an understanding about human life and the human condition.[2] The study of literature in particular, so often derided by utilitarian philistines, engaged young people's capacities for moral understanding and fostered insight, imagination and emotional sensitivity to help understand other people. A liberal education pursued truths through the study of history, mathematics or science. And through the study of literature, music and art it promoted an understanding of the national culture and a respect for past civilisations, as well as a celebration of achievement and a sense of the capacities of the human spirit. Through such an initiation, it managed to transmit the moral values of the culture but in a way that treated young people as ultimately responsible for their own judgments. In other words, it taught them to think for themselves, to base their conclusions upon evidence and to acquire that critical sense which forms the basis of judgment.

What young people learn from parents, teachers, peer group and the surrounding culture provides the moral, spiritual, intellectual and aesthetic shaping of their personalities. If there are no strong cultural traditions in which to anchor themselves, the ensuing rootlessness and sense of abandonment spells trouble, both for the individuals themselves and for the wider society to which they belong. Everyone needs to feel he or she belongs. Everyone needs to be attached: to identify, to feel safe and at home.

But now, the cultural attachments once provided by a liberal education have been eroded. The culture appears no longer to wish to regenerate itself. The big issue of our times is surely not the absence of a set of common values in a multicultural society. It is rather a battle between people who believe in something and those who believe in nothing: not in knowledge, not in authority, not in moral absolutes, and above all, not in themselves.

Throughout the system, now, there is evidence of self-disgust by our intelligentsia. It's in the desire to rewrite the national biography through the censoring of the English canon, or in the subjectivity and cultural relativism of much history teaching. It's in the rejection of knowledge by those charged with its nurture and growth. It's in the revisionism over the nature of teaching itself, explicitly stated by key disciples as an attempt to relocate power in the children and thus help overturn the very structures of society. The traditional purposes of education were rejected by those from both the political left and the right who came instead to view education instrumentally, as the means of transforming Britain's social or economic profile into either an egalitarian society or a society of economic self-interest. Neither side was interested any longer in connecting young people to the past in order to secure their future. It was, after all, the Conservative Education Secretary Kenneth Clarke who turned history into an option that pupils could drop altogether at 14. To the utilitarians of both left and right, despite the differences in their purposes, the past had become a dead country.

One can see this sickness illustrated very clearly in the teaching of history itself which, despite the attempts to restore rigour through the National Curriculum, still betrays strong elements of the retreat from the culture, particularly at GCSE. The emphasis here is still very much on the teaching of skills rather than the teaching of facts. For a number of years, indeed, facts were significantly reduced by fiction – or at best, 'faction' – through 'empathy' work. This meant that pupils had to place themselves imaginatively in the position of characters from the past. It was fundamentally an exercise in the creative imagination. Nothing wrong with that, except that it had very little to do with the teaching of history and everything to do with making things up; and by substituting subjectivity for objectivity, it opened the way for outrageously blatant political propaganda to fill the vacuum that had developed.

Empathy is now being phased out, but source work remains a significant element of GCSE and A-level history. This involves pupils studying a variety of 'sources', including photographs, and assessing their reliability in constructing a historical narrative. The result is that pupils learn how to evaluate these sources without necessarily learning the narrative itself and therefore end up knowing very little about a particular historical topic other than what is revealed in the sources. It is a formulaic, mechanistic process that bears little relation to the 'skills' employed by real historians. Its overwhelming message to the pupil is: there is *no* historical truth at all – the only meaning that can accrue to the historical record is what any individual makes of it. Insofar as it teaches pupils a skill, it teaches them comprehension rather than history. But the moral and cultural message is more insidious. It is that the culture has no authority.

The British political and intellectual classes during the 1950s and 1960s wanted to transform the purpose of education because they wanted to transform British society. Astonishingly, since it was so soon after a war in which the country had played a noble role in defeating tyranny, they had come to entertain feelings of shame, guilt and even loathing for the kind of society that was then in place. They wanted to atone for the sins of imperialism, colonialism and class division. After the trauma of Suez in particular in 1956, there was a feeling that Britain no longer counted in the world and that it should remake itself anew. The Labour Party identified the stratified English class system as a major obstacle in the way of creating that new society on properly democratic principles – as well as blaming it for the failure of the Attlee government – and it equated democracy with egalitarianism.

To create a common culture, the establishment turned not to Europe but to the American model of the common school. This was a significant and fateful choice. The American comprehensive school was designed to build a nation out of people who were strangers to each other, to construct a society from scratch with no innate tradition to build on. Through its comprehensive schools, America was going to put together American culture *ab initio*. But Britain already *had* deeply rooted traditions. The model simply wasn't appropriate, unless Britain wanted to destroy those traditions. The fact was that the egalitarians were hopelessly confused. They wanted to achieve the classless society, but they didn't realise how much of the culture they would destroy in the process. They didn't realise that in making a bonfire of British traditions to achieve that classless society, the ensuing vacuum would provide a fertile breeding ground for revolutionary and nihilistic ideologies.

These ideologies were not, in the main, home grown but as we saw earlier they came largely from America and France. Why, though, did they achieve such resonance in Britain? The answer is undoubtedly complex. But in essence it was bound up with a profound loss of faith by the intelligentsia in the British national project itself which went back at least to the shattering impact of the First World War and the subsequent Depression. That was the time when faith in British leadership was destroyed for ever and as a result the élites lost their faith in themselves.

Englishness

It is common knowledge that, after the Second World War, Britain lost an empire without gaining a role. But Britain also lost respect for itself in the process. The Empire became a source of national shame and self-reproach. And although the Empire was British, England was the dominant culture within it. So Englishness became associated with all those oppressive features of the Empire that were causing so much guilt and embarrassment. But in any case, Englishness was problematic in itself, as it was an identity that was intimately bound up with class. Despite the working-class radical tradition of English liberty, the dominant strands within Englishness – the manners and lifestyle of the gentleman, the rural idyll, the public schools, the English way of life – were associated first with the aristocracy and then with the upper-middle classes.

But after the War, while Empire became associated with being beastly to the natives abroad, the English class system became associated with being beastly to the natives at home. So while Britishness fell into bad odour, it was really Englishness that was the target of disenchantment. And the culture was vulnerable. Englishness is so understated, so tacit and oblique that most people use it interchangeably with Britishness and some deny it exists at all. Unlike, say, French nationality, Englishness has never defined itself in terms of ideas. The result is not only that it is inchoate but that it lends itself to anti-intellectualism, which is part of the reason why the English don't value education.

But nationality and the intelligentsia are intimately related. The intellectual élites often create the idea of a nation. Doctors, lawyers, teachers, architects, engineers, technicians and many others disseminate that idea because it is in their interests to have a special place in that nation. At the same time, the concept of the nation they disseminate is furnished by the intellectuals, the philologists or grammarians or poets or historians who constantly reinterpret the idea of the nation from the ethnic past they have reappropriated and authenticated.[3] But if, as has happened in Britain, those intellectuals have repudiated the nation, and those professionals no longer have a proper sense of themselves either, then the culture is clearly in trouble.

This vulnerability of English culture has led some educationists to want to draw a line under the past and cut the connection with the present, denying that English identity now exists at all. One head teacher has written:

> The common culture of pre-1940 England, based on the canon of English literature, the Whig interpretation of history and the liturgy of the Church of England, has died. Masefield, Keats, Trafalgar, Waterloo and the English hymnal do not speak for or to the country most people alive recognise and understand. Worse, local or regional cultures like those of Hull fishermen or Welsh miners have crumbled into the seas of economic change. The explosion of experience since the Second World War has undermined even George Orwell's version of England and no successor has been found. Life and language have outgrown the confines of English belief, history and ethnicity.[4]

Presumably, if this head had been a teacher during the Industrial Revolution he would have obliterated all teaching about feudalism on the basis that England had unrecognisably changed. Quite apart from the fact that for all those people who still go to church the English hymnal does have meaning and that the poetry of Keats speaks to timeless and universal values, it is only by understanding the past that we can begin to understand our present and grasp how and why it has altered.

In any case, English identity does still exist. There remain significant points of common reference: a shared language, a dominant religion (because despite our apparent lack of religiosity, this country still adheres publicly to Christian values) the common law and – despite some key disagreements – a common morality. There is a deep concern for justice, law and freedom of speech; there is still a strong culture of individualism, empiricism and scepticism of authority. The English respect for law and the love of freedom can be traced back to the thirteenth century, distinguishing England from the peasant societies of medieval Europe and Asia. Debased as it may have become, it is still a singular inheritance.

In the eighteenth century, the political pamphleteer William Cobbett epitomised the very particular concept of English liberty, the idea of the free-born Englishman who was not a selfish individualist but promoted the idea of the common weal. A champion of ordinary people and of freedom and independence as the English birthright, Cobbett also believed that mastery of grammar was the key to clarity of expression and comprehension. So he wrote his *Grammar of the English Language* to provide the tools of liberation and personal development for everyone, but most especially for the poor – for soldiers, sailors, apprentices and plough boys.[5] Today he'd be run out of town as a reactionary for that alone. Certainly, he doesn't even get a mention in National Curriculum history, not even as an optional example. Yet without a knowledge of such characters, it is not surprising that people come to believe that Englishness either doesn't exist or if it does that it is 'oppressive'. There is no way they can understand what it actually is.

They certainly cannot understand that as much as anything it is a certain disposition of mind. As the political theorist John Gray has written: 'In the British case, vague but still powerful notions of fair play and give and take, of the necessity of compromise and of not imposing private convictions on others, are elements of what is left of the common culture, and they are essential if a liberal society is to survive in Britain.'[6]

But these English characteristics began to be denied for other reasons. The growing revulsion against the whole western bourgeois ethic (which had surfaced as long ago as 1911 when the Chief Inspector of Schools Edmond Holmes wrote his book about education), the guilt about class and Empire and the feeling that Britain had lost its way as a force in the world all coincided with the arrival in Britain of ethnic mi:orities from the Caribbean and Asia. Undoubtedly, they presented difficulties for Englishness. As opposed to Britishness, which is a political and administrative definition, English cultural identity depends to a large extent on a shared historical narrative, religion, customs and traditions as well as law and language. So while British nationality can be acquired reasonably quickly, English cultural identity takes much longer to assimilate.

This latter development led a number of educationists to decide that it was no longer possible or desirable to assume there could be any common identity any more, and that any attempt to teach English or British identity was nothing less than racist colonialism. Because not everyone could have English cultural identity overnight, *no-one* must be allowed to claim it. Educationists dismissed the idea that it *did* exist and could not be denied without destroying the society, or that there were aspects of it which were *worth* assimilating and to which ethnic minority children should be exposed in order to understand the culture in which they lived and to start the process of making it their own. Teaching children about the culture into which they wanted and needed to be accepted was redefined as a species of prejudice. A vacuum was thus created, into which stepped political ideologues posing as educationists.

One paper on the subject written in 1995 by two lecturers in education, for example, was a highly tendentious piece of political propaganda masquerading as dispassionate research. To Bruce Carrington of Newcastle University's Department of Education and Geoffrey Short of the University of Hertfordshire, 'Englishness' was not only characterised as monolithic, anachronistic and pernicious, but it was somehow wrapped up with teaching children how to read. This, of course, was clearly beyond the pale. 'In view of the New Right's injunctions about national identity,' they wrote, 'it is not surprising that pluralist or anti-racist initiatives in education have been vigorously attacked during the past decade. For example, John Major told delegates at the Conservative Party conference in 1992 that teachers-in-training "should learn how to teach children to read" and "not waste their time on the politics of gender, race and class."'

It would seem to such people a shocking thing to want to teach children how to read rather than to indoctrinate them with political propaganda. These lecturers' views, however, are hardly borne out by hundreds of ethnic minority families which have been forced to supplement their children's education in privately funded classes and schools because they are not being taught those very things these academics so despise.

The assumption made by these academics was that British society was endemically racist. This opinion was presented as unchallengeable fact. They implied that teaching ethnic minority children Standard English and introducing them to classic English texts was also racist. They appeared to believe that such children should not be introduced to such authors or taught Standard English because English culture was one to which they must be refused entry. It appeared that they wished to confine them to ethnic ghetto cultures and wanted moreover to deny the validity or existence of any common culture at all. And so they took a very dim view indeed of remarks by Dr Nicholas Tate of the curriculum authority to the effect that teaching children British history played a key part in helping society maintain its identity. To the self-hating English intelligentsia, such a thought was, of course, almost criminal. The country's identity had to dissolve instead into a rainbow of pluralistic fragments. It seemed that we must all belong to nothing greater than ourselves. The only reality that could be admitted to was what we created in our own image – unless, of course, that image was indigenous English.

Nevertheless, despite this perceived racism in the cultural restorationists' curriculum, the authors found to their surprise that there was a dearth of racist comments among the children they interviewed. Since these facts didn't fit their preconceptions, they came up – chillingly – with other 'evidence' that did fit. The children's innocent responses, they suggested, contained 'potential for racism'. The evidence for this appeared to be merely that they identified with their own particular cultures.[7] So in other words, *any* cultural identification was by definition racist. But if there was no longer any common culture, and if particular cultural identification was racist, where lay the necessary attachments for *any* British child?

The purpose of education had been changed. Instead of transmitting the national culture, it was now to be used to correct people's prejudices. The system was therefore to be turned from the repository of disinterested knowledge to a vehicle for political and ideological propaganda. Instead of fostering the development of decent human beings by teaching them to think critically and independently on the basis of evidence laid before them, it was now enjoined to produce or eliminate certain attitudes. Instead of teaching children *how* to think, it was to tell them *what* to think. To ensure that pupils emerged with the right attitudes, the cultural traditions had to be disavowed and supplanted by others.

True pluralism and tolerance were confused with the doctrine of 'anti-racism', which far from promoting tolerance is inimical to the principles of a liberal society because it tolerates no dissent from a set of subjective opinions masquerading as objective truths. Anti-racists, moreover, wish to force human nature into a pattern into which it does not fit. As John Gray has written, traditional liberals realise that although prejudice is an evil, it is not possible to banish it from human nature. By tolerating our differences – unless they have demonstrably harmful effects – we discover paradoxically what we have in common with each other. This encourages compromise, rather than a war to the death. We should tolerate each other's prejudices, he wrote, not least because there is no agreement on what they are.[8] For the zealots of cultural nihilism, however, such liberal sentiments are merely to be despised.

Ethnic minority children, no less than any others, need to be initiated into the culture they inhabit. Not to do so disenfranchises them. It leaves them unattached and insecure, when what they need is certainty and mastery of their environment if they are to participate as equals in the public sphere. It also exaggerates differences and diminishes what is common to all of us. For example, being British entails among other things being monogamous, forbidding incest, permitting freedom of speech and adhering to the rule of law. A common culture forges common bonds through such common rules. If no common culture is acknowledged, people become unattached and fraternity becomes impossible. Instead of creating a tolerant society forged upon civic bonds, they descend into factionalism and civil strife. As Gray wrote: 'A stable liberal society cannot be radically multicultural but depends for its successful renewal across the generations on an undergirding culture that is held in common. This common culture need not encompass a shared religion and it certainly need not presume ethnic homogeneity, but it does demand widespread acceptance of certain norms and conventions of behaviour and, in our times, it typically expresses a shared sense of nationality.'[9]

There are two fundamental mistakes underpinning the attempt to deny an English or British culture in the interests of pluralism. The first is to believe that England is now a multicultural society. It is not. America is a multicultural society because it is a nation of immigrants; England, by contrast, has a dominant, indigenous culture and a relatively small number of immigrants. The fact that they tend to concentrate in pockets around the country where they may form local majorities does not mean that England no longer has an indigenous culture, which continues to dominate and in large measure define the wider British culture. The second mistake is to believe that a common culture necessarily entails a ruthless homogeneity and a denial of diversity. It does not follow that one must possess only one identity and belong to only one grouping. It is perfectly possible to belong to a culture within a culture, and for each of them to have different characteristics and play different roles. A common culture does not deny diversity: it enables diversity to flourish.

As Jonathan Sacks has written, there needs to be a delicate interplay between our first language of common citizenship which relates to the public sphere, and our second language of cultural identity linked to families and particular ethnic traditions. If we recognise only the first, it means the disappearance of minorities; if we recognise only the second, we shall end up with a society of conflicting ghettos.[10] If we have only our group language but no common language, then we will always demand to have all our aspirations met with no room for compromise, creating tribalism and civil strife. It is only in a common culture that we can set limits to our several aspirations and develop that sense of mutual respect and responsibility that is the basis for a harmonious society. It is only within such a culture that we recognise the legitimate existence of others and develop a true sense of liberal tolerance.

A shared culture is crucial to forge those common bonds with each other to produce a sense of a joint enterprise. Far from fearing it, minorities feel safest where the majority culture is strong and self-confident. It's when people feel unattached and insecure or when they are filled with self-hatred that they are likely to turn upon their most vulnerable neighbours. Minority faiths know that their freedom is guaranteed – paradoxically – by the Church of England remaining the established faith. The idea that a secular society is invariably a tolerant society is simply not true. Those squeamish members of the majority culture who think that by deconstructing it they will be kind to minorities seem to be motivated not so much by the need to be liberal and tolerant – they themselves tend to be some of the most illiberal and intolerant people around – but by their own self-disgust.

They also misunderstand the nature and purpose of British culture. Professor Brian Cox, whose impact on the curriculum has been discussed earlier, asks in his book *Cox on Cox*: 'Should we insist that children spend all their time with a literature whose main non-white representatives are Othello, Man Friday in *Robinson Crusoe* and the savages in Conrad's *Heart of Darkness*?' The answer surely depends on what one thinks teaching literature is for. Clearly for such people its main function is relevance, to teach children from texts which are about people like them. But theirs is a culture of narcissism. Education should teach children about *other* people; it should expand, not shrink, the child's mind.

To enter imaginatively into the lives of other people is to expand moral as well as aesthetic horizons and to equip one to be tolerant. Only in that way can we properly become social beings and only in that way can we learn about other people's prejudices and indeed about the universality of prejudice. Sanitising literature is tantamount to attempting to airbrush evil out of the child's cultural picture. There are worse things in life than hurting someone's feelings. It's also remarkably selective. One might as well say that a curriculum exposing children to Shylock, Fagin or the 'protozoic Jewish slime' of T. S. Eliot should be censored too. Strangely, though, that argument never seems to be advanced by those who are so anxious to promote multicultural education to spare the feelings of Britain's ethnic minorities. Jews aren't seen as ethnic minorities, partly because they are viewed as powerful – a differentiation that illustrates not merely some nasty prejudices lurking beneath the tolerant surface but also that these multicultural definitions owe as much to politics as they do to ethnicity.

But there is also a misunderstanding of the characteristics of English culture. Far from being exclusive, it is actually inclusive and welcoming to minorities. This is because it was itself pluralist in origin and consists to a considerable degree of internal criticism and questioning of values within the recognised limits of the culture. As the teacher and education writer Frank Palmer has observed: 'To transmit the particular culture of Britain to the children of Britain is not simply to indoctrinate them in beliefs and values which they are not to question. The child brought up in the British way of doing things is encouraged to question and criticise, to seek fair play and impartial judgment, and to receive as doctrine only that which he has independent reason to believe is true. A child brought up in such a culture does not need that presentation of "alternatives" which so many educationists wish to foist upon him.'[11]

This instinct for questioning authority is very clearly on display in the reviled canon of English literature. Far from being the way it is caricatured, the canon is not an establishment monolith but a collection of dissidents: Dickens; Swift; Orwell; Bunyan, who was imprisoned for his views; and Milton, who argued for a republic and supported the execution of a monarch. And because people disagree in their interpretation of literature, teaching the canon produces a questioning of values rather than obedience. The purpose of teaching literature, as the Oxford don John Carey has observed, is to make one think and criticise, to confront one's own prejudices by understanding others and to produce a habit of mind of applying rational standards to arguments. If what was taught was tailored to reflect students' own experiences it would merely confirm them in their prejudices rather than challenge them, and they would fail to grow as intellectual or emotional beings. 'I've heard the complaint: "But it isn't relevant to what I do, it's not like my experience",' said Carey. 'I'm afraid the answer is, of course, that's the point. You *don't* know it, and that's why you learn it. It's the same with any discipline, say mathematics or chemistry or physics. Of course you don't know it before you learn it; that's the point, you're learning, you're growing.'[12]

But the alarming fact now is that there's no 'of course' any more; far from it, the assumption now is that our children *do* know it and don't have to be taught it, and that what they are taught must relate to their own experience. As a result, they're not learning; and they're not growing. In both the formal education in the schools and the informal education within the family, there appears to be an unconscious drive to prevent children from growing and thriving. The conditions which enable them to grow – the parental relationship in which children are dependent for succour upon their parents and teachers – are being withheld from them. Instead, they are being treated as mini-adults who know it all without being taught, who are expected to fend for themselves and who are expected to construct their own understanding of the world out of their own experience. If ever there was a mechanism to stop children growing, to keep them trapped indefinitely in a state of childhood or social and intellectual disadvantage, this is it. And if the children don't grow, then the culture will not thrive either. It will implode; or it will be overrun by tyranny; or it will just wither and die. There appears to be no desire any more to ensure that this culture renews itself from generation to generation. There appears to be a profound and terminal despair, not just about England and the national project but about western culture, about the very idea of growth or progress on which it is based. We are on a self-destruction trajectory. We appear to have lost faith in the human project itself.

*   *   *

It is our cultural habits, above all, that are being lost in this deconstruction of our heritage. Our moral sense, our critical understanding of the world, our sense of duty and responsibility towards each other are learned as habits of heart and mind that become integral to our personal sense of identity. If these habits of the heart don't develop during our upbringing within the family, we cannot function as moral and social beings. If these habits of mind are not transmitted through the schools, we cannot play our proper part in the wider world because we will have no means of making it intelligible to ourselves. We are learning instead the wrong habits: to look after ourselves before our dependents, to tear up our traditions and to be indifferent to the social and moral ecology upon which the world depends after our own lives have ended. Not only are we losing these essential habits of mind and of heart, we are rapidly losing the knowledge of what we have lost.

Nietzsche believed that the decay of culture meant the decay of man. In the last century, de Tocqueville gave a warning that individualism would derive from egalitarianism and destroy the individual's sense of connection to the past and the future. The modern distrust of commitments of any kind, the destruction of attachments to the past, the non-binding and transitory relationships within the family, all now suggest a collapse of self-confidence and a profound reluctance to trust to the future.

The American political philosopher Hannah Arendt wrote more than thirty years ago: 'Education is the point at which we decide whether we love the world enough to assume responsibility for it and by the same token save it from that ruin which, except for renewal, except for the coming of the new and young, would be inevitable. And education, too, is where we decide whether we love our children enough not to expel them from our world and leave them to our own devices, nor to strike from their hands their chance of undertaking something new, something unforeseen by us, but to prepare them in advance for the task of renewing a common world.'[13]

We have indeed expelled our children from our world and from that relationship with us on which renewal depends. We have to act now to repair the damage before it becomes irreparable.

Chapter Thirteen

A PROGRAMME FOR SURVIVAL

Politics of Attachment

Our liberal democracy is in danger and we must take swift and clear-sighted action to save it. The extent of the damage already done and the danger that looms are simply not generally appreciated. We have slid into this state of affairs almost without knowing. We take for granted the advantages of our modern culture; many of us are unaware of how it is undermining what we cherish but treat so cavalierly. This is because our intelligentsia, who mediate our culture to us, are principally implicated in its undermining. For a variety of reasons, our élites have been losing faith progressively in the national project since the last century. Now, their terminal self-disgust has infected our understanding of our common culture and the common good, which they are at pains to deny has any entitlement to exist at all.

We actually have more in common with each other than we think. The inherent decency of the majority emerges when we are confronted by a shared horror: the murder of the south London head teacher Philip Lawrence by a schoolboy in 1995; the mass slaughter of 16 tiny children and their teacher by a paedophile at a Dunblane primary school in March 1996; or the murder of the Liverpool toddler James Bulger by two 10-year-old boys in 1993. We discover at that point that we want the same kinds of things for ourselves and our families. The measure of our alienation is such, however, that it takes such appalling events to make us realise our common humanity; and in that moment of recognition, we also realise that we have lost that everyday sense of a shared society.

We also see, in these thankfully still most rare and unusual events, our deepest fears and insecurities made flesh: the random and murderous rage that out of nowhere suddenly strikes down a headmaster who stood up for goodness and truth; or mows down a class of infant children being cared for in the safety of their school gymnasium; or that causes two 10-year-old boys to smash a toddler's skull open and leave him to die on a railway track. We grieve; but we find it difficult to acknowledge the reasons these things are happening even though they are often staring us in the face. The word 'evil' is sprayed around as people turn to metaphysics for explanation; but they choose to ignore those clearly obvious though more banal connections between the killer's murderous rage and his invariably violent, abusive, neglectful or otherwise inadequate family background. An increasing number of us are no longer caring properly for our children, but we are reluctant to admit it because that would make life too inconvenient.

Most parents instinctively want to nurture their children. Most teachers want desperately to teach their pupils so that they develop their potential and grow. But both parents and teachers find it increasingly difficult to do so because the surrounding culture has conspired to knock away all their props. There are many parents who exercise authority with love. There are many adults who understand what commitment to each other and to their children really means. There are many good teachers who understand the nature and importance of their own role as the transmitters of knowledge and cultural values. But these parents and teachers are swimming against a cultural tide. A frightening and increasing number no longer even understand what commitment or authority mean because they themselves were so inadequately parented or taught. Both commitment and authority have been stripped away by a nihilistic élite which, despite marching quite out of step with the vast majority of ordinary, decent people, nevertheless exercises a huge influence on the culture and in turn corrodes codes of conduct and helps corrupt general behaviour. In a world of bewildering change, with globalised, impersonal economies and deepening insecurity and flux, we are knocking away those fundamental elements of stability – our attachments to each other through family and culture – which we need more than ever to help us withstand such earthquakes. Left to define ourselves, we cannot. We have freedom, but no longer know what it is for.

The result is that we are in danger. There's the danger of creating two nations, not along lines of class or poverty or geography, but between those who come from backgrounds of emotional stability and strong intellectual and moral attachments, and those who do not and are therefore spinning uncontrollably in emotional and intellectual chaos. There's the danger of tribalism, with society broken into competing interests which, in the absence of a common culture, all demand their several 'rights' and brook no compromise. And there's the danger of fascism, since the calamitous loss of authority by the political institutions and the process of politics itself leaves the way wide open to totalitarianism.

We need as a matter of the utmost urgency to put back as much authority and as many solid structures and certainties as we can. The more vulnerable the individual, the family or the group, the more certainties they need to be given. We need to develop a good parenting model for the state as well as for families themselves. That means abolishing the unfettered market as the model for economic and social life. But it does not mean returning to the old discredited model of state control. We have got to understand that safeguarding a liberal democracy is not a job that can be hived off to politicians who, once elected, are left to get on with it. We're all in this together. This must be a properly national project, to reinstitute both in the private and public spheres the principles of a parenting culture: care, control and development.

In other words, parental authority must be exercised by the state but in a way that promotes the independence of the individual. That means establishing a reciprocal sense of duty and responsibility. Just as a parent giving a child everything it wants prevents that child from acquiring an adult sense of responsibility, so adults have become infantilised because they have been able to slough off responsibility onto the state which has to grant them ever more 'rights'. Conversely, the minimal state is a disaster since a properly responsible state cannot withdraw from promoting public welfare any more than parents can withdraw from their children without messing up their development. We've tried out both the state and the free market model. Both have failed us because they are in fact but the two sides of the same individualist coin. Now we need a completely different model. We need to create a new politics of attachment.

We live in a constant state of risk and insecurity, affecting us at every level. Nothing remains the same; everything is in a state of flux and change. We are insecure because our jobs may disappear from underneath us, we live from contract to contract and may have to uproot our families in pursuit of work. Normal relationships of trust between employees and employers have disappeared beneath the weight of downsizing and rationalisation. Instead of feeling attached to institutions like hospitals or structures such as the criminal justice system, we feel they are remote, ever more bureaucratic and even adversarial. A culture of attachment would go with the grain of human nature, instead of setting itself up as an adversary to it. It would also restore proper liberal principles to a society which has come to mistake the debased version for the real thing. Classical liberal principles produce freedom within order. Instead, we have growing disorder in which liberty has been confused with licence and tolerance with indifference. The effect has been to produce an increasingly intolerant society made up of grasping individualists. To produce a really inclusive society that exists in more than political rhetoric, we have to re-impose that self-discipline which flows from the paradox that constraints are vital for liberty to thrive. Attachment presupposes that there is something to which we can attach ourselves.

Creating an inclusive society must be a challenge primarily for the liberal left in politics. The Tories can't do it. Under the Thatcher/Major regime, the Conservatives became debased liberals and promoted a political culture of atomised and irresponsible individualism which has done untold damage to our society. But if they were to revert to what they once were, they would not be, as is often suggested, 'one nation Tories'. Most of them never were interested in 'one nation'. They would revert instead to what they were, a party representing the sectional interests of land and money, indifferent to the needs of the whole country.

The ideals of the left also became debased, after it swallowed an agenda of levelling egalitarianism and moral relativism which has done such damage to education and family. But its roots in English ethical socialism were noble. They sprang from a unique English understanding of liberty and from a fundamentally religious and moral sense that other people mattered. It is because the left owns that ancestry of the moral high ground that its current amoral pronouncements bamboozle so many people. Its ancestry means that its constant claims to be acting in others' best interests tend to be believed, especially when the other side so patently is not.

The left claims 'liberal values' for its own; but in fact they have been debased. Tolerance on the left has come to mean tolerating the intolerable, since moral judgment has been suspended. So its task must be to rediscover its own noble traditions. It must reclaim morality from the right for its own discourse, and rediscover in the process its own ethical roots. It must rescue liberal democracy from itself.

At present, both main political parties are locked together in the fatal embrace of individualist and relativist politics. But the challenges they face cannot simply be solved by soundbite politics in which the political agenda is marketed ruthlessly for short-term advancement. The customary small change of political discourse is utterly inadequate to tackle a cultural slide on such a scale. Calling for the sacking of poor teachers, for example, is almost irrelevant when no-one is grappling with what it is that's made those teachers so poor or, indeed, by what criteria such inadequacy might be defined. Every profession has its bad apples, of course; but the fact is that the rot in education has made victims of the teachers, paralysing them by a nihilistic ideology which captured the universities and higher reaches of the education establishment and filtered down.

Similarly, calling for parenting classes or more counselling to shore up the family is neither here nor there when no-one is tackling the values that would be transmitted in such classes or counselling sessions. If they were merely to pump out the relativist doctrine that everyone must do what makes them feel good and that their behaviour or lifestyle must be supported and never challenged, such interventions would be worse than useless. These doctrines in education or family have been pushed along by a hard core of ideologues whose influence has been grossly disproportionate to their numbers. But enormous numbers of middle-of-the-road people have gone along with them because they are too hamstrung and careworn to protest; or because they are careerists on the make; or because they are too timid to put their heads above the parapet and shout an alarm for fear of being denounced as 'intolerant' or 'reactionary'; and perhaps most important of all, because no-one has ever shown them what a trail of havoc these doctrines leave in their wake and that there are better ways to organise things.

On the left, nihilism is still very much in fashion. There is a common celebration of disorder and irresponsibility by the media or the universities which don't have to pick up the pieces. At a seminar on the future of the welfare state in June 1995 organised jointly by the South Bank University, the Institute for Public Policy Research and the *New Statesman*, the star speaker was Anthony Giddens, professor of sociology at Cambridge and an influential writer on modernity and identity. He came up with some remarkable assertions. To him, it appeared, everything was a social construct. Not only was everyone now out for him or herself, but this didn't seem to be a bad thing.

> We should be very suspicious of the idea that society is in a state of moral decay and disintegration caused by individualism. That is an idea of the right. We've got to be very cautious about the idea that we've got too many rights and not enough responsibilities. We should take the view that rights create responsibilities rather than destroy them. I take a very jaundiced view of the new Labour agenda around duty and obligation. We have to go on the left with the new individualism in a world where tradition and nature no longer determine our lives and we have to deal with the anxiety and uncertainty that now dominate our lives.

This declaration drew a protest from Bob Holman, a former professor of sociology who has had the moral courage to put theory into practice and who for the past 20 years has lived and worked in Easterhouse, a very poor area of Glasgow. Giddens's theories might be all very well for discussion in an Oxbridge common room, he said, but they bore no relation whatsoever to the lives of the people he lived with for whom individualism had destroyed employment, community and family life. This cut no ice, of course, with Giddens who said he had been misunderstood. And Holman's protest seemed to leave the rest of the meeting cold. They nodded approvingly at Giddens's remarks. The fact that there is little to distinguish Giddens's post-modernist dystopia from the Thatcherite deconstruction of society didn't seem to have occurred to them at all.

Institutions

Rights do not create responsibilities: they usurp them. But despite the new rhetoric of responsibility emanating from the Labour Party leadership, the fact is that for most of the left tribalism now rules, with its sights firmly fixed on the creation of more and more rights. Indeed, the left's analysis of what is wrong with British society rests on the perception that rights have been so badly trampled upon that the situation can only be remedied by root-and-branch constitutional reform including a written constitution and a Bill of Rights. This is very wide of the mark indeed. Our malaise has arisen through a surfeit of rights, not a shortage. It is duty and responsibility that have taken a nose-dive. It is said that constitutional reform is necessary because Mrs Thatcher destroyed consensus and the institutions. So she did; but then it is clearly an absurd proposal further to destroy those institutions. In any event, Mrs Thatcher drove a coach and horses through the institutions not because of constitutional weakness but through market forces predicated upon a 'democratic' levelling down. No written constitution can be proof against that kind of assault. One only has to look at America, where a written constitution and a Bill of Rights preside over a dangerously divided society made tribal by market individualism, to see that. Constitutional reform would merely transfer power from one section of the élite to another without challenging any of the fundamental assumptions that really have contributed to the breakdown of social cohesion.

To replace the culture of individualism by the politics of attachment, we need to shore up such institutions as Parliament and the monarchy which have come under such attack during the Thatcher/Major years, not further emasculate them or tear them down completely. It is the institutions which embody and represent to us a common culture, the shared national project. Destroy the institutions and all that is left is empty space to be filled by atomised individuals and the market. That is why the cultural revolutionary Mrs Thatcher was so hostile to them. They may need to adapt, of course; there are few things that don't. But they are essential mechanisms to carry the message that British society is bigger than the sum of its parts – the message Mrs Thatcher attempted to deny. To destroy them would be to reinforce the culture of individualism from which we are all suffering and which has itself been a prime cause of these institutions' problems.

* * *

The inappropriate instinct on the left to go for the constitutional jugular is fuelled by two impulses. One is sheer frustration: because it has been unable for so long to get its hands on the levers of power, it has decided to destroy those constitutional arrangements that have made this state of affairs possible. The other is the false identification of democracy with egalitarianism. It is simply not the case that democracy is best served by attempting to remove all hierarchies. The attacks upon the monarchy that have been unleashed within left-leaning circles, courtesy of the unbridled exploitative and commercial excesses of the press, are based on specious reasoning. It is said that the point of having a monarchy is vanishing because of the bad behaviour of members of the Royal Family. But the point of the monarchy is not above all to provide a set of role models for the nation. The point lies in its constitutional position as an embodiment of the state which is above temporal politics and which embraces everyone regardless of political preference. The argument that this constitutional position promotes social divisions is a fallacy: look at the Scandinavian monarchies, presiding over the most egalitarian and advanced welfare states in the world.[1]

Then it is argued that the monarchy allowed the Tory attacks on our 'rights' because we are subjects not citizens. But this is also nonsense. The Royal Prerogative didn't stop any reforms from taking place, because the only thing that's Royal about the Prerogative is its name. It is actually vested in politicians. The undoubted erosion of parliamentary power by the executive that has occurred was the result of politicians corrupting their own process; only they can put that right and they could do so tomorrow if they wanted to, regardless of the Royal Prerogative.

The attacks on 'rights' took place because the people didn't want to stop them. Local government, for example, was emasculated without protest largely because, so far from enjoying popular support, it had come to be seen as remote, bureaucratic, corrupt and inimical to people's interests. In the overall balance of rights, very few were actually taken away. The uncomfortable truth is that however much one might deplore such developments, those rights which were removed in the civil liberties arena, such as the right to silence or certain rights to free association, were taken away with the tacit or even enthusiastic consent of the majority. It was the élites who complained and set up through the media and political circles mutually reinforcing networks of the excluded intelligentsia to protest against policies that enjoyed 'democratic' support.

The constitutional reform agenda is not about giving power back to the people at all. It is rather a means of creating through patronage channels of influence of which the excluded intelligentsia has been stripped. It is a false assumption that by definition any body of equals is a good in itself. If all overarching authority is removed, there can be a tyranny of the majority. That's why there's a need for institutions commanding authority beyond egalitarianism, such as the monarchy or the Church. The paradox of liberty is that structures that enclose and constrain also protect and allow freedom; when such structures are removed, there follows not greater freedom but instead anarchy and the war of all against all. What holds true for families and the upbringing of children, which need firm boundaries if they are to thrive, holds for civic organisation too.

It is a modern fallacy to assume that by ending all distinctions between human beings they will all suddenly find a unanimity of purpose and start co-operating with each other. The precise opposite is the case. This kind of egalitarianism is a form of social Darwinism which gives everyone an equal right to muscle in on everyone else's territory, a situation in which those with the sharpest elbows will survive and the weakest will be pushed out. The political left needs to reclaim its old belief in meritocracy, under which everyone should be provided with an equal opportunity to make the best of themselves. This does *not* mean that everyone ends up with the same outcome. On the contrary, a meritocracy presupposes that outcomes will be different. It presupposes a hierarchy not just of achievement but also – crucially – of values, distinctions between good and bad, right and wrong. Fairness lies in providing equal opportunities to achieve good things or behave well; imposing equal results on the other hand, with all discriminating judgments of value forbidden, ends in totalitarianism. That was the dreadful mistake made by the left when it abandoned equal opportunity and embraced egalitarianism instead. And that was why it connived at the individualist ethic which it nevertheless found so distasteful when the Tories adapted it in the economic sphere. Its cries of dismay were all the sharper because it was being forced to look at its own distorted reflection in the mirror. Treating everyone as if they were the same is merely a higher species of indifference, and it has led to politicians across the political divide conspiring to reduce education and family life to a non-judgmental bear-pit as public life has washed its hands of responsibility for a range of conduct which individuals, including children, are now expected to decide for themselves.

There is little sign, however, that the political left has recognised its terrible mistake and the abandonment of its historic ethical principles of public duty and private responsibility. Its preoccupation with issues such as constitutional reform are yet another distraction upon which it has alighted to divert itself from accepting the much more difficult and painful challenge of rethinking its own first principles. It cannot expect to reform education unless it addresses the fundamental flaw in its own thinking; that although Tony Blair has said he wants to promote a meritocracy, the ideological underpinnings of the party he leads and the wider swathes of opinion are still wedded to egalitarian aims.

Meritocracy

For it is not just the political left that is anti-meritocratic but the right as well. Indeed, the fundamental assumptions about the aims of the education system are profoundly anti-meritocratic and are largely responsible for the mess it is in. The Tories historically were never interested in individual advancement. Indeed, they weren't interested in education at all until they became alarmed, along with the former Labour Prime Minister James Callaghan, that the education system was failing the country on economic grounds and causing it to be left behind in international competitiveness. But both parties and the entire establishment then came to the same misguided conclusion, that the measure of our international competitiveness and of the success of the education system was the number of young people who went to university. It followed, therefore, that the ideal situation would be one in which *everyone* went to university; or, since this was demonstrably unachievable, at least in very large numbers. But this was a philosophical absurdity, just as Labour's original 'grammar schools for all' was a piece of impossibilist utopianism. By definition, one cannot expect everyone to attain the highest level of achievement. If they all achieve it, it can no longer be the highest level. The very idea of the highest level presumes that there are lower levels. If everyone gets there, then the concept of the highest level vanishes. Just like 'grammar schools for all', 'higher education for all' was similarly self-destructive. Because the 'prize' of a university education was held to be the proof of success, everyone had to win the prize.

But the inevitable result of that was that the 'prize' became worthless. Subjecting the highest level of attainment to a process of egalitarian levelling progressively destroyed its claim to be anything worth striving towards. Since not everyone was able enough to reach that highest level, but since everyone had to be seen to have that ability, the inevitable outcome was that reaching it was made artificially easy. The result was the abolition of the two-tier structure of higher education: polytechnics were turned into universities simply by declaring them to be so. This led in turn to the creation of deeply unrigorous degree courses, to maintain the pretence that students at these new universities – who were actually not up to the original degree standards, whatever their paper qualifications – were able to obtain degrees of equal merit. It reinforced the corruption of the paper qualifications themselves – particularly the GCSE, the exam that quintessentially offered prizes to all – in order to maintain the pretence of ever-increasing education standards. And it corrupted the older universities, which were forced to adjust their own standards downwards in order to accommodate the disintegration throughout the whole system.

At the heart of this whole catastrophe lay the deeply-rooted belief, held right across the political spectrum, that it was of overwhelming importance for young people in the lower reaches of the ability range to believe that they were successful. The GCSE was thus a good exam because it enhanced their 'self-esteem'. The priority therefore was not what they actually were achieving but whether they felt good about themselves. This rotten corruption of the concept of self-esteem lies at the heart of the destruction of our relationship with our children. It is rotten because it is built on telling them a series of lies. The education system conspires to make them believe they have achieved when they have not. It sponsors individual self-delusion and national myopia.

Of course self-esteem is crucial. But self-esteem does not emerge from lies and self-delusion. It has to be based on something that is worth esteeming. The contemporary corruption of self-esteem merely panders to the supreme self-indulgence of the age, that the way individuals rate themselves is to be equated with whether they have been given everything they demand. So if they want 'academic' success, they have to be given it, otherwise their self-esteem will suffer. This ridiculous non-sequitur has meant that anything that causes effort or pain is forbidden and that failure is a banished concept altogether. But children are not stupid. They know when they are being sold a pup. Their alienation from the schools, their feeling that education just isn't worth the effort, the consistent spoken and unspoken question: 'What's the point?' all bear eloquent testimony to the havoc that has been wrought by a system that has eradicated challenge and demotivated effort.

What is more, it has produced a spiral of madness. The obsession with a university education as the hallmark of value has meant that people who don't get to university feel they are failures to an extent that never used to be the case. The perverse result is not only that they attempt to squeeze themselves into a model that simply doesn't fit them, but that the excellence they really could aspire to, in crafts or practical skills, has no chance of being achieved because neither they nor the culture appear to rate it. The more that self-esteem is promoted as the sole aim of the system, the more it erodes it. The result is 'vocational training' that tests 16-year-olds' ability to identify a bird feeding itself from a picture.

But it doesn't have to be like this. If one looks at those countries against whose standards we fall so lamentably behind, none of them suffers from this monstrous national delusion. No European country attempts to force its children into an identical mould. In Switzerland, for example, a country whose economic success we can merely gasp at, only 12 per cent go to university. This is not held to be a national disaster, because there university is not identified as the one and only criterion of excellence.

Educational Reform

We must take a leaf from the continental book and alter the fundamental assumption of our whole education system. We should make education fit children's different needs, not force children into one homogenising system. We need different types of schooling along the lines of the German or Swiss models, in which children would be separated – maybe at 11, maybe later at 14 – in order to be educated along different pathways. This should not be left to evolve by itself through market forces or parents' wishes. Instead, the government should take responsibility for establishing a structure which is appropriate to meet children's needs and the common good. There should be maximum flexibility, so that children can move from one pathway to another as their abilities and aptitudes develop. Employers should be induced to put their money where their mouth is and invest significantly – following the German example – in vocational education, at the end of which they should guarantee apprenticeships. Only by investing significantly in vocational education as a distinct route to achievement, by resourcing it well and by providing jobs at the end of the process will we crack the destructive assumption that a university degree is the only qualification which matters.

Such vocational education must itself be excellent. In Switzerland, one reason why university is no big deal is that a reverence for knowledge is not confined to the university but brought into everyone's life and certainly into vocational training. So the Swiss have a precision industry because their craftsmen are educated to carry out precise calculations and are taught in detail about the tools and materials they use. Like the Swiss, we must pay very much more attention to the building blocks of education. There is very little point teaching technology or science, for example, if pupils have not properly mastered mathematics. In Switzerland, much less time is spent on science and much more time is spent on maths than it is in Britain, which means more help is provided to a very much larger number of children who go on to develop vocational skills as a result.[2] We must rethink vocational education from such first principles, instead of taking refuge in flannel such as the review by Sir Ron Dearing of post-16 qualifications, published in 1996, which simply sought to blur the differences between academic and vocational education to provide an illusion of egalitarian progress instead of addressing the chronic weakness and inadequacies at the heart of vocational teaching.

We should rethink our National Curriculum from scratch. Given the damage it has done in institutionalising some of the worst aspects of relativised, child-centred education, there is a case for saying that it should be scrapped completely. However, it has been of some use in forcing schools to focus their attention on what teachers are teaching and how it is being taught, particularly in the primary schools. It is the primary schools which still need the most attention since it is there that so much of the damage is done. Secondary schools, however well organised, cannot be expected to remedy this damage. Moreover, since there is no public examination at the end of primary schooling there is no benchmark by which primary schools can measure their performance unless there is a curriculum and associated tests. If secondary schooling were differentiated, there would be an opportunity to scrap the GCSE exam which has itself done so much to lower standards and expectations throughout the system. But there would still be a need to ensure that secondary schools were equipping their pupils with a core of essential knowledge.

The National Curriculum should be re-drafted to concentrate solely on English, Maths and History. These are the absolute essentials without which pupils simply cannot take their proper place in society. Other desirable and necessary subjects such as the sciences or foreign languages cannot be taught properly unless these foundations are secure. If the National Curriculum were confined to these core subjects, this would give schools back the flexibility they have lost from their timetables while ensuring that every child is literate and numerate and knowledgeable about the culture he or she inhabits. In addition, the curriculum for these core subjects should be rewritten to be far more rigorous, with the emphasis shifted away from experiment and discovery and relativist values and towards teaching a body of knowledge instead.

But here the nature of the problem immediately becomes all too apparent. Reform cannot be piecemeal. To halt the deterioration in education, as in family relationships, there has to be a fundamental cultural shift. The central problem, however, is how to change the culture when by definition it is resistant. The history of education in the last few years has illustrated the awesome capacity of a relativist establishment to frustrate and even deform a set of founding objectives. Might not any attempt to change direction be subverted in precisely the same way? This is a serious problem and one for which there is no easy solution. There is a number of ways, however, in which the opportunities for such sabotage could be reduced.

The first priority is to safeguard the National Curriculum, school tests and the public examination system from interference by the Department for Education and Employment. Both the independence and the parliamentary accountability of the School Curriculum and Assessment Authority should be strengthened. At present, it is a quango with a chairman and chief executive appointed by the Department for Education and Employment. That should change, and its head – following the example of the Chief Inspector for Schools – should become a Crown servant appointed by the Queen on the advice of the Prime Minister. Like the Chief Inspector, the head of the curriculum body would be accountable to Parliament through the Commons Select Committee on Education. The curriculum body would thus become accountable to Parliament directly rather than through the Education Secretary, who at present if challenged can slough off responsibility to this quango. And it would be Parliament which would ultimately decide on the content of the curriculum through the Curriculum Orders laid before it. Appointments to the curriculum body, both its staff and its lay committees, would be made by its chief executive and thus become protected from interference either by the Department for Education and Employment or the Prime Minister. The curriculum would thus be handled at arm's length from the Department for Education and Employment, whose role would shrink accordingly. The next priority must be to rescue the teaching profession from the slough of low expectations and underachievement into which it has fallen. The principal deformation in the schools has arisen out of ignorance. Teachers are unaware of what they or their pupils are capable of achieving because they have little first-hand knowledge with which to combat the risible ideologies disseminated through teacher training colleges and the universities. Yet so far the government has done next to nothing to correct this situation. The damage done by teacher training has been all but ignored, as the government has chosen to allow the market to correct the faults of the system. But the market cannot possibly do that.

Teacher training colleges should be shut down in order to eradicate the influence of university-based educationists whose contribution to the ability of teachers to teach properly has been almost wholly negative and whose influence on the National Curriculum has been a disaster. Teaching should be recognised for what it is, a craft, and teachers should be trained accordingly in carefully selected schools from both the state and the independent sector.

Above all, teachers should learn by example from each other. The Teacher Training Agency should institute a rolling programme in which teachers from successful schools, including teachers from abroad, would explain and illustrate their teaching practices and show how they achieve the results that they do. The majority of primary teachers, for example, are simply unaware how quickly even the most disadvantaged child can be taught to read through systematic phonics. Similarly, they tend to disbelieve that whole class teaching can benefit the whole class, quite apart from making their own task infinitely easier than individualised teaching or group work. So they need to meet teachers who can show them the successes of these methods in order to challenge the prejudices which have been allowed to build up through ignorance. Where this has happened – as with teachers from east London who have been introduced to the Swiss methods of teaching maths – teaching practice in Britain has been transformed for the better, to the incalculable benefit of the pupils *and* of the teachers themselves.

Restoring Connections

Transforming the culture of individualism means making connections which have been allowed to atrophy. It means restoring a sense of personal and group responsibility, which means a new relationship between the individual and the state, and an enhanced role for the intermediate institutions of the voluntary sector. We need a state that enables people to take better care of each other. That means restoring the sense of the personal to structures which have become de-personalised, such as the criminal justice system for juveniles who commit the overwhelming majority of crimes. At present, most young offenders simply do not relate to what happens to them. For a start, hardly any are caught and even if they are virtually no-one forces them to confront what they've done in a way that makes a difference to them. The system positively discourages any serious engagement with them, not least because of the extreme reluctance of all involved to apply anything that might resemble punishment until, when they are older, they get sent to prison. What has been largely removed by this impersonal approach is the element of shame. Yet shame is a crucial social control.

We should follow the example of New Zealand which has introduced the concept of 're-integrative shaming' through its youth courts. These bring young offenders face to face with their victims in the company of members of the offenders' families, police, youth advocates, social workers and others in a family group conference. It is this group that decides what sanction should be imposed, unless the offence is serious enough to be referred to a judge. Usually the sanction is a plan of action: apology, reparation in money or work for the victim, community work, a curfew or other initiatives. The emphasis is thus on exposing offenders to the powerful responsibility of seeing and hearing the consequences of their actions upon their victims, and righting the wrong by making reparation. The system transfers power from the state to the community and makes possible a sharing of responsibility between the youth and those who have brought him up. The youth is therefore much more likely to feel shame when confronted directly with family members who make it plain he has let them down. It tells the youth that what he has done has not just hurt the victim but let down a wider community. It thus binds him into that community rather than giving him the status of a pariah, which is only likely to encourage him into further anti-social activity. It also associates his family with the responsibility for his continuing care, control and development which have by definition gone wrong.[3]

Promoting a culture of social responsibility means replacing a presumption of rights with a presumption of reciprocal duties. Only in this way, as the writer David Selbourne has argued, can we restore our lost ideal of civic responsibility. It means bringing an end to the rigid separation between the public and private spheres. It has to be recognised that state activity has some influence on private behaviour and that conversely private behaviour has an impact on the public good. We have to institute networks of reciprocal responsibility between individuals and the state. That means changing the welfare system to one that links benefits to behaviour, in the manner envisaged by the founding fathers of the welfare state before it became distorted into a species of demanding individualism.

Political Leadership

In the last analysis, the state cannot change behaviour by public fiat. Politicians cannot unilaterally change the culture. That can only be done by the people themselves. But the state has a part to play in shaping the public sphere in such a way that it is either conducive or hostile to socially productive or anti-social behaviour. It has, for example, a limited but crucial role to play in the fraught arena of family relationships.

Its first role is rhetorical. The only way to shift a culture is to change the way people think about themselves and what they do. That can only be done if they become aware of the effects upon others of their behaviour. Politics should be about leadership, and properly responsible leadership entails ensuring that people are aware of the consequences of their actions and that certain forms of behaviour and lifestyle are more socially responsible than others. We need a campaign of public education in which politicians have an important part to play. They should start treating the public as grown-up by telling them the truth. They should point out, for instance, the moral responsibility to pay income tax and the effects upon individuals if they pay very little. Politicians have a corresponding duty to spend such money in accordance with their public commitments and in a way that promotes transparency and accountability; duty and responsibility is a two-way street between the individual and the state.

Politicians should demonstrate to the public, by reference to the *Headstart* family support programmes in America, that by investing taxpayers' money in proactive and targeted infant support programmes, delinquency and other social ills can be prevented and much larger amounts of taxpayers' money can be saved. Politicians should say unequivocally that marriage is an important civic and social good and that it is better for individuals and for society if, wherever possible, children are brought up by both their natural parents. They should not be knocked off this course by ministerial scandal, any more than the church would feel obliged to stop preaching its doctrines if individual priests were discovered to be paedophiles. Hypocrisy has been promoted as the worst vice of society but it is not. It is essential that public ideals are promoted despite the frailties of individuals.

This campaign of public education should be backed up by policy which should promote the two-parent family as an explicit social good. Accordingly, the tax and benefit structure should at the very least not discriminate against married couples with families. The age of consent should be raised to 18, the divorce law should embody the concept of fault and local authority social services departments should be restructured so that child services become specialised, preventative and enabling, using parents to help each other by getting successful parents or grandparents to show others how to exercise loving authority over their children. Much more use should be made of the voluntary sector, Burke's 'little platoons', to stand as the intermediate tier between the individual and the state in ways that help bind us together as a mutually dependent and responsible community.

But here, the fundamental problem immediately reasserts itself. The voluntary sector at present is characterised by the worst excesses of moral relativism. Many children's charities promote the 'self-esteem on demand' approach which provides no boundaries for children but excuses their behaviour, while much so-called marriage guidance in effect promotes the destruction of marriages and sees no need to challenge individualistic behaviour. Yet the voluntary sector is potentially an invaluable resource. So it should be encouraged to change its attitudes. The bulk of the grants go to the big charities which tend to be most detached from people's real needs, while the most enlightened work displaying tough love and moral courage is done by tiny charities which get virtually no money at all. This perverse state of affairs should change, with government grants going explicitly to those voluntary bodies that reject the relativist approach and have a clear sense instead of the common good. That in turn might engender some clear moral leadership, the arrival on the voluntary scene of figures in the mould of Wilberforce or Shaftesbury or Rowntree in the last century, social reformers whose activities remoralised a nation by tempering the harshness of individualism with altruism.

But the idea of the common good is intensely problematic. It is rejected precisely because it repudiates the individualist ethic in which everyone makes up their own idea of what is good. In such an individualistic age, the idea of the common good is thought to be authoritarian because it asserts a set of virtues beyond our personal, subjective value system. To impose any values at all is deemed to be beyond the pale. But this ignores the fact that our culture is not value-free. It currently imposes a doctrine of moral relativism which has done us untold damage. And there are many moral aims which we do share, if we believe in the survival of liberal democracy: concepts of human rights and universal dignity and what is conducive to human well-being; what has been called the promotion of human flourishing.

It is notable that this hostility to the notion of the common good is suspended, however, when we come to judge the environmental movement, which is characterised by an evangelical belief in its possession of unchallengeable truths which must be acted upon to force limits upon individualistic behaviour. Environmentalists believe we cannot carry on as we are without destroying our environment. This may well be true; but it is no less true of the moral ecology than of the physical ecology. The reasons for this discrepancy surely lie in the fact that environmentalism shifts the blame onto recognisable and distant villains – big business and science – and involves problems remote from our personal experience. Since these are extremely difficult to solve because of their remoteness and complexity, we can salve our consciences by protest which is unlikely to result in any serious challenge to our personal behaviour. The breakdown of the moral ecology, by contrast, which is more immediate and accessible, requires us to take responsibility for ourselves and change our behaviour in ways that might cause us serious inconvenience. But we could make a significant difference here. Bringing up children in families where there are stable attachments to their parents promotes human flourishing, as does teaching children that there are limits beyond themselves which should be observed, and traditions which make sense of such boundaries.

The notion of the common good acknowledges the central paradox of individualism: that left unbridled it is self-undermining and leads to tyranny, and that freedom is only possible within constraints.

The great liberal turn-of-the-century thinker Leonard Hobhouse understood this well. A political theorist, he reformulated the principles of liberalism to recognise the claims of community, establish the centrality of welfare rights and legitimise an activist democratic state. As a result, he embodied the bridge between liberalism and ethical socialism. In his classic essay *Liberalism*, published in 1911, he wrote that individual rights must not conflict with or exist apart from the common good.

> The argument might seem to make the individual too sub-servient to society. But this is to forget the other side of the original suppositions. Society consists wholly of persons. It has no distinct personality separate from and superior to those of its members. It has, indeed, a certain collective life and character. The British nation is a unity with a life of its own. But the unity is constituted by certain ties that bind together all British subjects, which ties are in the last resort feelings and ideas, sentiments of patriotism, of kinship, a common pride, and a thousand more subtle sentiments that bind together men who speak a common language, have behind them a common history, and understand one another as they can understand no-one else. The British nation is not a mysterious entity over the forty odd millions of living souls who dwell together under a common law. Its life is their life, its well-being or ill-fortune their well-being or ill-fortune. Thus, the common good to which each man's rights are subordinate is a good in which each man has a share.[4]

Hobhouse's words stand today as the stern rebuke to Mrs Thatcher's belief that there is no such thing as society. The authentic liberal voice reaffirms that there is a need for fusion between the individual and society for civilised values to flourish, and in doing so reveals the neo-liberal to be the peddler of a debased and corrupted ideology.

The old political division between right and left is obsolete. The division that matters now is between selfish individualists and people with a moral sense which puts others' interests above their own. It is between people who, rooted in the traditions of the past, believe in the future; and those who don't. Liberal values do not depend on individualism. The urgent task is to rescue liberalism from its debased inheritance and restore the concept of freedom within order. We need a new politics of attachment that lays down a structure of reciprocal commitments, from the state to the individual and from individuals to each other, which allows us to nurture and be nurtured and which restores those virtues of fidelity, trust, loyalty and integrity that must be an integral part of the common good.

This is not nostalgia but a survival agenda. It is not an attempt to recreate some mythical golden age of the past, or undo the great movements of women's emancipation or freedom of expression. It is rather an attempt to recognise those virtues that have somehow been lost along the way and without which the liberal project is in danger of undermining its most precious assets. It means voluntarily agreeing to set limits for ourselves by discriminating between good and bad and deciding no longer to tolerate the intolerable. We are facing a crisis of western values. Liberal democracy has not brought us into the land flowing with milk and honey. It has given us freedom, but we have misused it and no longer have any idea what we want it for. Liberation from superstition and tyranny gave us liberty, but it also led to the unparalleled excesses in this century of oppression, terror and fanaticism. Unless we recognise the ambiguous nature of our inheritance we will surrender the freedom we so dearly prize.

The loss of authority, wrote Hannah Arendt, is the loss of the groundwork of the world.[5] Of course we cannot recreate the authority that characterised the past; nor would we wish to. But we can rediscover our common humanity, and decide that our commitment to human flourishing and to the future is greater than our love of ourselves. It doesn't have to be like this. The choice lies with us.

NOTES

Chapter One

1 Sheppard, Richard: 'Getting Down to Brass Syntax: German Teaching and the Great Standards Debate', in *German Teaching No. 8* (Association for Language Learning, December 1993)

2 *Study Skills, Transferable Skills; A Guide to Essay Writing, Oral Presentation, Grammar, Punctuation and Related Matters* (School of English and American Studies, University of East Anglia, 1995)

3 'Another important consideration has been to make as much provision as is politically possible for coursework executed in formats other than the traditional essay . . . More contentious, perhaps, is our bid to provide an optional space for non-verbal responses . . . Why should a student not explore Hardy's landscapes in paint or interpret a poem by setting it to music?' Hackman, Sue: 'A New A-Level Proposal – An Examination Capable of Stimulating Change', from *A-Level English – Pressures for Change* (National Association for the Teaching of English, 1990)

4 *Daily Telegraph* (7 September 1995)

5 Green, A. and Steedman, H.: *Educational Provision, Educational Attainment and the Needs of Industry; A Review of Research for Germany, France, Japan, the USA and Britain* (National Institute of Economic and Social Research, 1993)

6 *Daily Telegraph* (28 November 1995)

7 *ibid.*

8 Figures from the School Curriculum and Assessment Authority

9 *Independent* (19 February 1996)

10 *Guardian* (5 July 1990)

11 Conversation with the author, 1995

12 Gordon, J. C. B.: 'Kannitverstan' in *Languages Forum 1*, No. 4 (Institute of Education, London, April 1995)

13 *ibid.*

14 *ibid.*

15 Conversation with the author, 1995

16 *ibid.*
17 *ibid.*
18 *ibid.*
19 *Guardian* (7 July 1990)
20 *The Times* (21 September 1990)
21 London Mathematical Society, Institute of Mathematics and its Applications and the Royal Statistical Society: *Tackling the Mathematics Problem* (London Mathematical Society, 1995)
22 Evidence to Conservative backbench education committee, February 1995
23 *The Times Higher Education Supplement* (24 February 1995)
24 Conversation with the author, 1995
25 Gardiner, Tony: *Observed Effects of Recent Changes in English School Mathematics on Those Entering Universities at Age 18* (printed for private circulation, Birmingham, 1994)
26 *ibid.*
27 Her Majesty's Inspectors of Schools
28 Gardiner, Tony: *op. cit.*
29 London Mathematical Society *et al, op. cit.*
30 Conversation with the author, 1995
31 Conversation with the author, 1996
32 London Mathematical Society *et al, op. cit.*
33 *Dispatches: Making the Grades*; transmitted on Channel Four (11 October 1995)
34 Draft report: 'Consistency of Awarding and Standards of Attainment in GCSE Mathematics Examinations' (Centre for Innovations in Mathematics Teaching, Exeter University, November 1995)
35 Conversation with the author, 1995
36 Taylor, Gillian and Edwards, David: *Étoiles* (BBC/Longman, 1992)
37 Conversation with the author, 1995
38 Conversation with the author, 1995
39 Conversation with the author, 1996

Chapter Two

1 Conversation with the author, 1995
2 Conversations with the students by the author, 1995
3 *Access and Achievement in Urban Education* (Ofsted, HMSO 1993)
4 *The Annual Report of Her Majesty's Chief Inspector of Schools, 1993/94* (HMSO, 1995)
5 *The Annual Report of Her Majesty's Chief Inspector of Schools, 1994/95* (HMSO, 1996)
6 *The Teaching of Reading in 45 Inner London Primary Schools* (Ofsted, 1996)

7 Interview with Jim Rose, head of inspection at Ofsted, and Sinclair Rogers, head of the inspection team for the reading standards report, May 1996

8 *Evening Standard* (31 January 1996)

9 *The Annual Report of Her Majesty's Chief Inspector of Schools, 1993/94* (HMSO, 1995)

10 *The New Teacher in School: A Survey by Her Majesty's Inspectors in England and Wales 1987* (HMSO, 1988)

11 Conversation with the author, 1995

12 A series of books published by Routledge and the Open University

13 Pollard, Andrew and Bourne, Jill, eds: *Teaching and Learning in the Primary School* (Routledge/Open University, 1994)

14 Brindley, Susan, ed.: *Teaching English* (Routledge/Open University, 1994)

15 Slinger, Michelle, ed.: *Teaching Mathematics* (Routledge/Open University, 1994)

16 Bourdillon, Hilary, ed.: *Teaching History* (Routledge/Open University, 1994)

17 *Guardian* (4 December 1990)

18 Steiner, George: *No Passion Spent*, p. 15 (Faber and Faber, 1996)

19 C. T. Lewis and C. Short: *A Latin Dictionary*, p. 627 (Oxford University Press, 1963). Hence *educit obstetrix, educat nutrix:* the midwife brings forth, the wet-nurse rears.

20 Conversation with the author, 1996

21 Cox, Brian: *The Great Betrayal*, p. 21 (Chapmans, 1992)

22 Gipps, Caroline: 'Policy Making and the Use and Misuse of Evidence' *British Educational Research Journal*, Vol. 19, No. 1 (1993)

23 Barker, Bernard: *Anxious Times: The Future of Education* (Stanground College, Peterborough, 1995)

24 Conversation with the author, 1996

25 Bierhoff, Helvia and Prais, S. J.: *Schooling as Preparation for Life and Work in Switzerland and Britain* (National Institute for Economic and Social Research, 1995)

26 *Investigations in Science Education Policy: Sc.1 in the National Curriculum for England and Wales* (Centre for Policy Studies, University of Leeds, 1993)

27 *Daily Mail* (15 December 1993)

28 *Sunday Times* (10 March 1996)

29 Letter to the *Observer* (18 December 1995)

30 Letter to the *Observer* (12 December 1995)

31 *Report of the Consultative Committee on the Primary Schools – Hadow report* (HMSO, 1931)

32 *Children and their Primary Schools – Plowden report* (HMSO, 1967)

33 Galton, Maurice: *Crisis in the Primary Classroom*, p.6 (David Fulton, 1995)

34 *op. cit.*, pp. 11–12

35 Alexander, Robin, and Rose, Jim, and Woodhead, Chris: *Curriculum Organisation and Classroom Practice in Primary Schools* (DES, 1992)

36 Alexander, Robin: *Primary Education in Leeds* (University of Leeds, 1991)

37 Alexander, Rose and Woodhead, *op. cit.*

38 At the press conference, in response to a question from the present author who asked him to quantify the teaching which had suffered from these 'highly questionable dogmas', Alexander replied: 'These questionable dogmas are far less widespread than may be thought.' Reported in the *Guardian* (14 February 1992)

39 Conversations with the author, 1996

40 'Primary Matters: A Discussion on Teaching and Learning in Primary Schools' (Ofsted 1994)

41 Conversation with the author, 1996

42 Woodhead, Chris: Annual Lecture of Her Majesty's Chief Inspector of Schools (26 January 1995)

43 Woodhead, Chris: speech to Girls' Schools Association (9 November 1994)

44 Oakeshott, Michael: *The Voice of Liberal Learning* (Yale University Press, USA, 1989)

Chapter Three

1 Hoyles, Martin, ed.: *The Politics of Literacy*, 1977; cited in the paper by Peter Traves delivered to the conference on English at Ruskin College, Oxford, 1991

2 Brooks, Gorman, Kendall and Tate: *What Teachers in Training are Taught About Reading* report to the Reading Review Working Group of the Council for the Accreditation of Teacher Education (National Foundation for Educational Research, 1991)

3 *The Basic Skills of Young Adults* (Adult Literacy and Basic Skills Unit, March 1994)

4 *Evening Standard* (29 January 1996)

5 *GCSE: General Certificate of Secondary Education* (Institute of Economic Affairs, 1990)

6 *Mail on Sunday* (3 March 1991)

7 *The Times Educational Supplement* (29 March 1991)

8 *Inspection Report on Culloden School, Tower Hamlets* (HMI report 1991, available from Ofsted)

9 Styles, Morag, and Drummond, Mary Jane, eds.: *The Politics of Reading*; University of Cambridge Institute of Education and Homerton College, Cambridge, 1993. Chapter by Helen Bromley: 'At Last He

Is Looking At The Words', pp. 41–46

10 Visit by the author, 1996

11 Conversation with the author, 1996

12 Turner, Martin: *Sponsored Reading Failure* (Education Unit, Warlingham Park School, 1990)

13 Conversation with the author, 1995

14 Conversation with the author, 1996

15 *Annual Report of Her Majesty's Chief Inspector of Schools, 1994/95* (HMSO, 1996)

16 HMI: *The Teaching and Learning of Reading in Primary Schools* (DES, 1990)

17 *Standards of Reading in Primary Schools* (House of Commons Education, Science and Arts Committee, 1991)

18 Gorman, Tom: *Reading in Recession* (National Foundation for Educational Research, 1992)

19 Author's sources

20 Stierer, Barry and Maybin, Janet: *Language, Literacy and Learning in Educational Practice*, pp. 128–138 (Multilingual Matters/Open University, 1994)

21 *Education* (28 September 1990)

22 *Education* (5 April 1992)

23 *Guardian* (6 July 1990)

24 *Sunday Times* (10 October 1993)

25 *Guardian* (20 March 1990)

26 Chall, Jeanne: *Learning to Read: The Great Debate* (McGraw-Hill, USA, 1967)

27 Turner, Martin, *op. cit.*

28 Blumenfeld, Samuel: *Trojan Horse in American Education* (The Paradigm Company, USA, 1984)

29 Jager Adams, Marilyn: *Beginning to Read*, p. 19 (MIT Press, USA, 1990)

30 Turner, Martin, *op. cit.*

31 Willinsky, John: *The New Literacy*, pp. 6–24 (Routledge, 1994)

32 Mathieson, Margaret: *The Preachers of Culture: A Study of English and its Teachers* (Allen & Unwin, 1975). Cited in *The New Literacy*, p. 12

33 Chew, Jennifer: paper submitted to the Campaign for Real Education conference, 1995

34 *Reading Words* (National Association for the Teaching of English, 1994)

35 Letter to the *Guardian* (4 July 1990)

36 Brubacher, John: *History of the Problems of Education* (McGraw-Hill, USA, 1966)

37 Jager Adams, *op. cit.*

38 Goodman, Kenneth S.: 'I Didn't Found Whole Language' from *The*

Reading Teacher (International Reading Association, 1992)

39 Stierer and Maybin, *op. cit.*

40 Gorman, Tom: *What Teachers in Training Read About Reading* (National Foundation for Educational Research, 1989)

41 Brooks, Gorman, Kendall and Tate, *op. cit.*

42 Waterland, Liz: *Read With Me: An Apprenticeship Approach to Reading* (Thimble Press, 1985)

43 Beard, Roger, and Oakhill, Jane: *Reading by Apprenticeship* (National Foundation for Educational Research, 1994)

44 Redfern, Angela, and Edwards, Viv: *How Schools Teach Reading* (Reading and Literacy Information Centre, University of Reading, 1992)

45 Styles and Drummond, *op. cit.*, pp. 10, 3–4, 7

46 *The Times Educational Supplement* (7 September 1990)

47 *The Times Educational Supplement* (19 April 1991)

48 *Daily Mail* (30 April 1991)

49 Dombey, Henrietta: *Words and Worlds: Reading in the Early Years of School* (National Association of Advisers in English, 1992)

50 Evidence by Brian Cox to the Committee of Inquiry into Reading and the Use of English – Bullock committee (1972)

51 Smart, Nicholas, ed.: *Crisis in the Classsroom*, p. 25 (Hamlyn/Daily Mirror Book, 1968)

52 Conversation with the author, 1995

53 *The Times Educational Supplement* (25 September 1992)

54 *Guardian* (9 October 1992)

55 Private correspondence with the author, 1992

56 Private correspondence with the author, 1992

57 *The Times* (8 April 1996)

Chapter Four

1 Stubbs, Michael: *Language, Schools and Classrooms* (Methuen, 1976)

2 King, Lid and Boakes, Peter: *Grammar!* (Centre for Information on Language Teaching and Research, 1994)

3 *ibid.*

4 Mason, Mary; Mason, Bob and Quayle, Tony: 'Illuminating English: How Explicit Language Teaching, Improved Public Examinations in a Comprehensive School' from *Education Studies,* Vol. 18, No. 3 (1992)

5 Conversation with the author, 1996

6 *Junior Education* (March 1994)

7 Steiner, George: *In Bluebeard's Castle* (Faber & Faber, 1994)

8 Conversation with the author, 1996

9 Cox, Brian: *Cox on Cox: An English Curriculum for the 1990s*, pp. 78–79 (Hodder & Stoughton, 1991)

10 Evidence by Brian Cox to the Bullock committee, 1972
11 *Report*: Association of Teachers and Lecturers' Magazine (January 1995)
12 Conversation with the author, 1994
13 Stierer and Maybin, *op. cit.* p. 30
14 *Guardian* (4 July 1994)
15 *Sunday Telegraph* (6 August 1995)
16 Dixon, John: *A Schooling in 'English'* (Open University, 1991)
17 Edmundson, Mark: from an adaptation of *Literature Against Philosophy, Plato to Derrida* published in *Harper's* (August 1995)
18 *Commentary*, Vol. 100, No. 3 (September 1995)
19 Lyotard, Jean-François: *The Post-Modern Condition: A Report on Knowledge* translated by Bennington, G. and Massumi, B., *Editions de Minuit* (University of Minnesota, USA, 1984)
20 Blake, Nigel: 'Truth, Identity and Community in the University' from *Curriculum Studies*, Vol. 3, No. 3 (1995)
21 *In Bluebeard's Castle*
22 Hewison, Robert: *Culture and Consensus: England, Art and Politics Since 1940*, p. 116 (Methuen, 1995)
23 *op. cit.*, p.97
24 Hoggart, Richard: *The Way We Live Now* (Chatto & Windus, 1995)
25 *op. cit.*, p. 57
26 *op. cit.*, pp. 58–59
27 *op. cit.*, p. 172
28 McRobbie, Angela: *Magazine of Cultural Studies*, No. 4 (1991)
29 Bloom, Harold: *The Western Canon*, p. 519 (Macmillan, 1994)
30 Cox, *op. cit.*, pp. 77–78

Chapter Five

1 Private correspondence with the author, 1992
2 Private correspondence with the author, 1990
3 Private conversation with the author, 1996
4 Private conversation with the author, 1995
5 *Hansard* (House of Lords, 21 July 1989)
6 *The Times Educational Supplement* (2 June 1995)
7 Cox, Caroline, and Marks, John: *The Insolence of Office: Education and the Civil Service* (Claridge, 1988)
8 *ibid.*
9 *The Times* (8 December 1983)
10 Cox and Marks, *op. cit.*
11 *Guardian* (29 November 1983)
12 Cox and Marks, *op. cit.*
13 Conversation with the author, 1996

14 Conversation with the author, 1996
15 Conversation with the author, 1996
16 Private conversation with the author, 1995
17 Private conversation with the author, 1996
18 Author's sources
19 Marsland, David and Seaton, Nick: *The Empire Strikes Back* (Campaign for Real Education, 1993)
20 *The Times Educational Supplement* (1 July 1988)
21 *Newcastle Evening Chronicle* (9 July 1987)
22 *National Curriculum Consultation Document* (Newcastle City Council, 1 September 1987)
23 *Newcastle Evening Chronicle* (17 February 1989)
24 *Newcastle Evening Chronicle* (8 February 1989)
25 *Education: Putting the Record Straight* (Network Educational Press, 1992)⁻
26 *ibid.* These references to statements by NCC and SEAC members are cited in *The Empire Strikes Back*
27 Graham, Duncan and Tytler, David: *A Lesson For Us All: The Making of the National Curriculum*, p.1 (Routledge, 1993)
28 *op. cit.*, p. 37
29 *Curriculum Guidance 3, The Whole Curriculum* National Curriculum Council, 1990. Cited in *The Empire Strikes Back*
30 Graham and Tytler, *op. cit.*, p. 23
31 *The Empire Strikes Back*
32 *The Times* (2 February 1993)
33 *Proposals for Modern Languages for Ages 11–16: Proposals of the Secretary of State for Education and the Secretary of State for Wales* (HMSO, 1990)
34 Conversation with the author, 1995
35 *English for Ages 5–11* (HMSO, 1988)
36 Graham and Tytler, *op. cit.*, p. 49
37 *The Great Betrayal*, p. 4
38 Cox, Brian: *Cox on Cox*, p. 57 (Hodder & Stoughton, 1991)
39 *Daily Mail* (2 June 1982)
40 *Education: Putting the Record Straight*
41 *Daily Mail* (2 June 1982)
42 *Daily Express* (3 February 1993)
43 *English Today* (1990)
44 Evidence to the Bullock committee (1972)
45 *The Times Educational Supplement* (28 September 1990)
46 Letter from Brian Cox to Jennifer Chew (19 June 1992)
47 Cox, Brian: *Cox on the Battle for the English Curriculum*, p. 36 (Hodder & Stoughton, 1995)
48 *Cox on Cox*, p. 4

Chapter Six

1 Author's sources
2 Author's sources
3 Serving civil servants contacted by the author during 1996 declined to comment
4 Letter to the author (14 February 1996)
5 Speech by Sir Geoffrey Holland to the Foundation for Manufacturing and Industry (4 October 1995)
6 Baker, Kenneth: *The Turbulent Years*, p. 168 (Faber & Faber, 1993)
7 *The Times Educational Supplement* (7 December 1988)
8 Baker, *op. cit.*, p. 167
9 *Sunday Telegraph* (27 June 1993)
10 Graham and Tytler, *op. cit.*, p. 12
11 Author's sources
12 Author's sources
13 Graham and Tytler, *op. cit.*, p. 15
14 Graham and Tytler, *op. cit.*, p. 19
15 Conversation with the author, 1995
16 Conversation with the author, 1995
17 Conversation with the author, 1996
18 Conversation with the author, 1996
19 Letter to Sir Ron Dearing from J. M. M. Vereker, Deputy Secretary at the Department for Education, 10 June 1993. The letter said: 'It is perfectly possible to redefine the attainment targets to accommodate a different approach to graduating achievement without amending the primary legislation . . . In short, our advice is that the primary legislation allows a significant measure of flexibility.'
20 Conversation with the author, 1995
21 Conversation with the author, 1996
22 Conversation with the author, 1996
23 Author's sources
24 *The National Curriculum* (DFE, HMSO, 1995)
25 *The National Curriculum* (DFE, HMSO, 1995)
26 Conversation with the author, 1996
27 Conversation with the author, 1995
28 McGovern, C. J. M.: *The SCAA Review of National Curriculum History: A Minority Report* (Campaign for Real Education, 1994)
29 *The National Curriculum* (DFE, HMSO, 1995)
30 *Cox on the Battle for the English Curriculum*, p. 168
31 *The Times Higher Education Supplement* (11 June 1993)
32 *A Language for Life* report of the Bullock committee (HMSO, 1975)
33 Willinsky, *op. cit.*, p. 37
34 HMI: *English from 5–16* (HMSO, 1984)
35 Conversation with the author, 1995

36 *Daily Telegraph* (20 June 1991) All the quotations from the Ruskin meeting are taken either from that article or from notes taken by observers who were present at the meeting.

37 Kimberley, Keith: 'The Third Limb: Assessment and the National Curriculum' from *New Readings: Contributions to an Understanding of Literacy*, Kimberley, Keith, Meek, Margaret, and Miller, Jane, eds. p. 91 (A&C Black, 1992)

38 Author's sources

39 *Guardian* (25 March 1991)

40 *The Times Educational Supplement* (25 December 1992)

41 *English in Education*; National Association for the Teaching of English (Spring 1993)

42 London Association for the Teaching of English *Bulletin* (March 1993)

43 National Association for the Teaching of English *News* (Spring 1994)

44 *Cox on the Battle for the English Curriculum* (p. 21)

45 *Review of Qualifications for 16 to 19-year-olds* (SCAA, 1996)

Chapter Seven

1 I am grateful to Professor Robert Pinker of the London School of Economics for access to his paper on the Enlightenment

2 Kant, Immanuel: *What Is Enlightenment?* 1784; translated by Lewis White Beck (University of Chicago Press, USA, 1950)

3 Porter, Roy: *The Enlightenment*, p. 3 (Macmillan, 1990)

4 Chadwick, Owen: *The Secularisation of the European Mind in the Nineteenth Century* (Cambridge University Press, 1975)

5 Locke, John: *Essay Concerning Human Understanding*, 1689; Nidditch, P. H., ed. (Oxford University Press, 1975)

6 I am grateful to Professor Anthony O'Hear of Bradford University for access to his paper on the philosophy of education

7 Rousseau, Jean-Jacques: *Émile*, 1762 (Everyman, 1993)

8 O'Hear, *ibid.*

9 Rousseau, Jean-Jacques: *The Social Contract*, 1762 (Penguin, 1968)

10 De Tocqueville, Alexis: *Democracy in America*, Vol. 1, 1835, Vol. 2, 1840 (Modern Library College Editions, 1981)

11 *Democracy in America* discussed in Ian Hampsher-Monk: *A History of Modern Political Thought*, p. 352 (Blackwell, 1992)

12 Wilson, James Q.: *The Moral Sense*, p. 216 (The Free Press/ Macmillan, 1993)

13 Sacks, Jonathan: *The Persistence of Faith*, p. 14 (Weidenfeld & Nicolson, 1991)

14 Briggs, Asa: *A Social History of England*, pp. 279–80 (Penguin, 1983)

15 Nietzsche, Friedrich: *Twilight of the Idols*, 1888; discussed in *The Demoralisation of Society* by Gertrude Himmelfarb, p. 188 (Institute of Economic Affairs, 1995)

16 Thomson, David: *England in the Nineteenth Century*, p. 107 (Penguin, 1950)

17 Himmelfarb, *op. cit.* (p. 189)

18 Morley, John: *On The Study of Literature* (Macmillan, 1887)

19 Houghton, Walter E.: *The Victorian Frame of Mind, 1830–1870* (Yale University Press USA, 1957)

20 Lecky, W. E. H. : *The Map of Life: Conduct and Character* (Longmans, 1899)

21 Conversation with the author, 1995

22 Holmes, Edmond: *What Is And What Might Be*, pp. 4, 123, 50, Preface V, VII (Constable, 1911)

23 Selleck, R. J. W.: *English Primary Education and the Progressives 1914–1939*, p. 24 (Routledge and Kegan Paul, 1972)

24 Selleck, *op. cit.*, p. 55

25 Wooldridge, Adrian: *Measuring the Mind*, p. 55 (Cambridge University Press, 1994)

26 Cited in A. Davis: 'Piaget, Teachers and Education', from *Learning to Think: A Reader*, Light, P., Sheldon, S. and Woodhead, M., eds. (Routledge/Open University, 1991)

27 *ibid.*

28 Wooldridge, *op. cit.*, p. 166

29 Crook, David: 'Edward Boyle, Conservative Champion of Comprehensives?' in *History of Education* Vol. 22, No. 1 (March 1993)

30 Wooldridge, *op. cit.*, pp. 296–297

31 *op. cit.*, p. 306

32 Crosland, Anthony: *The Future of Socialism* (Jonathan Cape, 1956)

33 Conversation with the author, 1995

34 Crosland, Susan: *Tony Crosland*, p. 148 (Jonathan Cape, 1982)

35 I am grateful to Phil Gardner of Cambridge University for sharing these insights gathered from his research into the oral history of education

36 O'Hear, Anthony: *Father of Child Centredness* (Centre for Policy Studies, 1991)

37 Dewey, John: *Reconstruction in Philosophy* (Boston, USA, 1948)

38 Dewey, John: *Democracy and Education*, 1916 (Free Press, USA, 1966)

39 O'Hear, *op. cit.*

40 Dewey, John: *Education and Experience* (Collier/Macmillan USA, 1938)

41 *Democracy and Education*

42 Hofstadter, Richard: *Anti-Intellectualism in American Life* (Random House, USA, 1962)

43 Conversation with Fred Naylor, former Schools Council administrator, 1996

44 Rogers, Carl: *Freedom to Learn for the '80s* (Charles E. Merrill Publishing Co., Columbus, Ohio, USA, 1983)

45 Rogers, Carl: 'Interpersonal Relationships: USA 2000', from *The Journal of Applied Behavioural Science*, 4 (3) (1968)

46 Kirschenbaum, Howard: *Advanced Value Clarification* (La Jolla University Associates, USA, 1977)

47 Greenberg, Jerrold: 'Health Education as Freeing' *Health Education* March–April, 1978

48 Coulson, William: 'OBE: Outcome-Based Education' in *Free World Journal*, Vol. 1, No. 1 (Spring 1994)

49 Stenhouse, Lawrence: 'The Discussion of Controversial Value Issues in the Classroom' (Schools Council/Nuffield Foundation Humanities Curriculum Project, 1969)

Chapter Eight

1 Bloom, Allan: *The Closing of the American Mind*, p. 25 (Penguin, 1987)

2 *Sunday Times* (24 July 1994)

3 *Guardian* (23 March 1993)

4 Mill, John Stuart: *On Liberty* (Cambridge University Press, 1989)

5 Himmelfarb, Gertrude:, *op. cit.*, p. 26

6 Gray, John: *Enlightenment's Wake*, p.19 (Routledge, 1995)

7 *The Times* (11 December 1995)

8 Lasch, Christopher: *Haven in a Heartless World*, p. 99 (W. W. Norton and Co., USA, 1977)

9 Sykes, Charles J.: *A Nation of Victims*, pp. 56–60 (St Martin's Press, USA, 1992)

10 Hobhouse, L. T.: 'Liberalism' 1911. In L. T. Hobhouse: *Liberalism and Other Writings* (Cambridge University Press, 1994)

11 Sacks, Jonathan: *Faith in the Future*, p. 66 (Darton, Longman and Todd, 1995)

Chapter Nine

1 Cited in Norman Dennis, *Who's Celebrating What?* (Christian Institute, 1995)

2 Halsey, A. H.: *Children and Society* (Nuffield College, Oxford, 1993)

3 *Marriage and Divorce Statistics 1993* (Office of Population, Censuses and Surveys, HMSO, 1995)

4 *Population Trends 83* (OPCS HMSO, 1996)

5 Wellings, Field, Johnson and Wadsworth: *Sexual Behaviour in Britain* (Penguin, 1994)

6 Rutter, Michael and Smith, David J., ed: *Psychosocial Disorders in Young People*, pp. 783–785 (Wiley, 1995)

7 *Daily Telegraph* (17 June 1994)

8 *Daily Mail* (4 January 1995)

9 *ibid.*

10 *Daily Mail* (10 June 1994)

11 *Daily Telegraph* (31 May 1995)

12 Creighton, Susan J. and Russell, Neil: *Voices from Childhood* (NSPCC, 1995)

13 Whelan, Robert: *Broken Homes and Battered Children* (Family Education Trust, 1994)

14 Conversations with the author, 1993

15 Conversations with the author at Deerbolt Young Offender Institution, 1995

16 Conversation with the author, 1995

17 Family Policy Studies Centre *Bulletin* (December 1991)

18 Wallerstein, Judith, and Blakeslee, S.: *Second Chances: Men, Women and Children a Decade After Divorce* (Ticknor and Fields, N.Y., 1989)

19 *Atlantic Monthly* (April 1993)

20 Conversation with the author, 1995

21 Peter Laslett's authoritative work, *Household and Family in Past Time*, published by Cambridge University Press, 1972, demolished the myth that the small, nuclear family had arisen during the Industrial Revolution and that prior to that time families had consisted merely of loosely extended kinship networks. The extended networks existed, but grouped around the nucleus of the mother, father and children.

22 *Atlantic Monthly, op. cit.*

23 Coote, Anna, and Franklin, Jane: 'In Place of Moral Panic' *IPPR Bulletin*, Autumn 1995

24 Conversation with the author, 1991

25 Cockett, Monica and Tripp, John: *The Exeter Family Study: Family Breakdown and its Impact on Children* (University of Exeter Press, 1994)

26 *Guardian* (6 January 1996)

27 *ibid.*

28 Dennis, Norman: 'Crime and the Family' paper delivered to Institute of Economic Affairs conference (29 June 1993)

29 Shorter, Edward: *The Making of the Modern Family*, pp. 3–5 (Collins, 1976)

30 Author's interview with Roy Porter, 1995

31 Cited in Mount, Ferdinand: *The Subversive Family* (Free Press, 1992)

32 Lasch, Christopher: *Haven in a Heartless World*, pp. 85–96 (W. W. Norton and Co., USA, 1977)
33 Laing, R. D.: *Reason and Violence: The Politics of Experience* cited in *The Abolitionists: The Family and Marriage Under Attack (see below)*
34 Fletcher, Ronald: *The Abolitionists: The Family and Marriage Under Attack*, pp. 38–39 (Routledge, 1988)
35 *op. cit.*, pp. 93–106
36 Greer, Germaine: *The Female Eunuch*, pp. 235–6 (Paladin, 1971)
37 Stone, Lawrence: *Road to Divorce*, p. 51 (Oxford University Press, 1992)
38 Shorter, *op. cit.*, p. 280
39 Halsey, A. H.: *Children and Society*
40 *Guardian* (13 February 1996) The British Medical Association said surrogacy was acceptable as a last resort for infertile couples.
41 Lasch, *op. cit.*, Preface, p. XVI
42 Winn, Marie: *Children Without Childhood*, p. 98 (Penguin, 1984)
43 *op. cit.*, pp. 99–101
44 Steed, David and Lawrence, Jean: *Come Back Spock – All Is Forgiven! Disruptive Pupil Behaviour in the Primary School* (Goldsmiths College, London, 1986)
45 Lasch, *op. cit.*, p. 126
46 *Sunday Mirror* (21 January 1996)
47 Krasny Brown, Laurene and Brown, Marc: *Dinosaurs Divorce*; cited by Barbara Dafoe Whitehead in *Atlantic Monthly* (April, 1993)
48 Elkind, David (Eliot-Pearson Department of Child Study, Tufts University, Medford, Mass., USA): 'School and Family in the Post-Modern World' in *Phi Delta Kappan* (September, 1995)
49 Harris, John: 'Liberating Children' p. 138 in *The Liberation Debate*, Michael Leahy and Dan Cohn-Sherbok, eds. (Routledge, 1996)
50 Wolkind, Stephen: letter to the *Association of Child Psychiatrists and Psychologists Review and Newsletter*, 1993
51 Wilson, James Q.: *The Moral Sense*, p. 200 (The Free Press/ Macmillan, 1993)
52 Todd, Emmanuel: *The Explanation of Ideology* (Blackwell, 1988)
53 Bloom, Allan: *The Closing of the American Mind*, p. 120 (Penguin, 1987)

Chapter Ten

1 *Guardian* (14 October 1995)
2 Reiner, Robert: 'Crime and Control: An Honest Citizen's Guide' *LSE Magazine* (Spring, 1994)
3 *ibid.*
4 *The Times* (10 July 1995)

5 *Guardian* (18 December 1995)
6 *Guardian* (11 May 1995)
7 *Daily Mail* (17 June 1995)
8 *Guardian* (19 August 1995)
9 *Evening Standard* (12 December 1995)
10 Hoghughi, Masud: *Adolescent Crime: Surviving at the Margins*. To be published in 1996 by Bower Dean, Oxford.
11 Conversation with the author
12 *Guardian* (4 October 1995)
13 *Guardian* (8 November 1995)
14 *Guardian* (15 February 1995)
15 Field, Simon: *Trends in Crime and their Interpretation* (Home Office Research Study, HMSO 1990)
16 Rutter and Smith, *op. cit.*, p. 464
17 *op. cit.*, p. 464
18 Lasch, *op. cit.*, p. 103
19 Young, Jock: *The Oxford Handbook of Criminology* (Oxford University Press, 1994)
20 Duff, R. A. and Garland, D.: 'Thinking About Punishment' pp. 1–34; also Morris, Herbert: 'A Paternalistic Theory of Punishment' pp. 95–111; from *A Reader on Punishment*, Duff and Garland, eds. (Oxford University Press, 1994)
21 *ibid.*
22 Conversation with the author, 1996
23 Conversation with the author, 1995
24 *Children and Violence* report of the Commission on Children and Violence by the Gulbenkian Foundation, p. 177 (Calouste Gulbenkian Foundation, 1995)
25 *op. cit.*, p. 32
26 *Guardian* (2 May 1995)
27 *Daily Telegraph* (28 November 1995)
28 *Guardian* (15 July 1995)
29 *Daily Mail* (21 June 1995)
30 *The Times* (10 July 1995)
31 *Young People and Crime* Home Office Research Study (Home Office, 1995)
32 Farrington, David P.: 'The Challenge of Teenage Antisocial Behaviour' in Rutter, Michael, ed.: *Psychosocial Disturbances in Young People: Challenges for Prevention*, pp. 101–105 (Cambridge University Press, 1995)
33 Hoghughi, *op. cit.*

Chapter Eleven

1 Tawney, R. H.: 'History and Society' in *Essays by R. H. Tawney;* discussed in *English Ethical Socialism*, p. 253 (see below).

2 Dennis, Norman and Halsey, A. H.: *English Ethical Socialism,* pp. 254–255 (Clarendon Press, Oxford, 1988)

3 Himmelfarb, Gertrude, *op. cit.*, p. 256

4 *Woman's Own* (31 October 1987)

5 The Health Secretary, Stephen Dorrell, announced a package of measures to improve emergency and intensive care services; *The Times* (7 March 1996)

6 *Report of the Inquiry into the Export of Defence Equipment and Dual-Use Goods to Iraq and Related Prosecutions* Sir Richard Scott (HMSO, February 1996)

7 *Guardian* (7 October 1993)

8 Under cross-examination in the *Spycatcher* hearings Sir Robert Armstrong, the Cabinet Secretary, explained that the difference between telling an untruth and giving a misleading impression lay in 'being economical with the truth'. *Guardian* (19 November 1986)

9 *Standards in Public Life* Lord Nolan (HMSO, May 1995)

10 The Public Accounts Committee reported an unprecedented level of corruption and squandering of public money. The creation of so many executive agencies and quangos, it said, had brought into government people who were used to taking short cuts. *Financial Times* (28 January 1994)

11 Conversation with the author, 1996

12 *Sunday Telegraph* (25 February 1996)

13 *Guardian* (20 January 1994)

14 *Daily Telegraph* (19 January 1994)

15 'Rising unemployment and the recession have been the price that we've had to pay for getting inflation down. That is a price well worth paying'. Norman Lamont, *Independent* (18 May 1991)

16 Smith, Adam: *Lectures on Jurisprudence*; available in Liberty Classics (1982)

17 Tawney, R. H.: *The Acquisitive Society*, 1921; discussed in Dennis and Halsey, *op. cit.*, pp. 234–235

18 Gray, John, *op. cit.*, pp. 101–102

19 *op. cit.*, pp. 87–89

20 Selbourne, David: *A Civic Politics/Movement for Christian Democracy Occasional Papers* No. 5 (Christian Democrat Press, 1996)

21 This point has been elaborated upon by Professor David Rose of Essex University; *Observer* (2 July 1995)

22 *Sunday Times* (10 March 1996)

23 Conversation with the author, 1995

24 Conversation with the author, 1995

25 *Tackling the Mathematics Problem*

Chapter Twelve

1 Warner, Geoffrey: 'The Anglo-American Special Relationship' from *Diplomatic History*, Vol. 13, No. 4 (1989)
2 Arnold, Matthew: *Literature and Dogma*; Preface to 1873 edition
3 These influences were discussed by Professor Anthony Smith of the London School of Economics in an interview with the author, 1995
4 Barker, Bernard, *op. cit.*
5 Dennis and Halsey, *op. cit.*, p. 31
6 Gray, *op. cit.*, p. 24
7 Carrington, Bruce and Short, Geoffrey: 'What Makes a Person British? Children's Conceptions of their National Culture and Identity' *Educational Studies*, Vol. 21, No. 2 (1995)
8 Gray, *op. cit.*, p. 27
9 *op. cit.*, pp. 23–4
10 Sacks, Jonathan: *The Persistence of Faith: Religion, Morality and Society in a Secular Age*, p. 67; (Weidenfeld & Nicolson, 1991)
11 Palmer, Frank, ed.: *Anti-Racism: An Assault on Education and Values* (Sherwood Press, 1986)
12 Interview with the author, 1995
13 Arendt, Hannah: *Between Past and Future*, p. 196; 1961 (reissued in Penguin, 1993)

Chapter Thirteen

1 Bogdanor, Vernon: *The Monarchy and the Constitution*, p. 300 (Oxford University Press, 1995)
2 Prais, Sig: *Economic Performance and Education: The Nature of Britain's Deficiencies*; Keynes Lecture in Economics (British Academy, 28 October 1993)
3 McElrea, Fred: 'Justice in the Community: The New Zealand Experience' pp. 93–103 in *Relational Justice*, Burnside, Jonathan and Baker, Nicola (Waterside Press, 1994)
4 Hobhouse, L. T.: *Liberalism and Other Writings*, James Meadowcroft, ed., p. 61 (Cambridge University Press, 1994)
5 Arendt, *op. cit.*, p. 95

SELECT BIBLIOGRAPHY

Arendt, Hannah: *Between Past and Future* (1961; reissued in Penguin, 1993)

Baker, Kenneth: *The Turbulent Years* (Faber and Faber, 1993)

Bloom, Allan: *The Closing of the American Mind* (Penguin, 1987)

Bloom, Harold: *The Western Canon* (Macmillan, 1994)

Blumenfeld, Samuel: *Trojan Horse in American Education* (The Paradigm Company, USA, 1984)

Bogdanor, Vernon: *The Monarchy and the Constitution* (Oxford University Press, 1995)

Bourdillon, Hilary, ed.: *Teaching History* (Routledge/Open University, 1994)

Briggs, Asa: *A Social History of England* (Penguin, 1983)

Brindley, Susan, ed.: *Teaching English* (Routledge/Open University, 1994)

Brubacher, John: *History of the Problems of Education* (McGraw-Hill, USA, 1966)

Burnside, Jonathan and Baker, Nicola, eds.: *Relational Justice* (Waterside Press, 1994)

Chadwick, Owen: *The Secularisation of the European Mind in the Nineteenth Century* (Cambridge University Press, 1975)

Chall, Jeanne: *Learning to Read: The Great Debate* (McGraw-Hill, USA, 1967)

Cox, Brian: *The Great Betrayal* (Chapmans, 1992)

Cox, Brian: *Cox on Cox: An English Curriculum for the 1990s* (Hodder & Stoughton, 1991)

—: *Cox on the Battle for the English Curriculum* (Hodder & Stoughton, 1995)

Cox, Caroline and Marks, John: *The Insolence of Office: Education and the Civil Service* (Claridge, 1988)

Crosland, Anthony: *The Future of Socialism* (Cape, 1956)

Crosland, Susan: *Tony Crosland* (Cape, 1982)

De Tocqueville, Alexis: *Democracy in America* (Vol. 1, 1835, Vol. 2, 1840; Modern Library College Editions, 1981)

Dennis, Norman and Halsey, A. H.: *English Ethical Socialism* (Clarendon Press, Oxford, 1988)

Dewey, John: *Reconstruction in Philosophy* (Boston, USA, 1948)

—: *Democracy and Education* (1916; Free Press, USA, 1966)

—: *Education and Experience* (Collier/Macmillan USA, 1938)

Dixon, John: *A Schooling in 'English'* (Open University, 1991)

Duff, R. A. and Garland, D. eds.: *A Reader on Punishment* (Oxford University Press, 1994)

Fletcher, Ronald: *The Abolitionists: The Family and Marriage Under Attack* (Routledge, 1988)

Galton, Maurice: *Crisis in the Primary Classroom* (David Fulton, 1995)

Graham, Duncan and Tytler, David: *A Lesson For Us All: The Making of the National Curriculum* (Routledge, 1993)

Gray, John: *Enlightenment's Wake* (Routledge, 1995)

Greer, Germaine: *The Female Eunuch* (Paladin, 1971)

Gulbenkian Foundation: *Children and Violence* (Report of the Commission on Children and Violence; Calouste Gulbenkian Foundation, 1995)

Hampsher-Monk, Ian: *A History of Modern Political Thought* (Blackwell, 1992)

Hewison, Robert: *Culture and Consensus: England, Art and Politics Since 1940* (Methuen, 1995)

Himmelfarb, Gertrude: *The Demoralisation of Society* (Institute of Economic Affairs, 1995)

Hobhouse, L. T.: *Liberalism and Other Writings* (Cambridge University Press, 1994)

Hofstadter, Richard: *Anti-Intellectualism in American Life* (Random House, USA, 1962)

Hoggart, Richard: *The Way We Live Now* (Chatto & Windus, 1995)

Hoghughi, Masud: *Adolescent Crime: Surviving at the Margins* (to be published in 1996 by Bower Dean, Oxford)

Holmes, Edmond: *What Is And What Might Be* (Constable, 1911)

Houghton, Walter E.: *The Victorian Frame of Mind, 1830–1870* (Yale University Press USA, 1957)

Jager Adams, Marilyn: *Beginning to Read* (MIT Press, USA, 1990)

Kant, Immanuel: *What Is Enlightenment?* (1784; translated by Lewis White Beck, University of Chicago Press, USA, 1950)

Kimberley, Keith; Meek, Margaret; and Miller, Jane, eds: *New Readings: Contributions to an Understanding of Literacy* (A&C Black, 1992)

Kirschenbaum, Howard: *Advanced Value Clarification* (La Jolla University Associates, USA, 1977)

King, Lid and Boakes, Peter: *Grammar!* (Centre for Information on Language Teaching and Research, 1994)

Lasch, Christopher: *Haven in a Heartless World* (W. W. Norton and Co., USA, 1977)

Laslett, Peter: *Household and Family in Past Time* (Cambridge University Press, 1972)

Leahy, Michael and Cohn-Sherbok, Dan, ed.: *The Liberation Debate* (Routledge, 1996)

Lecky, W. E. H.: *The Map of Life: Conduct and Character* (Longmans, 1899)

Lewis, C. T. and Short, C.: *A Latin Dictionary* (Oxford University Press, 1963)

Light, P., Sheldon, S., and Woodhead, C., eds.: *Learning to Think: A Reader* (Routledge/Open University, 1991)

Locke, John: *Essay Concerning Human Understanding*, 1689; ed. P. H. Nidditch (Oxford University Press, 1975)

Lyotard, Jean-François: *The Post-Modern Condition: A Report on Knowledge* translated by Bennington, G., and Massumi, B., *Editions de Minuit* (University of Minnesota, USA, 1984)

Mathieson, Margaret: *The Preachers of Culture: A Study of English and its Teachers* (Allen and Unwin, 1975)

Mill, John Stuart: *On Liberty* (Cambridge University Press, 1989)

Morley, John: *On The Study of Literature* (Macmillan, 1887)

Mount, Ferdinand: *The Subversive Family* (Free Press, 1992)

Oakeshott, Michael: *The Voice of Liberal Learning* (Yale University Press, USA, 1989)

Palmer, Frank, ed.: *Anti-Racism: An Assault on Education and Values* (Sherwood Press, 1986)

Pollard, Andrew and Bourne, Jill, ed.: *Teaching and Learning in the Primary School* (Routledge/Open University, 1994)

Porter, Roy: *The Enlightenment* (Macmillan, 1990)

Rogers, Carl: *Freedom to Learn for the '80s* (Charles E. Merrill Publishing Co., Columbus, Ohio, USA, 1983)

Rousseau, Jean-Jacques: *Émile*, 1762 (Everyman, 1993)
—: *The Social Contract*, 1762 (Penguin, 1968)
Rutter, Michael and Smith, David J., ed.: *Psychosocial Disorders in Young People* (Wiley, 1995)
Rutter, Michael, ed.: *Psychosocial Disturbances in Young People: Challenges for Prevention* (Cambridge University Press, 1995)
Sacks, Jonathan: *Faith in the Future* (Darton, Longman and Todd, 1995)
—: *The Persistence of Faith* (Weidenfeld and Nicolson, 1991)
Selleck, R. J. W.: *English Primary Education and the Progressives 1914–1939* (Routledge and Kegan Paul, 1972)
Shorter, Edward: *The Making of the Modern Family* (Collins, 1976)
Slinger, Michelle, ed.: *Teaching Mathematics* (Routledge/Open University, 1994)
Smart, Nicholas, ed.: *Crisis in the Classsroom* (Hamlyn/Daily Mirror Book, 1968)
Smith, Adam: *Lectures on Jurisprudence;* available in Liberty Classics, 1982
Steiner, George: *In Bluebeard's Castle* (Faber and Faber, 1994)
—: *No Passion Spent* (Faber and Faber, 1996)
Stierer, Barry and Maybin, Janet: *Language, Literacy and Learning in Educational Practice* (Multilingual Matters/Open University, 1994)
Stone, Lawrence: *Road to Divorce* (Oxford University Press, 1992)
Stubbs, Michael: *Language, Schools and Classrooms* (Methuen, 1976)
Styles, Morag and Drummond, Mary Jane, eds.: *The Politics of Reading* (University of Cambridge Institute of Education and Homerton College, Cambridge, 1993)
Sykes, Charles J.: *A Nation of Victims* (St Martin's Press, USA, 1992)
Taylor, Gillian and Edwards, David: *Étoiles* (BBC Longman, 1992)
Thomson, David: *England in the Nineteenth Century* (Penguin, 1950)
Todd, Emmanuel: *The Explanation of Ideology* (Blackwell, 1988)
Wallerstein, Judith, and Blakeslee, S.: *Second Chances: Men, Women and Children a Decade After Divorce* (Ticknor and Fields, N.Y., 1989)
Waterland, Liz: *Read With Me: An Apprenticeship Approach to Reading* (Thimble Press, 1985)

Wellings, Field, Johnson and Wadsworth: *Sexual Behaviour in Britain* (Penguin, 1994)

Willinsky, John: *The New Literacy* (Routledge, 1994)

Wilson, James Q.: *The Moral Sense* (The Free Press/Macmillan, 1993)

Winn, Marie: *Children Without Childhood* (Penguin, 1984)

Wooldridge, Adrian: *Measuring the Mind* (Cambridge University Press, 1994)

Young, Jock: *The Oxford Handbook of Criminology* (Oxford University Press, 1994)

INDEX